2nd International Conference on Material Strength and Applied Mechanics (MSAM 2019)

IOP Conference Series: Materials Science and Engineering
Volume 629

Kiev, Ukraine
27 – 30 May 2019

ISBN: 978-1-5108-9833-2
ISSN: 1757-8981

Printed from e-media with permission by:

Curran Associates, Inc.
57 Morehouse Lane
Red Hook, NY 12571

Some format issues inherent in the e-media version may also appear in this print version.

Institute of Physics
Dirac House, Temple Back
Bristol BS1 6BE UK

Phone: 44 1 17 929 7481
Fax: 44 1 17 920 0979

techtracking@iop.org

Additional copies of this publication are available from:

Curran Associates, Inc.
57 Morehouse Lane
Red Hook, NY 12571 USA
Phone: 845-758-0400
Fax: 845-758-2633
Email: curran@proceedings.com
Web: www.proceedings.com

Table of contents

Volume 629

The 2nd International Conference on Material Strength and Applied Mechanics

27–30 May 2019, Kiev, Ukraine

Accepted papers received: 27 August 2019
Published online: 8 October 2019

Preface

Papers

IOP Conf. Series: Materials Science and Engineering **629** (2019) 011001 doi:10.1088/1757-899X/629/1/011001

MSAM2019 Preface

The 2nd International Conference on Material Strength and Applied Mechanics (MSAM 2019) was held from May 27-30, 2019 in Kiev, capital of Ukraine. Sponsored by Pisarenko Institute for Problems of Strength, National Academy of Sciences of Ukraine, MSAM 2019 represented a broad forum for all aspects of material strength to promote our fundamental understanding of material strength under different conditions. MSAM 2019 also provided an opportunity to present and discuss recent analytical, experimental and computational results in the field of Applied Mechanics to bridge the gap between physical theory and its application to technology. The conference had parallel sessions with dedicated selected subjects related to Material Strength and Applied Mechanics. The technical program consisted of 4 international keynote speakers, 68 oral presentations, and 18 poster presentations.

The present issue of IOP Conference Series: Materials Science and Engineering (MSE) records the proceedings of MSAM2019 and contains 38 specially selected manuscripts submitted to the MSAM2019 conference. The electronic submission and handling of manuscripts via the conference website, including the selection of reviewers and evaluation of manuscripts, were identical to the procedures applied to manuscripts submitted as regular contributions for international per-reviewed publications.

The organization of this conference and the preparation of proceedings volume would have been impossible without the tremendous efforts and dedication of many individuals, especially from Ms. Li Chen, who oversaw the organization of the conference and the program, and a large team of reviewers with their timely submission of quality reports. We express our sincere thanks to all authors and presenters for their contributions. We also thank very much our sponsors for their generous support.

Prof. Nao-Aki Noda
Technical Program Committee Chair MSAM2019
Kyushu Institute of Technology
Kitakyushu, Japan, July 2019

IOP Conf. Series: Materials Science and Engineering **629** (2019) 011002 doi:10.1088/1757-899X/629/1/011002

Peer review statement

All papers published in this volume of *IOP Conference Series: Materials Science and Engineering* have been peer reviewed through processes administered by the proceedings Editors. Reviews were conducted by expert referees to the professional and scientific standards expected of a proceedings journal published by IOP Publishing.

IOP Conf. Series: Materials Science and Engineering **629** (2019) 012001 doi:10.1088/1757-899X/629/1/012001

Influence of additional elements (Si, Ti and B) on the microstructure, mechanical properties and castability of aluminum alloys (A201)

S A E Majid[1,2], M Bamberger[1] and A Katsman[1]

[1]Technion – Israel Institute of Technology, Haifa 3200, Israel

E-mail: eng.suzan26@gmail.com

Abstract. Improving the castability, mechanical properties and modifying the microstructure of the A201 alloy (Al - 4.97wt. % Cu – 0.56 wt. % Ag) were done by adding Si, Ti and B elements. The four alloys of A201 with additions of: (1) Si, (2) Ti and B, (3) Si, Ti and B, (4) non, were investigated in the as-cast, solution treated at 550°C for 20 hr and aged at 170°C up to 32 days conditions. Combination of different microscopes (HRSEM/EDS, TEM) and microhardness tests were used to investigate how the alloy properties influenced by the precipitation. Additions of Si resulted in the grain coarsening and an increase in the amount of near-grain boundary secondary phases. Additions of Ti and B resulted in the grain refinement, prevented hot tears, but caused a large amount of Al_3Ti phase which makes the melt more viscose, and less fluid and castable. The mechanisms of the influence were recognized as facilitating the GP zones nucleation: Si atoms provide more probable homogeneous nucleation of the $<001>_\alpha$ GP zones transforming to θ'' phase during aging; TiB_2 particles serve as the sites for heterogeneous nucleation of $<111>_\alpha$ GP zones which then transform to Ω phase. The precipitation sequence during aging was the following: SSSS → GP zones → θ''+ Ω → θ' + Ω → θ' → θ. The maximum microhardness was achieved with two kinds of semi-coherent precipitates in the structure: θ' (Al_2Cu- tetragonal) and Ω (Al_2Cu-orthorhombic). Therefore, joint additions and corresponding thermal treatment can provide optimal microstructure and mechanical properties of the A201 based alloys.

1. Introduction

A201 alloy used in aerospace and automobile parts due to its good properties and easy of production [1-6]. Solution treatment followed by aging cause the decomposition of Supersaturated Solid Solution (SSSS) to different precipitates: GP zones → θ'' → θ' → θ and it is depending on the degree of super-saturation and the aging temperature [1-3, 7-12]. The process and/or kinetics of precipitation can be changed by adding a small amount of alloying elements [1, 2, 9, 13].

For the Al alloys, it was investigated the effect of different elements additions on the mechanical properties [13-17]. The addition of Si controls the amount of eutectics structure of Al-Si and also influences the precipitation sequence. Additions of Ti and B cause the grain refinement of the primary Al phase. However, the cross and mutual influence of alloying elements is not fully understood.

In the present work, the followinf alloys were investigated: (1) Al201 (referred hereafter as Base-alloy – BA), (2) BA+1 wt. % Si, (3) BA+1.33 wt. %B= 3.17 wt. % Ti and (4) BA+1 wt. % Si +1.33 wt. %B+ 3.17 wt. % Ti al. These compositions were selected to achieve improved properties and castability. They were investigated by a combination of high resolution scanning electron microscopy (HRSEM), energy dispersive X-ray spectrometry (EDS) and transmission electron microscopy (TEM).

Microhardness measurements were performed by a Vickers hardness tester to reveal the effect of precipitation on the mechanical properties of the alloy.

2. Experimental

The chemical composition of the A201 alloy (alloy I) is given in table 1. The nominal compositions of the prepared alloys (II, III and IV) are given in the table 1.

Table 1. The compositions (in wt. %) of the investigated alloys.

Alloy	Nominal composition
I	BA: 4.97wt. % Cu-0.56wt. % Ag- 0.33wt%Mn-0.24wt. %Mg
II	BA with addition of 1 wt. % Si
III	BA with additions of 1.33 wt. %B and 3.17 wt. % Ti
IV	BA with additions of 1 wt. % Si, 1.33 wt. %B and 3.17 wt. % Ti

For detailed experimental procedure refer to [13].

3. Experimental results and discussion

Castability and mechanical properties of the alloys after ST and aging were investigated together with comprehensive investigation of the alloy's microstructure.

3.1. Castability

To check the fluidity and castability of the alloys, a metal die with three plates having different thicknesses was used. The ability of the cast to fill in was used to evaluate the castability. The additions of Si to A201 alloy have been shown to increase castability of the alloy [13]. However, samples with addition of 1wt. % Si contained a lot of cracks (figure 1(b)). The additions of Ti and B allowed to avoid the presence of cracks (figures 1(c) and 1(d)), but made the melt more viscous, and as a result, the fluidity and castability of the as-cast alloys decreased.

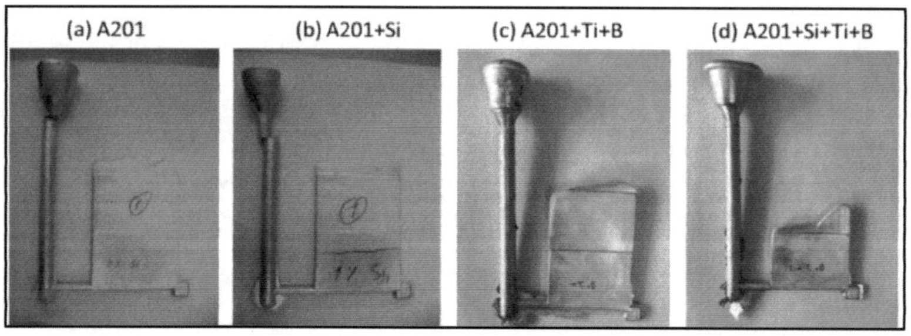

Figure 1. Castability tests for the four alloys cast into a metal die with three plates of different thicknesses: (a) A201, (b) A201-1 wt. % Si, (c) A201-1.33 wt. %B- 3.17 wt. % Ti and (d) A201-1 wt. % Si -1.33 wt. %B- 3.17 wt. % Ti.

3.2. Age hardening behavior

The hardeness Vs time curves of the four investigated alloys during isothermal aging at 170°C for up to 32 days are presented in figure 2. It was decided that the aging temperature should be 170°C, since it was mentioned by Gergi [18], that the early stages precipitates do not appear at higher temperatures (above 180°C) of aging.

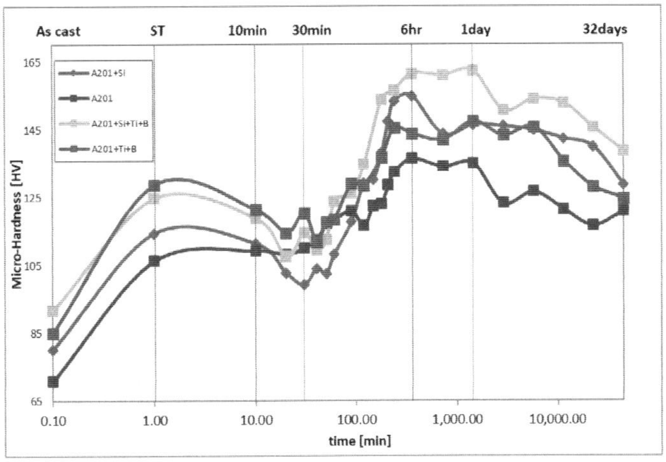

Figure 2. Hardness Vs time curve of the four alloys aged at 170°C; the error bars of 5 HV.

For all investigated alloys microhardness reached a maximum value after a certain aging time, and then further aging led to its decrease (overaging). For alloys I and II (A201 and A201- 1 wt. % Si) the maxima (~138 HV and ~155 HV, respectively) were reached after ~ 6 h of aging; For alloys III and IV containing Ti and B the maxima (~147 HV and ~162 HV, respectively) were reached after 1 day of aging.

3.3. Microstructure

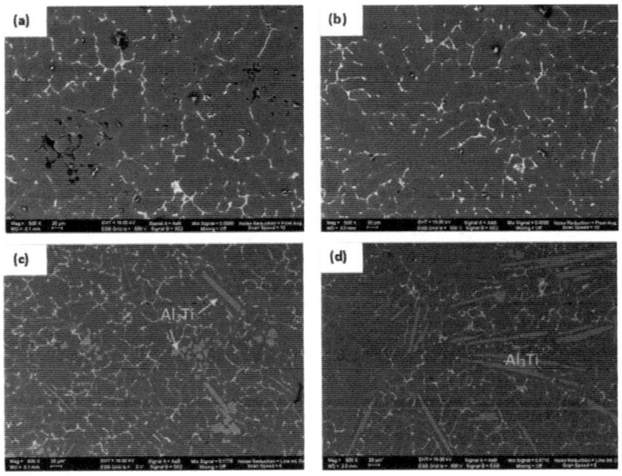

Figure 3. Backscattering (AsB detector) HRSEM micrographs of the as-cast alloys; (a) BA, (b) BA+1 wt. % Si,(c) BA+1.33 wt. %B+ 3.17 wt. % Ti and (d) BA+1 wt. % Si +1.33 wt. %B-+3.17 wt. % Ti alloys.

3.3.1. As-cast alloys. As can be seen from HRSEM micrographs (figure 3), the as-cast alloys consisted mainly of the α-Al phase as a matrix and eutectic structures (α-Al/ θ-Al2Cu) in the near grain boundary regions. Alloys III and IV, with Ti and B additions, (figures 3(c) and 3(d)), contain also hundred microns size particles having two different shapes: needle-like and round-like particles. The composition of both the needle-like and round-like particles, was measured by EDS HRSEM analysis, and was found to be close to Al3Ti.

3.3.2. Aged alloys: microstructure and phase composition. Several samples of the aged alloys were selected for detail microstructure investigation: after aging periods of 30 min and 3 hr (under- aged), 6 hours and 1 day (maximum peak-aged).

- Under-aged samples (first peak after 30 min)

As seen from the HAADF STEM images of the alloys II and III in the under-aged condition (figure 4), the microstructures contained very thin disc-shape precipitates with diameters about 2–10 nm, elongated along two perpendicular directions, which are identified as GP zones. Two types of GP zones were recognized: (i) elongated along <001>α directions and (ii) elongated along <111>α directions.

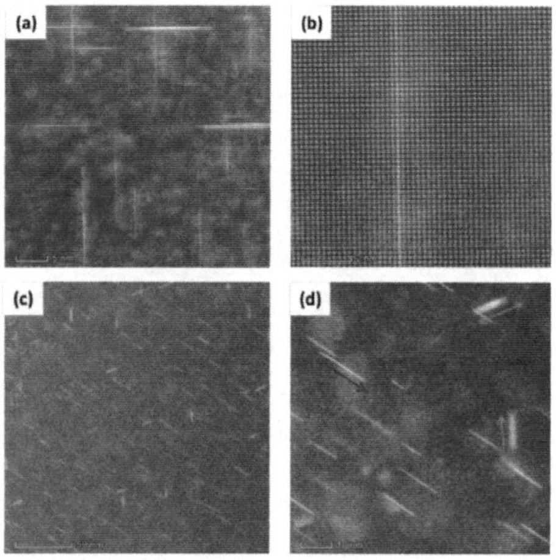

Figure 4. (a) HAADF STEM image of alloy II (A201-1wt% Si) aged for 30 min at 170°C, with Z.A.$[001]_{\alpha}$ showing GP zones. (b) Magnification of (a); (c) HAADF STEM image of alloy III (BA+1.33 wt. %B+ 3.17 wt. % Ti) aged for 30 min at 170°C, with Z.A. $[101]_{\alpha}$ showing GP zones alongated along $<001>_{\alpha}$ (red arrow) and $<111>_{\alpha}$ (green arrow); (d) Magnification of (c).

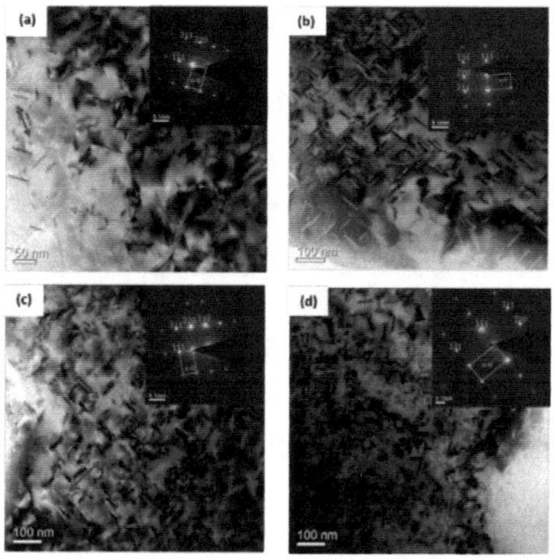

Figure 5. TEM image and corresponding SAED pattern of (a) BA, (b) BA+1 wt. % Si, (c) BA+1.33 wt. %B+ 3.17 wt. % Ti and (d) BA+1 wt. % Si +1.33 wt. %B+ 3.17 wt. % Ti alloys aged for 6hr (a, b) and for 1day (c, d), with Z.A. $[112]_{\alpha}$.

- Maximum peak-aged samples (6 hours / 1 day)

Typical TEM image and corresponding SAED pattern of the four investigated alloys aged for 6 hr (alloys I, II) and for 1 day (alloys III, IV) are presented in figure 5, with zone axis $[112]_{\alpha}$. As seen from the images, the microstructure in the peak-aged condition contained two kinds of precipitates which can be recognized as θ' and Ω phases. The platelet-like θ' (Al_2Cu, tetragonal, semi-coherent) precipitates are extending along $<001>_{\alpha}$ direction, while Ω (Al_2Cu, Orthorhombic, semi-coherent) precipitates are extending along $<111>_{\alpha}$.

4. Discussion

In order to understand how the Si addition influences the precipitation process, the EDS analysis was performed on the GP zone in the alloy II (A201-1wt. % Si) after 30 min aging, it was found that increased amount of Si atoms is present along the GP zone.

The alloying of the BA with Ti and B increases the Ω phase precipitates. This result can be explained by heterogeneous nucleation of <111>α GP zones on the TiB2 particles. As was shown by [19], the activation energy for the Ω phase nucleation was greatly reduced by the addition of TiB2 and silver, and the TiB2 particles were even more effective than Ag.

The θ' precipitates are thermodynamically more favorable than θ" and Ω phases. That is why they grow during following aging in the expense of existed θ" and Ω precipitates.

5. Conclusion

The additions of silicon, titanium and boron to the A201 alloy can improve the alloy properties by proper modification of the microstructure. The Si addition alone may improve castability and increase the microhardness, however, it results in grain coarsening and increase amount of the secondary phases in the grain boundary regions. On contrary, the additions of Ti and B resulted in the grain refinement, prevented hot tears, but caused a large amount of Al_3Ti phase making the melt more viscose, less fluid and castable. The mechanisms of the influence were recognized as facilitating the GP zones nucleation: Si atoms provide more probable homogeneous nucleation of the $<001>_\alpha$ GP zones transforming to θ'' phase during aging; TiB_2 particles serve as the sites for heterogeneous nucleation of $<111>_\alpha$ GP zones which then transform to Ω phase. The precipitation sequence during aging of the (A201+Si+Ti+B) alloy was the following: SSSS → GP zones → θ''+ Ω → θ' + Ω → θ' → θ. The maximum microhardness was achieved with two kinds of semi-coherent precipitates in the structure: θ' (Al_2Cu- tetragonal) and Ω (Al_2Cu-orthorhombic). Therefore, joint additions and corresponding thermal treatment can provide optimal microstructure and mechanical properties of the A201 based alloys.

References

[1] Ringer S P and Hono K 2000 Microstructural evolution and age hardening in aluminium alloys: Atom probe field-ion microscopy and transmission electron microscopy studies *Mater. Charact.* **44** 101-31
[2] Davis J R 1993 *Aluminum and Aluminum Alloys* (ASM Specialty Handbook)
[3] Straink M J and Zahra A M 1997 Mechanisms of combined GP zone and θ' precipitation in an Al - Cu alloy *J. Mater. Sci. Lett.* **6** 1613-5
[4] Kapranos P 2012 Thixoforming A201 aluminium alloy: Is there a future in aerospace applications *Metal. Ital.-n* 25-9
[5] Tiryakioglu M and Campbell J 2009 Ductility, structural quality, and fracture toughness of Al-Cu-Mg-Ag (A201) alloy castings *Mater. Sci.Tech.* **25** 784-9
[6] Blanco, Azpilgain Z, Lozares J, Kapranos P and Hurtado I 2010 Rheological characterization of A201 aluminum alloy *T. Nonferr. of China* **20** 1638-42
[7] Konno T J, Kawasaki M and Hiraga K 2001 Direct imaging of Guinier-Preston zones by high-angle annular detector dark-field scanning transmission electron microscopy *J. Electron Microsc* **50** 105-11
[8] Kabir M S, Minhaj T I, Ashrafi E A and Islam M 2014 The influence of aging time and temperature on the structure and properties of heat treated A201. 0 aluminum alloy *Int. J. of Rec. Tech. and Eng.* **3** 78-83
[9] Polmear I J 2004 Aluminium alloys – A century of age hardening *Mater. Forum* **28** 1-14
[10] Palmerin J M, Rosales H J D, Hirata V M L, Castro N C, Velázquez J L G and Ramirez A D J M 2009 Hardening behavior in aged Al-4%Cu-0.3%Mg alloys with 0.5 and 2%Ag additions *Mate. Trans.* **50** 2785-9
[11] Hirano K 1989 Studies on age-hardenable aluminium alloys by atom-probe field-ion microscope *Bull. Mater. Sci.* **12** 313-23
[12] Möller H, Masuku E P, Curle U A, Pistorius P C, Knutsen R D and Govender G 2011 The influence of Mn on the tensile properties of SSM-HPDC Al-Cu-Mg-Ag alloy A201. *J. South. Afr.Inst. Min.Metal.* **2011** 27-9
[13] Abd El Majid S, Bamberger M and Katsman A 2018 Influence of additional elements (Si, Ti AND B) on the castability, corrosion and mechanical properties of A201 alloys *Light Met.2018* 259-66
[14] Caceres C H, Svensson I L and Taylor J A 2003 Strength-ductility behaviour of Al-Si-Cu-Mg casting alloys in T6 temper *Int. J. Cast Metals Res.* **15** 531-43
[15] Backerud S L, Chai G and Tamminen J 1990 Solidification characteristics of aluminium alloys: foundry alloys. Skanaluminium – aluminum alloys

[16] Couture A 1981 Iron in aluminum casting alloys: A literature survey *AFS Inter. Cast Metals* **6** 9-17
[17] Sabatino M D and Arnberg L 2009 Castability of aluminium alloys *T. Indian I. Metals* **62** 321-5
[18] Gergis A Z 2014 On the performance of low pressure die – cast Al-Cu based automotive alloys: Role of additives. MSc Thesis University of Quebec 2014
[19] Men H 2016 Effects of TiB_2 particles and Ag on the activation energy of Ω phase in Al alloys *Mater. Sci. Eng.* **117** 1-6

Deformation and energy distribution of metal tubes under internal explosive loading

L N Feng[1,3], D Li[2], J D Tian[1], Z F Zhang[1] and Z X Hu[1]

[1]Beijing Institute of Astronautical Systems Engineering, Beijing 100076, China
[2]China Academy of Launch Vehicle Technology, Beijing 100076, China

E-mail: flnhit@126.com

Abstract. Deformation and energy distribution of two kinds of metal tubes under internal explosive loading have been studied by means of experiments and simulations in this paper. The velocity and displacement-time history of characteristic points on the flat metal tubes were measured by Photonic Doppler Velocimetry (PDV). Ls-dyna fluid-solid coupling algorithm was used to simulate the experiment. Simulation results agree well with the experimental results. On this basis, the expansive properties of the round tube in the same condition were obtained, the round tube directly expands to the maximum diameter, does not like the expansion and contraction fluctuant process of the flat tube. The simulation result shows that the flat tube will output more energy in the form of kinetic energy, while the round tube tends to store energy in the form of plastic strain energy.

1. Introduction

Explosives can output high energy in a short period of time, and work quickly, which has an important application in areas that require rapid response [1,2]. In practical applications, explosives are often coated with metal tubes or containers outside, previous studies on metal shells under internal explosive load include two types, one is the damage problem of the metal tube, most of which are directed to the metal tube in the weapon warhead. The metal tube is expected to damage so that high-speed fragments can produce to damage the goal. Therefore, factors such as damage and scattering velocity [3-7] are often considered. The Gurney Formula [8] and related corrections [9] can predict the fragment velocity. Another type of the research is about explosion protection, the research object is mainly explosive containers [10-13]. The deformation of thin-walled shells under explosive loading has a certain research basis, most of which focused on the response of thin-walled shells under the internal explosive loading of concentrated charges. In the aspect of theoretical research, there were researchers simplified the pressure time curve generated by the explosion into an ideal triangular pressure pulse which can simplify the problem [10,11]. In addition, Fouad and Lowell used the single-degree-of-freedom analysis and energy method to predict the response of the shell wall, the theoretical analysis agrees well with the finite element analysis results [13]. Under the action of explosion loading, the structure will undergo large deformation, which causes the traditional finite element analysis no longer applicable, there will be problems such as element distortion. ALE technology can deal with the locking phenomenon caused by the distortion during the large deformation [14]. Corresponding finite element analysis software Ls-dyna [15] and Autodyne [16] can simulate such experiments. In terms of experimental research, Test equipment requires high frequency response due to the rapid explosion response. Related experimental equipment includes X-ray photography, high-speed photography, electric probes, dynamic pressure gauges, strain wires, fiber displacement interferometer etc. [10,14,17-21].

In the above studies, the energy generated by the explosion of the explosive is transmitted to the metal directly or through the air, and the research objects are mainly weapons or explosive containers. The research object of this paper is the flat tube components (flat tube, filler, and lead detonation cord), the strong impact generated by the explosive does not directly affect the external structure, but a certain attenuation is generated by the filler, and evenly acts on the metal tube. The metal tube produces a large expansion, causing the external structure to produce the expected damage. At the same time, the explosive product is enclosed inside the metal tube without cracking and therefore no pollution [19]. In this paper, the expansion process of the metal tubes under the internal explosive load was obtained by means of experiment and simulation analysis. The velocity and displacement-time history curve and energy distribution were analysis, which has certain guiding significance for engineering practice.

2. Experiment design

The test specimens are the flat tube components. We used the Photonic Doppler Velocimetry(PDV) [20,21] developed by Yang Jun et al to monitor the change of velocity and displacement at characteristic points on the flat tube. The basic principle of PDV is to use a fiber laser to output laser light, the laser is divided into two beams. One laser is irradiated on the test point which is the test laser and another laser is used as reference laser. When the test laser is reflected back and processed with the reference laser, the displacement and velocity of the test point can be obtained. To prevent the overall displacement of the tube from affecting the test results, the ends of the test specimens were clamped. The test device and cross-sectional view of the test specimens are shown in figure 1. Through the test method, the expansion velocity and displacement history of the four points in the middle section of the flat tube were obtained. In fact, due to the symmetry of the structure, two points can already represent the motion and deformation characteristics of the tube. But the explosion experiment is not repeatable and the experimental results often have certain discreteness. In order to obtain as much effective data as possible in one experiment, four measured points were set, among which, points 1 and 3 are symmetrical points, the test results are basically the same, similar to points 2 and 4.

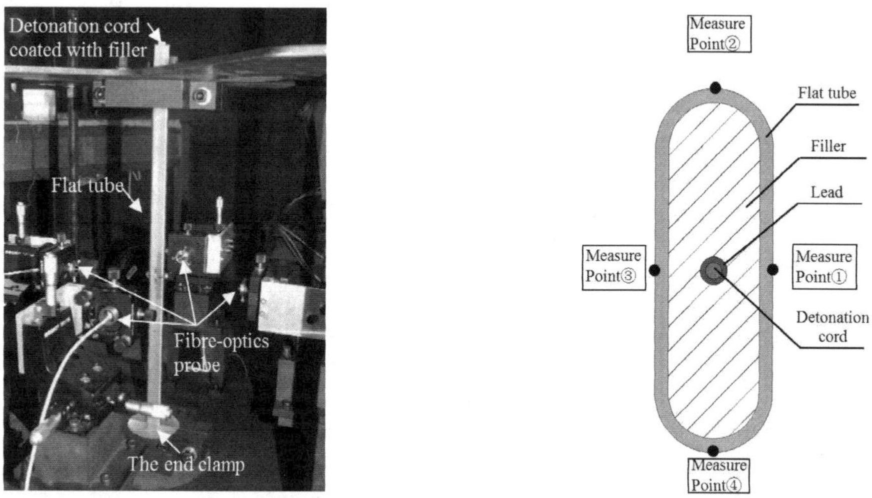

Figure 1. The test device and cross-sectional view of the test specimens.

3. Simulation analysis

We used Hypermesh to create and mesh the geometric model according to the actual size of the specimen and Ls-dyna fluid-solid coupling algorithm to simulate the experiment. The flat tube was in the form of Lagrange elements, detonation cord, lead, the filler and air were Euler elements. Its element models are shown in figure 2.

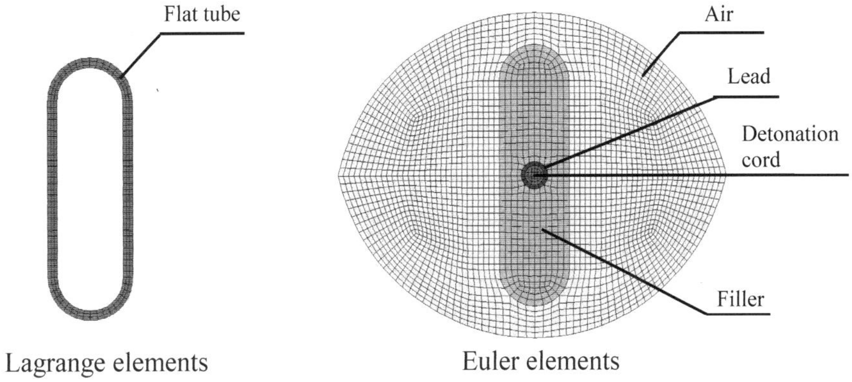

Figure 2. Finite element model.

The material parameters of the flat tube components refer to relevant references and research reports [22-24], the detonation velocity of the detonation cord is the average value used for this experiment provided by the manufacturer of the cords. The specific parameters are shown in table 1.

Table 1. Material and its parameters.

part	parameters
Detonation Cord(RDX)	ρ=1.5g/cm^3, D=7240m/s, P_{CJ}=23.11Gpa, A=611.3Gpa, B=10.65Gpa, R_1=4.4, R_2=1.2, ω=0.32, E_0=8.9GPa, V_0=1.0
Lead	ρ=11.34g/cm^3, C=2006m/s, S_1=1.429, γ_0=2.77
Filler(polymer)	ρ=0.95g/cm^3, E=3Gpa, v =0.47, σ_y=0.015Gpa
Air	ρ=1.29×10^{-3}g/cm^3, γ=1.4, E_0=2.5×10^{-4}Gpa, V_0=1
Flat tube (1Cr18Ni9Ti)	ρ=7.83g/cm^3, E=205Gpa, v=0.29; σ_y=0.34Gpa, k=1.34Gpa, n=0.34

The detonation cord, lead, and air were subject to the corresponding equation of state. The equation of state for the explosive was the JWL equation of state as equation (1).

$$P = A\left(1-\frac{\omega}{R_1 V}\right)e^{-R_1 V} + B\left(1-\frac{\omega}{R_2 V}\right)e^{-R_2 V} + \frac{\omega E}{V} \tag{1}$$

Where P and V represent the pressure and relative specific volume of the explosive respectively, E is the internal energy of the explosive per unit volume, and A, B, R_1, R_2, ω are the coefficients of the JWL equation of state, which are depended on the kind of the explosive. E_0 is the initial internal energy per unit volume of explosive and V_0 is the initial relative specific volume, which are the initial values of E and V before the detonation reaction.

The Gruneisen equation of state was used for lead. This equation defines the pressure of the material in the form of equations (2) and (3) when μ has different values.

$$P = \frac{\rho_0 C^2 \mu\left[1+\left(1-\frac{\gamma_0}{2}\right)\mu - \frac{a}{2}\mu^2\right]}{\left[1-(S_1-1)\mu - S_2\frac{\mu^2}{\mu+1} - S_3\frac{\mu^3}{(\mu+1)^2}\right]^2} + (\gamma_0 + a\,\mu)E \quad (\mu \geq 0) \tag{2}$$

$$P = \rho_0 C^2 \mu + (\gamma_0 + a\mu) E \ (\mu \leq 0) \tag{3}$$

Where C is the intercept of the v_s-v_p curve (shock wave velocity - post wave particle velocity curve). S_1, S_2, and S_3 are slope parameters of the v_s-v_p curve; γ_0 is the Gruneisen coefficient, a is the first-order volume correction to γ_0. μ is a parameter indicating the change of the material volume, $\mu=\rho/\rho_0-1$, ρ is the current density of the material, and ρ_0 is the initial density of the material.

The air was modelled with the gamma law equation of state as equation (4).

$$P = (\gamma - 1)\frac{\rho}{\rho_0}E \tag{4}$$

Where P is the pressure in the air, γ is the ratio of specific heats, ρ/ρ_0 is the ratio of current density to initial density, and E is the internal energy of air per unit volume. In the finite element analysis of the explosion, this equation of state is often used for the ideal gas [16,24].

4. Result and discussion

4.1. Comparison of simulation and experimental result

The velocity and displacement-time history of the feature points obtained by simulation analysis are compared with the experimental results. We did four experiments, of which points 1, 3 are structural symmetry points, so the results are the same, similar for points 2 and 4. Velocity and displacement time curves of points 1 and 3 are shown in figure 3. Curves of points 2 and 4 are shown in figure 4.

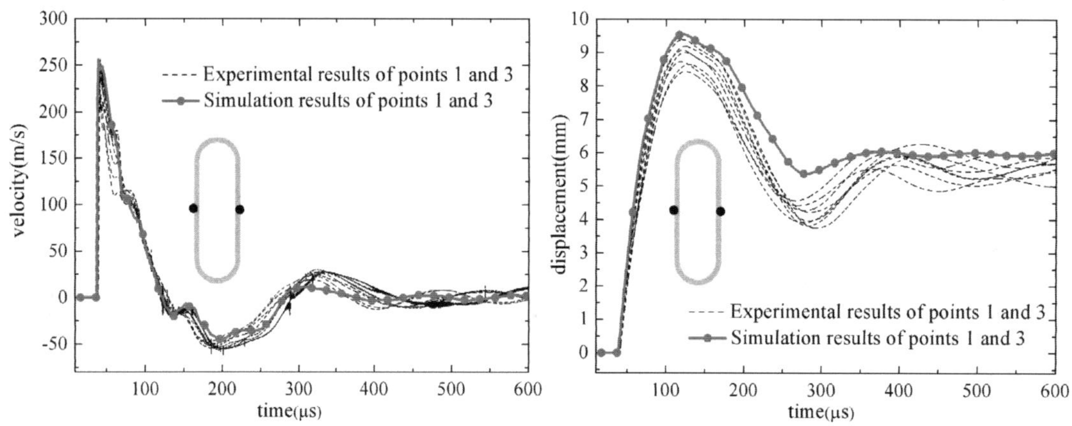

Figure 3. Experimental and simulation results of points 1 and 3.

From the experimental and simulation results, it can be seen that the points 1 and 3 of the flat tube experience the process of expanding firstly and then contracting and gradually becoming stable. The points 2 and 4 are slightly expanded, greatly contracted, expands outward and finally stabilizes gradually. The motion process is complicated. The experimental results of the maximum expansion velocity of points 1 and 3 are between 212.38-256.58 m/s, average value is 236.95 m/s, for points 2 and 4, the experimental value is between 46.84-158.65 m/s, average value is 101.21 m/s, the location of points 2 and 4 is narrow which is difficult to monitor, so the experimental results are somewhat discrete. The simulation result for points 1 and 3 is 255.86 m/s, and 127.26 m/s for points 2 and 4, which are all in the range of the experimental results. As for the maximum expanding displacement, the experimental results of point 1,3 is between 8.45 mm and 9.55 mm, 9.01 mm on average, for points 2,4, the results range from 0.26 mm to 0.63 mm, average value is 0.43 mm, correspondingly, the simulation displacement results are 9.55 mm and 0.48 mm on points 1,3 and 2,4 respectively. The data results show that the simulation and experimental results correspond well, which proves that the simulation results are reliable.

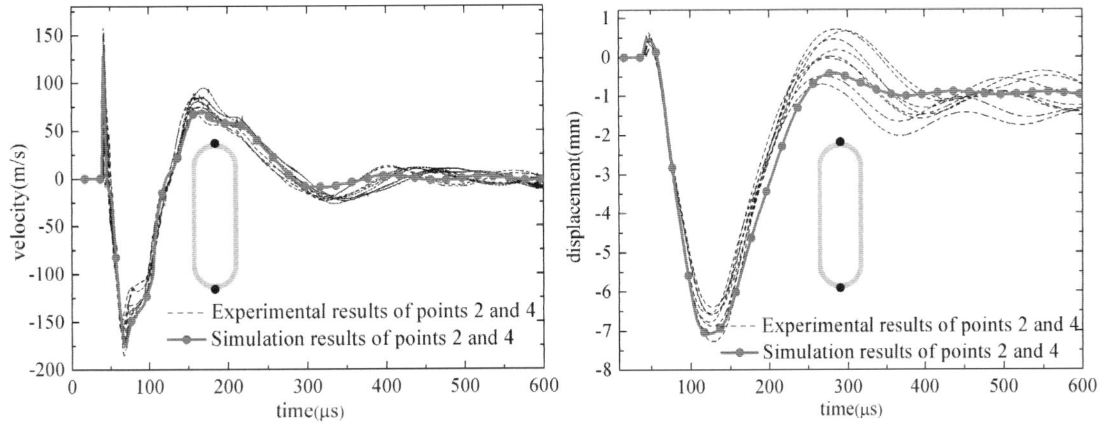

Figure 4. Experimental and simulation results of points 2 and 4.

4.2. Comparison of expansion velocity and displacement of round and flat metal tube

The same method, material parameters and detonation cord were used to simulate the expansion process of the round metal tube with the same quality as the flat tube, the wall thickness of the two kind of tubes and the perimeter of the section are the same. Therefore, geometrically, points 1 and 3 on the flat tube are closest to the detonation cord, points 2 and 4 are farthest, the distance between the points on the round tube and the detonation cord is between the above two. The geometric comparison figure and element model of round metal tube components are shown in figure 5.

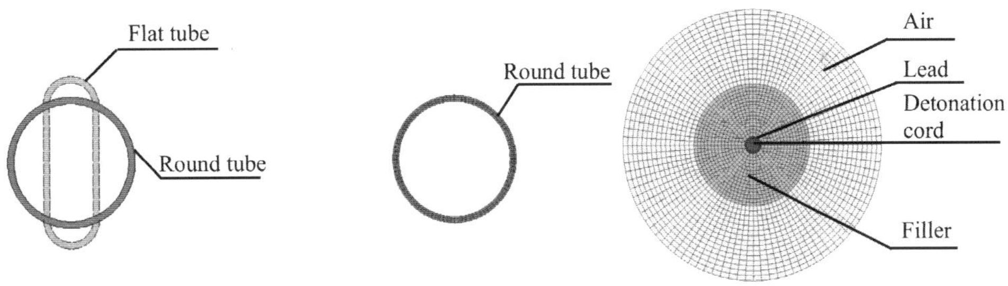

geometric comparison of two tubes   Lagrange elements   Euler elements

Figure 5. Geometry comparison of two tubes and finite element model of round metal tube components.

The expansion velocity, displacement of feature points on two tubes were obtained as shown in figure 6.

The results of the data show that the round tube does not have the large fluctuation process of expansion and contraction like the flat tube, but gradually expands to the maximum value to reach the stable stage. The maximum expansion velocity is 168.20 m/s, which is between the 1, 3 and 2, 4 points of the flat tube, and the displacement tends to be stable at 2.10 mm after expanding to the maximum value 2.14 mm. In order to understand the energy distribution of explosive energy on two kinds of metal tubes during the internal explosive load, more in-depth study was then carried out.

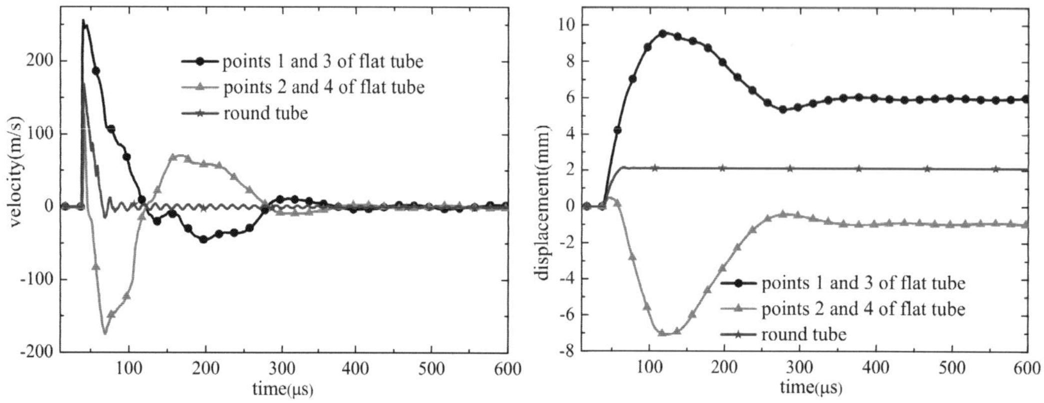

Figure 6. Comparison of expansion velocity and displacement of round and flat metal tubes.

4.3. Analysis of explosion energy distribution

To reduce the influence of boundary on the calculation of explosion energy distribution, the middle parts of the test specimens were extracted as independent parts shown in figure 7.

The total energy of the detonation cord, the kinetic energy and plastic strain energy of the metal tubes in the middle portion were output. The data was subjected to dimensionless treatment with reference to the initial total energy of the detonation cord in this middle potion. The energy change process is shown in figure 8. Since the energy of the detonation cord is released at the moment of explosion, its change with time is basically the same in both cases, so it is represented by the same curve in the figure.

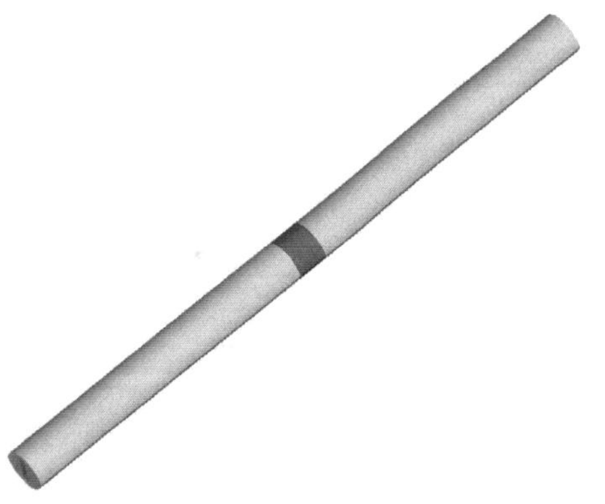

Figure 7. Model for the energy distribution analysis.

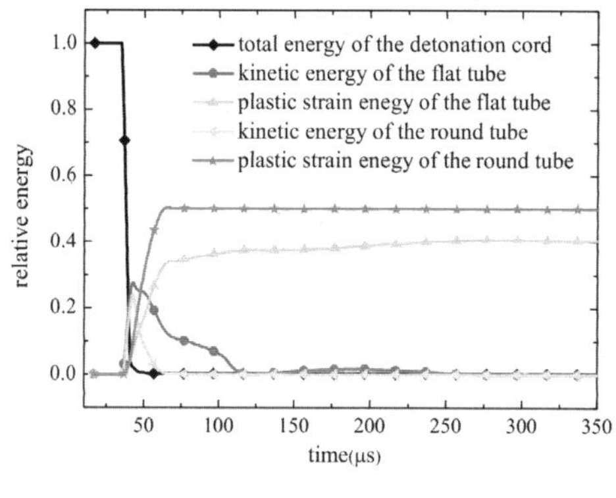

Figure 8. Energy changing process of the detonation cord and the metal tubes.

As shown in figure 8, the maximum kinetic energy of the flat tube accounts for 27.65% of the initial energy of the detonation cord, and the final stable plastic strain energy ratio is 40.43%. For the round tube, the extreme value of the kinetic energy is 24.10% and the final stable plastic deformation energy ratio is 50.03%. The initial energy of the detonation is consistent. The flat tube is more inclined to output the explosive energy in the form of kinetic energy. The extreme value of kinetic energy and the duration time that the kinetic energy greater than zero are longer than that of the round tube. Round tube tends to store energy in the form of plastic strain energy, and the final stable energy value is larger than that of the flat tube. This is due to the fact that the flat tube will have more energy loss during expansion and contraction.

5. Conclusion

In this paper, the motion and deformation process of the flat metal tube under internal explosive loading have been tested and simulated. The experimental and simulation results correspond well. The deformation characteristics of round tube and energy change history of both tubes were also analysed by the simulation results. The main conclusions are as follows.

- Under the internal explosive loading, the flat tube expands and then contracts in the direction of the short axis (points 1 and 3), in the direction of the long axis (points 2 and 4), it expands slightly, contracts, expands outward and then finally gradually stabilizes.
- Under the internal explosive loading, the round tube directly expands to the maximum deformation and stabilizes gradually.
- The maximum kinetic energy value of the flat tube accounts for 27.65% of the initial energy of the detonation cord, and the final stable plastic deformation energy ratio is 40.43%. The maximum kinetic energy value of the round tube is 24.10%, and the final stable plastic deformation energy ratio is 50.03%. For the flat tube, its duration time that the kinetic energy greater than zero are longer than that of the round tube.
- The flat tube tends to output energy outward in the form of kinetic energy, while the round tube stores energy in the form of plastic deformation energy. Therefore, it is preferable to use a flat tube for destroying the external structure, and round tube is better to protect the external structure, energy is stored in the form of its own plastic deformation energy.

Acknowledgments

Thanks to Zhang Dezhi, Yang Jun *et al* from the Northwest Institute of Nuclear Technology for their support of the experiments in this paper.

References

[1] Chen T, Guo L F, Zhao Z *et al* 2017 Dynamic model and experimental study of low-shock separation device *Astronaut Syst Eng Technol* **1** 44-53
[2] Liu Z S, Wang X J, Zhu X C *et al* 2012 *Aerospace Firework Device* (Beijing: China Aerospace Publishing House)
[3] Mu R D and He S M 2006 Research on initial velocity of in-explosive cylinder fragment *J Aeronaut Mater* **26** 140-3
[4] Li Y C, Li D H, Wei Z G *et al* 1999 Research on the deformation damage and fracture rules of circular tubes under inside-explosive loading *Acta Mech Sinica* **31** 442-9
[5] Ji C, Long Y, Fang X *et al* 2012 Numerical simulation of large-aperture steel pipe subjected to contact explosion loading *J Vib Shock* **31** 72-76
[6] Wang X Y, Wang S S and Ma F 2018 Experimental study on the expansion of metal cylinders by detonation *Int J Impact Eng* **114** 147-52
[7] Du Y, Ma L, Zheng J Y *et al* 2016 Coupled simulation of explosion-driven fracture of cylindrical shell using SPH-FEM method *Int J Pres Ves Pip* **139-140** 28-35
[8] Gurney R W 1943 The initial velocities of fragments from Bombs, Shell, Grenades *BRL Report No 405*
[9] Danel J F and Kazandjian L 2004 A few remarks about the gurney energy of condensed explosives *Propell Explos Pyrot* **29** 314-6
[10] Rushton N, Schleyer G K, Clayton A M *et al* 2008 Internal explosive loading of steel pipes *Thin Wall Struct* **46** 870-7
[11] Baker W E 1960 The elastic-plastic response of thin spherical shells to internal blast loading *ASME J Appl Mec*h **27** 139-44
[12] Simangunsong G M, Kubota S, Saburi T *et al* 2008 Dynamic response of a steel pipe to internal blast loading *Mater Sci Forum (Switzerland)* 29-34
[13] Fouad F and Lowell G 1988 Simplified analysis for impulsively loaded shells *J Struct Eng* **114** 885-99

[14] Zhang S W, Hua J S, Liu C L *et al* 2004 A numerical simulation of the metallic tube expansion induced by inside head-on hitting two detonation waves *Explo Shock Waves* **24** 219-25
[15] Zhao Y L, Xiao D S and Dai H B 2007 Effects on the velocity of metallic tube expansion by inside head-on hitting detonation waves *J Ordn Eng Coll* **19** 30-2
[16] Choi M, Lee J R and Kong C W 2014 Development of a numerical model for an expanding tube with linear explosive using AUTODYN *Shock and Vib* **2014** 1-10
[17] Qin X J, Zhang D Z, Yang J *et al* 2014 Electric probe measurement technique on deformation process of cylindrical steel shell under inside-explosion loading *Explo Shock Waves* **34** 115-9
[18] Li X M, Jin X G and Li D H 2007 Elementary measurement of free surface velocity of cylindrical metal tube under inward detonation *Acta Armament* **39** 1252-5
[19] Li J, Xiong C, Zhang D Z *et al* 2017 Strain wire measurement technique used in calibrating circumferential large deformation of cylindrical explosion vessels *Explo Shock Waves* **37** 976-81
[20] Yang J 2012 The development of fiber displacement interferometer and its applications (Hefei: University of Science and Technology of China)
[21] Yang J, Wang K Y, Xu H B *et al* 2013 Development of an optical-fiber displacement interferometer and its application in Hopkinson pressure bar experiment *Infrared Laser Eng* **42** 102-7
[22] Sun J and Yang Z G 2006 Simulation and optimization of explosive separation process for flat tube separator *Proc 9th Natl Conf Explo Safety* Shenyang pp 140-3
[23] Wang J and Li H 2011 Simulation on the separating performance of linear separation System *Struct Environ Eng* **38** 28-33
[24] Wen X J 2016 Separation security of fragments during linear explosive separation process (Changsha: National University of Defense Technology)

Component layout optimization design based on genetic algorithm

B Shen[1,2], Z W Pan[1], H Lin[1] and W Z Li[1]

[1]Beijing Institute of Astronautical Systems Engineering, Beijing, 100076, China

E-mail: 1192931044@qq.com

Abstract. When working, electronic equipment is often subjected to a mechanical environment dominated by vibration. The printed circuit board (PCB) in the equipment is prone to deformation and transmits vibration to the components on the board. As the weakest link, the component corner position has a large acceleration, which is one of the important reasons for the failure of components and the entire electronic equipment. Based on the finite element modeling of PCB, the interaction between the finite element analysis software NASTRAN and the programming software MATLAB is realized, and the application of genetic algorithm in component layout optimization becomes possible. The optimization results show that the method adopted in this paper can effectively reduce the acceleration value at the component corner and optimize the mechanical environment when the electronic equipment works.

1. Introduction

Electronic equipment generally contains many sophisticated electronic components. To meet special requirements, it needs to be installed on a special platform. When the electronic equipment is subjected to mechanical environment such as vibration input from the platform, the PCB undergoes large dynamic bending deformation, which is an important reason for the failure of the electronic equipment. If we can estimate the acceleration of the PCB under the vibration in the design, optimize the component layout and reduce the acceleration value of the component corners, the mechanical environment of the printed circuit board and the reliability of the product can be improved.

The traditional means of dynamic analysis through vibration test not only cost and time, but also difficult to grasp the structural characteristics of the internal components of electronic systems. But finite element simulation can effectively make up for the above defects. Therefore, verification is often performed through finite element simulation in the optimization. Zhu [1] analyzed the calculation formula of natural frequencies to obtain the conclusion that large and heavy components should be placed as close to the constraint as possible. Zhu introduced inertial forces into the proposed integrated layout optimization method designing the multi-component systems. A consistent material interpolation scheme between element stiffness and inertial load was presented to avoid the singularity of localized deformation, which was due to the presence of design dependent inertial loading when the element stiffness and the involved inertial load were weakened with the element material removal. The tested numerical example show the proposed methods extended the actual concept of topology optimization and were efficient to generate reasonable design patterns [2].

Due to the large number of components on the board and adjustable locations, the traversal optimization cannot be performed under normal circumstances. It is often done by calculating the results of models under a number of typical locations to compare and analyze, and then put forward suggestions for the optimization of its layout. However, due to the limited selection of typical locations, the optimization results are often not optimal. In recent years, intelligent optimization algorithms have gained attention, which can overcome the limitations of traditional optimization algorithms and achieve global optimization [3-5]. Krishna [6] presented a fixture layout optimization technique that used the genetic algorithm to find the fixture layout. The technique minimized the deformation of the machined surface due to clamping and machining forces over the entire tool path. Particle swarm optimization (PSO) intelligent algorithm was adopted to perform optimization, and obtained the optimal solution in the multidisciplinary optimization design for the section layout of umbilicals [7].

In this paper, based on the genetic algorithm and the finite element simulation, the component layout is optimized. In the solution of the optimization model, an improved genetic algorithm is proposed and the idea of parametric modeling is used to realize the interactive optimization of MATLAB and NASTRAN to achieve the purpose of optimizing the component layout and improving the mechanical environment.

2. Finite element simulation of printed circuit board

This paper takes the circuit board in an electronic device as the research object, and uses finite element software PATRAN to model it. The finite element model is shown in figure 1.

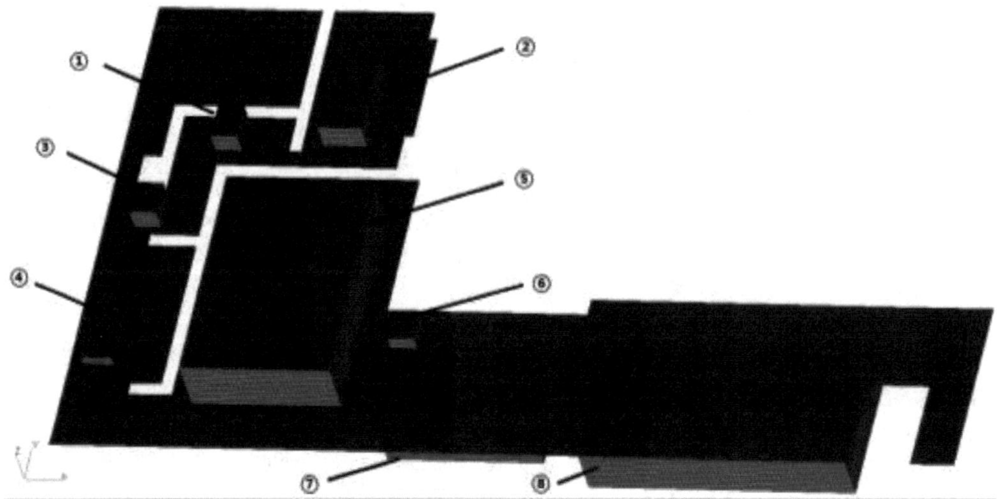

Figure 1. Finite element model.

There are 8 components on the board, and each component has been numbered in figure 1. Considering the limited installation space near the location of the component, this paper focuses on component No. 6 and improves its mechanical environment by optimizing its layout.

When constructing the finite element model, the PCB is constructed by surface, and the quadrilateral element (Quad4) is used to divide the mesh. The surface is cut according to the position of the components on the board, ensuring that the corner and the node on the board are coincident, and the shell element is stretched to modeling the components. At the end, common node processing is performed to achieve the unity of the cutting surfaces and the constraints between the components and the board. The total finite element model includes 65597 Hex elements, 13274 Quad4 elements and 78871 nodes.

The circuit board is mounted on the platform by bolts at the edges. Bolt connections limit the X, Y, Z movement in three directions. So the constraint is applied to the node on the model at the bolts position. According to actual working conditions, the load on the circuit board comes from the vibration of the platform. So, the acceleration in the Z direction is applied at the model by 1 g, where g is about 9.81 m/s^2. In the calculation process, the linear elastic constitutive equation model is adopted for the board and the components. The material parameters used in each part are shown in table 1.

Table 1. Material parameters.

	Elastic Modulus/Pa	Poisson ratio	Density/kg/m^3
PCB	1.40e10	0.12	1800
Component NO. 1	7.65e10	0.32	7500
Component NO. 2	7.65e10	0.32	7500
Component NO. 3	7.65e10	0.32	7500
Component NO. 4	7.65e10	0.32	7500
Component NO. 5	2.06e11	0.30	7800
Component NO. 6	1.86e11	0.32	16600
Component NO. 7	7.00e10	0.34	2700
Component NO. 8	2.06e11	0.30	7800

First, the modal analysis of the board model is performed. The results are shown in table 2 below.

Table 2. Results of modal analysis.

Modal Order	1	2	3	4	5
Frequency/Hz	485.16	1174.6	2315.0	2715.8	3233.4

According to modal analysis, the natural frequency of the circuit board model is 485.16 Hz. The sinusoidal frequency excitation of acceleration is applied to the circuit board and the interval is 20~2000 Hz. Initially set the modal damping ratio to 0.01. We get the frequency response curve of the average acceleration of component No. 6 in the current position, as shown in figure 2.

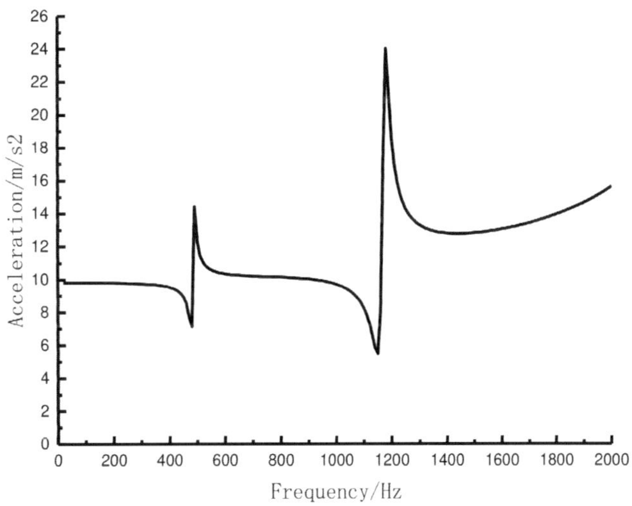

Figure 2. Frequency response curve of average acceleration (Z direction).

The average value of the acceleration at the component corners peaks at 1180 Hz with a peak value of 24.01 m/s^2. The average acceleration frequency response curve of the component has obvious resonance characteristics. The resonance frequency is close to the modal frequency of each order, and the response is the most obvious at the second-order modal frequency.

3. Genetic algorithm and improvement

3.1. Genetic algorithm

The generation of Genetic Algorithm (GA) was attributed to the pioneering work of Professor Holland of the University of Michigan in the United States in the 1960s and early 1970s [8]. Later, after further research by Jong and Goldberg, the genetic algorithm was more perfect [9]. It has been widely used to solve complex combinatorial optimization problems. It can be said that genetic algorithm is the most widely used and most successful modern optimization method.

Genetic algorithm mimics the evolutionary processes of living things. In the process of evolution, children maintain the characteristics of their parents to a certain extent, which is hereditary. At the same time, variations may occur during the generation of offspring. Both heredity and variation occur on the chromosome, reflecting changes in the gene on the chromosome. The entire evolutionary process is controlled by the competition mechanism of "survival of the fittest, survival of the fittest", which makes the most viable chromosomes survive with the greatest possibility, thus allowing the organism to evolve and develop.

The strategy of roulette is used to mimic the survival mechanism. The probability that each individual will survive is

$$p_i = \frac{F_i}{\sum_{j=1}^{N} F_j} \tag{1}$$

where F is the objective function value, N is the number of the individuals in the population. The purpose of the formula (1) is to allow individuals with higher objective function value to obtain higher survival chances, and to eliminate inferior solutions during the evolution process and optimize the objective function values of the population. Considering that the objective function value in this paper is as small as possible, the above formula is rewritten as

$$p_i = \frac{\sum_{j=1}^{N} F_j - F_i}{(N-1) \times \sum_{j=1}^{N} F_j} \tag{2}$$

Taking into account the above characteristics and advantages of the simulated annealing algorithm, this paper optimizes the component layout on the printed circuit board based on the simulated annealing algorithm. The optimization parameters of simulated annealing algorithm are shown in table 3.

Through the selection, crossover and mutation operations in the algorithm, the genetic algorithm can finally obtain a good solution with low objective function.

Considering the above characteristics and superiority of genetic algorithm, this paper optimizes the layout of components on the printed circuit board based on genetic algorithm. The optimization parameters of the genetic algorithm used are shown in table 3.

Table 3. Optimization parameters of genetic algorithm.

Name	Value
Cross rate	0.75
Mutation rate	0.05
Population size	10
Genetic algebra	200

3.2. Adaptability improvement

In order to make the genetic algorithm more suitable for the layout optimization of components, it has been improved adaptively in the programming, mainly in the following aspects.

- Considering that the program calls NASTRAN to perform the calculation for a long time, each calculation result is stored to avoid repeated calculations for the same state, and the iteration time can be shortened;

- Since each current state has a limited neighborhood, the searched neighborhoods are recorded during the optimization process, avoiding repeated searches to the same neighborhood, which can increase the convergence speed.

4. Optimization design of component layout

The acceleration value at the component corners is an important indicator of the mechanical environment of the printed circuit board. When the material properties meet the requirements, the reduction of the acceleration value means the increase of the life of the components, which is particularly important for the reliability of the circuit board and even the entire electronic equipment.

The most effective way to reduce the acceleration at the component corners is to adjust the component layout. Layout optimization is the design and redo of the circuit board model. In order to automate modeling, the idea of parametric modeling is used to update the model by updating the relevant data of nodes, units, etc. in the model file. And on this basis, interacting with the finite element calculation software NASTRAN to obtain the calculation results of the new model, so as to judge whether the acceleration value under the new layout has decreased.

4.1. Objective function

The output file of finite element calculation contains the real and imaginary values of the accelerations in three directions. Calculate the modulus length of accelerations of each of the three directions, and then calculate the total acceleration. The formula is as follows

$$a=\left(a_{11}^2+a_{12}^2+a_{21}^2+a_{22}^2+a_{31}^2+a_{32}^2\right)^{1/2} \tag{3}$$

where a_{i1} is the real value of the i-direction acceleration and a_{i2} is the imaginary value of the i-direction acceleration.

4.2. Restrictions

There are two restrictions as follows.

- The distance from other components is not less than 8mm to facilitate the installation of components.
- The distance from the circuit board boundary is not less than 3mm, to facilitate the connection between the circuit board and the platform.

4.3. Interaction methods between MATLAB and NASTRAN

In this paper, the joint programming of NASTRAN and MATLAB can not only use the program to realize the powerful search ability of the algorithm, but also realize the finite element calculation of the current state. The software interaction process is shown in figure 3.

The establishment of the optimization model and the reading of the calculation results are all implemented using MATLAB programming. At the same time, the finite element calculation is performed using NASTRAN and the calculation results are output. MATLAB programmatically controls the input parameters calculated by NASTRAN and reads the result of the calculation. Then the component position parameters of the next iteration are generated and output. Through the above loop iteration process, the optimization problem is ultimately solved.

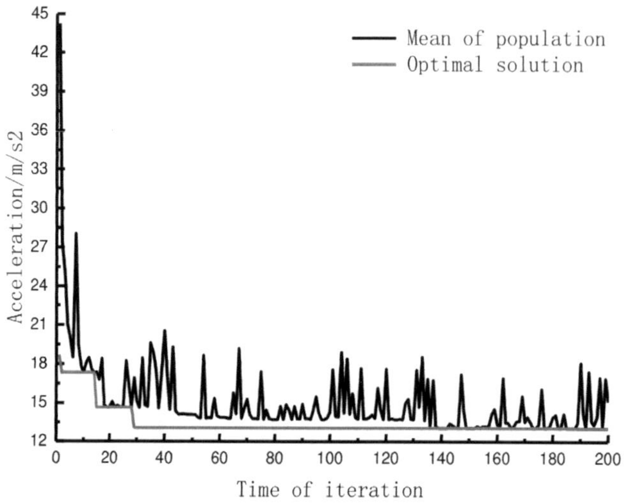

Figure 3. Convergence result of acceleration value optimization.

4.4. Optimization results

With the current mesh density, there are 840 adjustable positions for the target components. If traversing the search, the time required is 98h. The optimal solution obtained by traversal search is 13.08 m/s^2.

The optimization result of the objective function is shown in figure 3. The abscissa is the time of iterations, and the ordinate is the acceleration value. The red line represents the optimal solution found during the optimization process and the black line represents the average of the accelerations of all individuals in the population under the current iteration. As the number of iterations increases, the average acceleration value of the population decreases and the optimal solution is reduced to a minimum of 13.08 m/s^2.

To verify the stability of the program, repeat the operation 5 times, record the time and results of each run, as shown in table 4.

Table 4. Optimization results.

Num	1	2	3	4	5
Time/min	815	772	742	703	863
Result/m/s^2	13.08	13.08	13.08	13.08	13.08

The average program running time is 779 minutes. The length of the run time is related to the initial solution, which is the randomly selected initial positions of the component.

It can be seen that using the improved genetic algorithm for search optimization, the average time required is 779 min. Compared with the ergodic search, the time is shortened to less than 2/15 of the original, and the searched results are all global optimal solutions.

The finite element model of the components before and after optimization is shown in figure 4. The distance between the optimized position and other components is greater than 8 mm, and the circuit board boundary is greater than 3mm, which meets the restrictions.

The average frequency response curve of acceleration at the component corners after optimization is shown in figure 5. Compared with the pre-optimization, the curve trend is constant and the frequency at which the average value of the acceleration peaks is basically constant, which is because the optimized components have light weight and little influence on the overall natural frequency.

Figure 4. Model of printed-circuit board. (a) Before optimization and (b) after optimization.

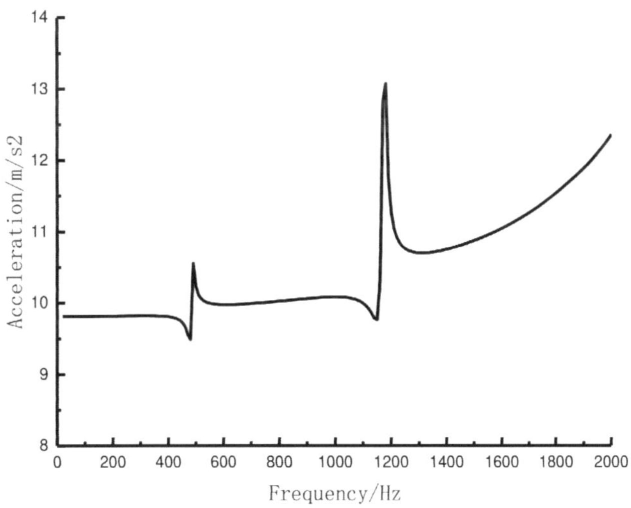

Figure 5. The mean acceleration response curve after optimization (Z direction).

5. Conclusion

Based on the finite element modeling analysis, this article uses the idea of parametric modeling to realize the interaction between MATLAB and NASTRAN, and realizes the application of the improved genetic algorithm on the component layout through programming. The acceleration value at the corners has been reduced, which improves the mechanical environment of the circuit board, prolongs the service life of the electronic device, and improves the overall reliability. The optimization results show that the component layout optimization method adopted is reasonable and feasible, and has practical value and promotion significance.

References

[1] Zhu J Y, Zhou D J and Wu Z H 2007 A study on vibration analysis technique of PCB and componet placement optimization technology *Electro-Mechanical Engineering* **23** 1-4
[2] Zhu J H, Beckers P and Zhang W H 2010 On the multi-component layout design with inertial force *J Comput Appl Math* **234** 2222-30
[3] Zhao S S, Zhang W, Li J and Zhong J H 2017 Mass optimization design of separation devices for pint-size load based on multi-island genetic algorithm *Missiles and Space Vehicles* **2017** 16-9
[4] Ren L, Shao Y Y and Han H 2008 Research and application of multidisciplinary design optimization technology based on iSIGHT *Hoisting and Conveying Machinery* **30** 45-8

[5] Wang L, Pan Z W, Lian Y Z *et al* 2018 Research on H∞ semi-active control of magnetorheological stewart vibration isolation platform *Astronautical Systems Engineering Technology* **2** 41-8

[6] Krishnakumar K and Melkot S N 2000 Machining fixture layout optimization using the genetic algorithm *Int J Mach Tool Manu* **40** 579-98

[7] Yang Z, Lu Q, Yan J *et al* 2017 Multidisciplinary optimization design for the section layout of umbilicals based on intelligent algorithm *J Offshore Mech Arct* **140** 1-12

[8] Holland J H 1975 *Adaptation in Natural and Artificial Systems* (Ann Arbor: University of Michigan Press)

[9] Goldberg D E 1989 Genetic algorithms in search, optimization, and machine learning (New Jersey: Addison-Wesley Professional)

Analysis of deformation characteristics of high-strength reinforced concrete beams under explosion load

Z Liao[1], D G Tang[1], L Z Shao[1], Z Z Li[1] and D P Yang[1,2,3]

[1]State Key Laboratory for Disaster Prevention & Mitigation of Explosion & Impact, The Army Engineering University of PLA, Nanjing 210000, Jiangsu, China
[2]Research Institute for National Defense Engineering of Academy of Military Science PLA, Luoyang 471000, Henan, China

E-mail: ydpheixiami@sina.com

Abstract. The dynamic response and failure characteristics of reinforced concrete beams under explosion loads are affected by various factors. To study the effect of reinforcement strength on the deformation properties of RC beams and to analyze the deformation differences of high-strength RC beams under different explosion loads, comparative explosion tests on two kinds of simply supported RC beams with longitudinal reinforcement yield strength of 400 MPa and 700 MPa were carried out respectively in this paper. The test results show that the deformation of the mid-span section concrete of the high-strength RC beam is basically the same as that of the ordinary RC beam when the peak overpressure value is small, and the strength advantage of the high-strength reinforcement has not been fully exerted. With the increase of explosion load, the maximum tensile strain and compressive strain of concrete at the mid-span section of high-strength RC beams are 46.6% and 43.5% of ordinary RC beams respectively. Under the action of high overpressure explosion load, the deformation of RC beams can be significantly reduced by high-strength reinforcement, which limits the development of cracks.

1. Introduction

The anti-explosion performance of reinforced concrete beams under explosion load has always been a research hotspot in the field of protective engineering [1-3]. To obtain structures with better anti-explosion performance, researchers at home and abroad have carried out a lot of research on a series of new materials and structural forms [4-6], such as ultra-high performance concrete [7,8], steel fiber reinforced concrete [9,10] and new reinforced concrete composite structure [11,12], and have made a lot of achievements. The reinforcement plays a skeleton role in the RC structure. Therefore, the strength index of the reinforcement is of great significance to the deformation and failure of the structure. At present, the use of high-strength reinforcement is an important development direction in the field of both military and civil construction engineering. In the field of civil architecture, some high-rise buildings have made extensive use of high-strength and high-performance reinforcement with yield strength above 600 MPa [13,14].

With the continuous development of RC structure to high resistance, it is difficult for low-strength reinforcement to meet the practical needs. High-strength reinforcement can not only improve the bearing capacity of the structure, but also reduce the consumption of steel and save resources. However, there are few studies on the anti-explosion performance of RC structures with high strength reinforcement under explosion load. Therefore, it is necessary to carry out further research on this issue. In this paper, explosion tests on RC beams with two kinds of strength reinforcement are conducted. The deformation characteristics of high-strength RC beams under different explosion loads are also studied so as to provide reference for the application of high-strength reinforcement in engineering anti-explosion field.

2. Explosion tests

2.1. Specimen design

Specimens of the two types have the same size, both of which are 1700 mm ×150 mm ×300 mm (length ×width ×height), as shown in figure 1. The compressive strength of the concrete is 40 MPa and the thickness of the protective layer of the specimen is 25 mm. The standard strength of stirrup is 400 MPa and the stirrup rate is 0.38%. The standard strength of longitudinal reinforcement of the two types of specimens are different, namely 400 MPa and 700 MPa respectively, but the reinforcement ratio is identical. Before the test, the flexural capacity of the RC beams under uniformly distributed explosive load was estimated, and the peak overpressure of the applied load was determined to be 0.25 MPa and 0.35 MPa. The test conditions are shown in table 1.

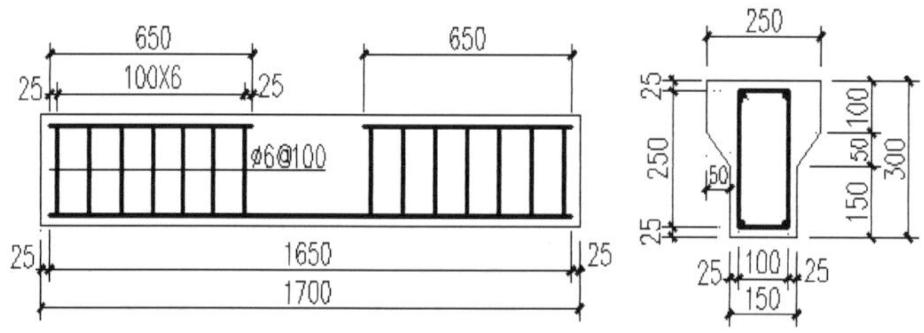

Figure 1. Diagram of specimen and reinforcement layout.

Table 1. Explosion test conditions.

Test	Specimen	Reinforcement strength (MPa)	Peak overpressure (MPa)
1	B400-1	400	0.25
	B700-1	700	
2	B400-2	400	0.35
	B700-2	700	

2.2. Test principle and sensor layout

The tests were carried out in a cylindrical explosion pressure simulator with an internal diameter of 1900 mm, as shown in figure 2(a). The explosion pressure simulator consists of explosion section, transition section, test section and control system, as shown in figure 2(b). The detonating cord is detonated in the explosion section to produce a large amount of detonation products. The detonation product acts on the surface of the specimens through a grid system arranged in the transition section to form a uniformly distributed blast wave. Due to the natural cooling and the leakage action of the reserved vaporous hole, the explosion overpressure in the simulator decreases, and finally forms a long positive duration blast wave. The pressure sensors are symmetrically arranged at both ends of the specimens. The collection and analysis of test data are completed by the DH5922N dynamic signal testing system, as shown in figure 3. It is composed of a variety of front-end measurement systems with different sampling rates, a unified post-processing and operation interface. The system also has good software interface and powerful data processing and analysis function, which provides great convenience for the analysis of test results.

(a)

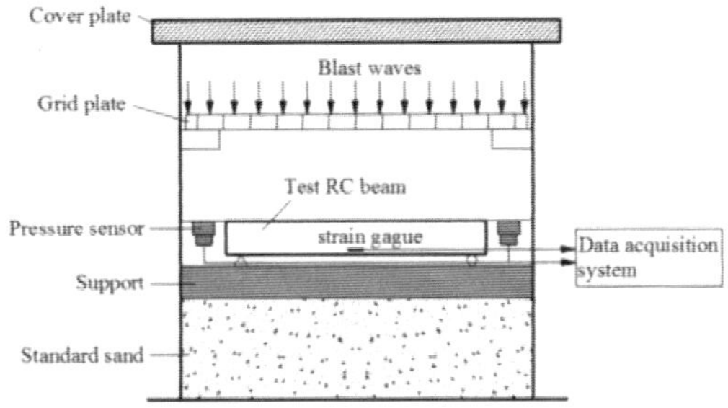

(b)

Figure 2. The diagram of test device. (a) Test setup and (b) Schematic diagram of test.

Figure 3. DH5922N dynamic signal testing system.

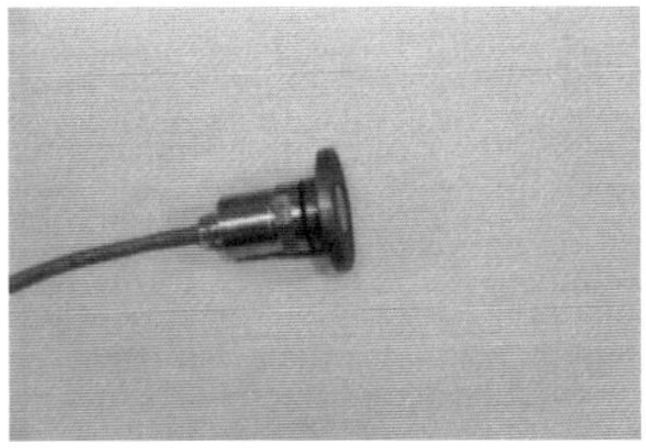

Figure 4. KH1122-PC12B pressure sensor.

The KH1122-PC12B pressure sensor manufactured by Nanjing Wotian Technology Co., Ltd. was used to measure the shock wave load on the specimen, as shown in figure 4. Two pressure sensors of this type were placed in each test and placed at the ends of the specimen. The sensitivity of the pressure sensor was 28.038 mV/MPa.

To study the effect of reinforcement strength on deformation parameters of RC beams under different explosive loads, the strain of concrete was measured with BX120-20AA strain gage, as shown in figure 5. There are altogether 7 strain gauges on the upper surface, side and bottom of the mid-span section of the specimen, which are numbered ε1~ε7. The detailed layout of the strain gauges is shown in figure 6.

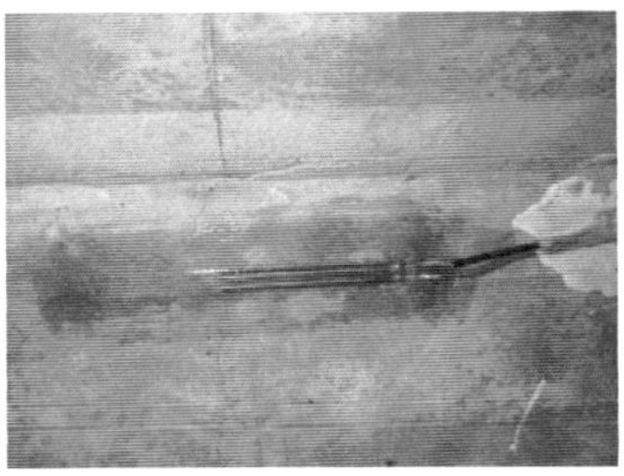

Figure 5. BX120-20AA resistance strain gauge.

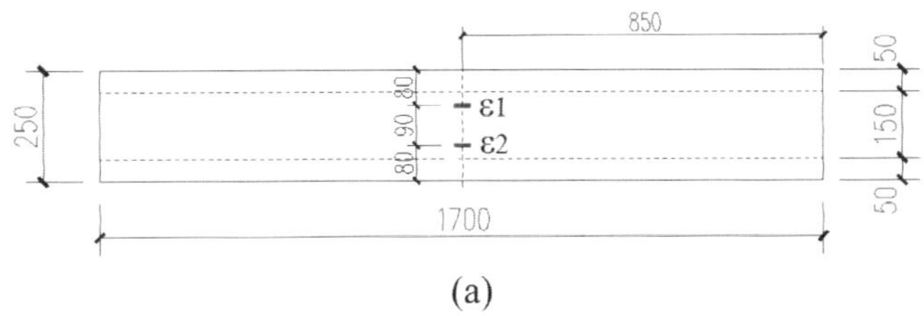

(a)

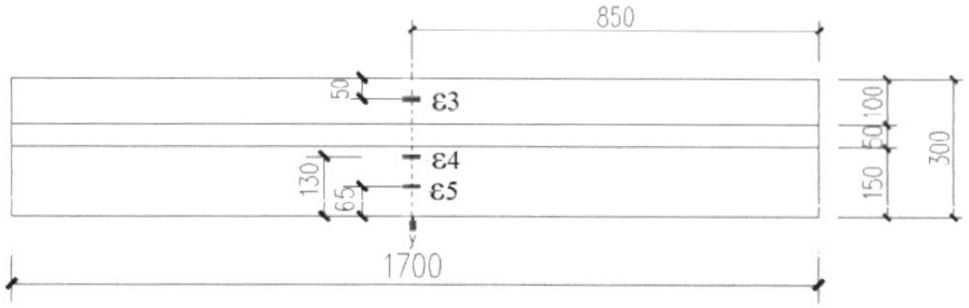

(b)

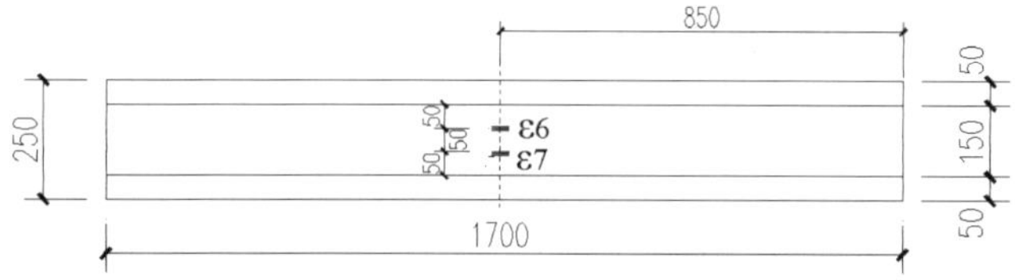

(c)

Figure 6. Layout of resistance strain gauges. (a) Layout of strain gauges on the upper surface of the specimen, (b) layout of strain gauges on the side surface of the specimen and (a) layout of strain gauges on the bottom surface of the specimen.

3. Test results and analysis

3.1. Explosion load

To get more accurate explosion load, the measured values of the two pressure sensors are averaged. The overpressure time-history curves of test 1 and test 2 are shown in figure 7. As shown in figure 7, the explosive loads acting on RC beams reaches the peak overpressure in an instant, and then decreases exponentially. The positive duration time of explosive loads is almost not affected by the peak overpressure value, and the peak overpressure corresponding to test 1 and 2 is 0.245 MPa and 0.372 MPa, respectively.

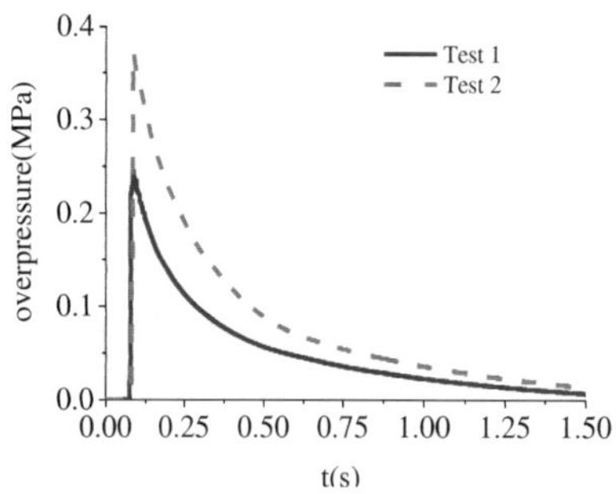

Figure 7. Overpressure time-history curves of explosion load.

3.2. Concrete strain

The RC beam is deformed under uniformly distributed explosive load, and the concrete strain value at the mid-span section directly reflects the deformation degree of the RC beam. Figures 8 and 9 show the strain time-history curves of RC beams at mid-span section with the peak overpressure values of 0.245 MPa and 0.372 MPa respectively. The corresponding maximum strain of the concrete is shown in table 2.

IOP Conf. Series: Materials Science and Engineering **629** (2019) 012004 doi:10.1088/1757-899X/629/1/012004

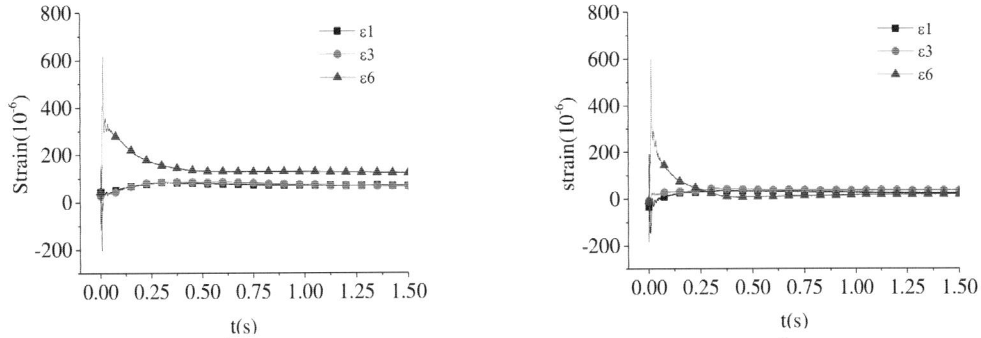

(a) (b

Figure 8. Strain time-history curves of concrete (0.245 MPa). (a) B400-1 and (b) B700-1.

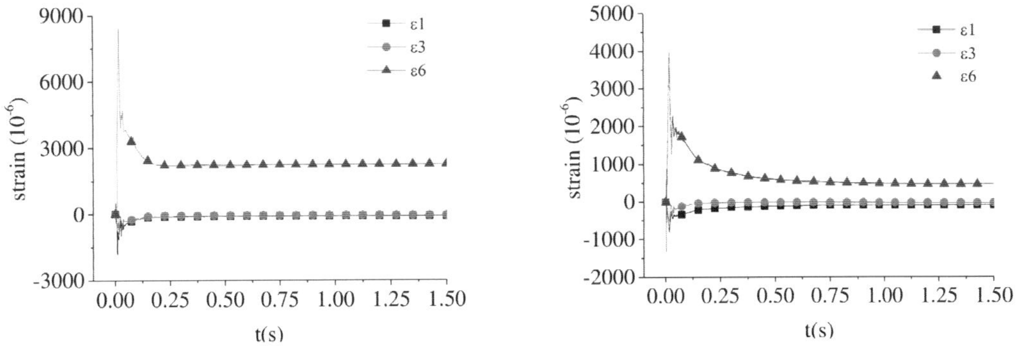

(a) (b)

Figure 9. Strain time-history curves of concrete (0.372MPa). (a) B400-2 and (b) B700-2.

Table 2. Maximum strain of concrete.

Measuring point	Specimen			
	B400-1	B700-1	B400-2	B700-2
$\varepsilon1/10^{-6}$	-202[a]	-147	-1806	-786
$\varepsilon3/10^{-6}$	-126	-121	-881	-520
$\varepsilon6/10^{-6}$	610[b]	594	8409	3921

[a] "-" denotes the compressive strain.
[b] "+" denotes the tensile strain.

It can be seen from table 2 that when the peak overpressure of explosion load is 0.245 MPa, the deformation of the mid-span section concrete of high-strength RC beam is basically the same as that of ordinary RC beam. This is because the deformation of the two RC beams is relatively small. Both the high strength reinforcement and ordinary reinforcement are in the elastic deformation at this time. When the peak overpressure increases to 0.372 MPa, the maximum concrete strain in the tensile zone on the bottom surface of high-strength RC beam is 3921 με, which is only 46.6% of ordinary RC beam (8409 με); The maximum concrete strain in the compressive zone on the top surface of high-strength RC beam is 786 με, which is only 43.5% of ordinary RC beam. The deformation of RC beam increases with the increase of explosive load. When the ordinary RC beams enter the yield stage and produce large deformation, the high-strength RC beams are still in the elastic stage due to the high yield strength of its longitudinal reinforcement. Therefore, high-strength reinforcement can effectively reduce the deformation of RC beams and improve the anti-explosion performance of the component.

4. Summary

In this paper, two kinds of simply supported RC beams with longitudinal reinforcement yield strength of 400 MPa and 700 MPa are tested and analyzed under different uniform explosion loads. The main conclusions are as follows:

- when the peak overpressure of explosion load is less than 0.245 MPa, the strength of longitudinal reinforcement has little influence on the deformation of RC beams, and the deformation characteristics of high-strength RC beam and ordinary RC beam are basically the same.
- when the peak overpressure of explosion load increases to 0.372 MPa, the maximum strain of concrete in tension zone and compression zone of high-strength RC beams is only 46.6% and 43.5% of that of ordinary RC beams, respectively. The concrete deformation of high-strength RC beam is significantly less than that of ordinary RC beam.

References

[1] Zhang D, Yao S J and Lu F 2013 *Eng. Fail. Anal.* **33** 173-214
[2] Acito M, Stochino F and Tattoni S 2011 *AMM* **82** 434-9
[3] Kalateh F 2018 *Front. Struct. Civ. Eng.* **3** 1-23
[4] Thomas R J, Steel K and Sorensen A D 2018 *Eng. Struct.* **168** 838-51
[5] Li J, Wu C and Hao H 2017 *Int. J. Impact. Eng.* **110** 242-54
[6] Stewart M G 2018 *Struct. Saf.* **71** 13-23
[7] Wu C, Oehlers D J and Rebentrost M 2009 *Eng. Struct.* **31** 2060-9
[8] Millard S G, Molyneaux T C K and Barnett S J 2010 *Int. J. Impact. Eng.* **37** 405-13
[9] Drdlová M, Buchar J and Krátký J 2016 *Struct. Concrete* **16** 508-17
[10] Li J, Wu C and Hao H 2016 *JSIM* **1** 51-9
[11] Fan J Q, Xin K and Song H M 2011 *Journal of Experimental Mechanics* **26** 329-36
[12] Yang R L and Guo H 2011 *AMR* **253** 2866-71
[13] Rong X, Duan W W and Wang H 2017 *J. Civil. Archi. Environ. Eng.* **39** 148-54
[14] Gong W, Hu K X and Wang Y D 2017 *Journal of Hebei University of Engineering (Natural Science Edition)* **34** 6-11

IOP Conf. Series: Materials Science and Engineering **629** (2019) 012005 doi:10.1088/1757-899X/629/1/012005

Investigation into the differences of pull-out resistance between normal and osteoporotic cancellous bone

C Liu[1], J B Guo[2] and Y H Yan[1,3]

[1]School of Mechanical Engineering & Automation, Northeastern University, Shenyang, Liaoning 110819, P. R. China
[2]School of Mechanical Engineering, Shijiazhuang Tiedao University, Shijiazhuang 050043, P. R. China

E-mail: yanyh@mail.neu.edu.cn

Abstract. Pedicle screw loosening or pullout has occurred from time to time in patients with poor bone quality. However, there have been no rigorous mechanical analysis of the pullout process, and the resistance region to pullout force remains under-investigated. The objective of the current study was to investigate the stress state of cancellous resistance to pullout in normal and osteoporotic conditions using plane finite element method. The pull-out model was simplified to a simpler two dimensional axisymmetric model by symmetry. A prescribed axial displacement of 2 mm was imposed on the screw head and a fixed boundary condition was set on edge nodes of the bone to simulate the pullout process. The results of numerical simulation showed that the high-stress area of normal bone was clearly larger than that of osteoporotic bone, and the normal bone provided the higher pullout force of 86.09 N, which was about 3 times than that for osteoporotic bone. From the study it was found that osteoporosis would diminish the active region of resistance to pullout around the screw and decrease the pullout strength about threefold. The results of the current study can be used for screw augmentation techniques.

1. Introduction

Spinal surgery has been significantly improved by using pedicle screws. These screws are placed into cancellous bone to provide rigid fixation and load transmission. Screw pullout strength is closely associated with material properties of cancellous bone, such as bone density, elastic modulus, plastic yield stress and strain, and so on. Therefore, pedicle screw loosening or pullout has occurred from time to time, especially in patients with poor bone quality [1]. Orthopaedic surgeons often face clinical situations where improved screw holding power in cancellous bone is needed. However, the enhancement of screw fixation strength in osteoporosis patients is challenging. It is essential to study the stress state during pullout process in cancellous with different bone properties.

Researchers have long been interested in the investigation of fixation strength of pedicle screw. The holding power of a screw is directly influenced by the number of threads and their depth and for a given length of the screw, and screws fail most commonly on the bone side of the bone screw interface [2,3]. Tests using polyurethane foam blocks with different mechanical properties reported by Ramaswamy *et al* [4] suggest that the fixation strength of screw is directly correlated to bone density, thread design and number of threads engaging the bone. To overcome difficulties of fixing screws in poor quality cancellous bone, the injectable bone cement was used in clinic. Various biomechanical tests and clinical studies have confirmed that an appropriate use of cements can generally significantly increase the pullout forces of bone screws and improve stability [5-8]. In addition, other studies have investigated the effects of screw orientation to the axis of pullout in osteoporotic bone [9,10]. To date, work on screw pullouts has focused mainly on pull-out tests using cadaveric and animal bone and synthetic materials. There have been no rigorous mechanical analysis of the pullout process, and the resistance region to pullout force remains under-investigated.

The objective of this study was to investigate the resistance behavior of cancellous bone to screw pullout. The maximal axial pull-out strength and stress distribution in cancellous bone with normal and osteoporotic properties were predicted and expected to help understand the mechanical behavior of the pullout process. The findings are intended to determine the resistance region of bone around the screws for application of screw augmentation techniques.

2. Methods

In order to accurately and visually determine the resistance region with stress distribution, plane finite element method was used to simulate the pullout process.

2.1. Simplified finite element model

The screw model is based on the universal lumbar pedicle screw manufactured by Weigao Orthopaedic Materials Co. LTD (5.5 mm major diameter, 32 mm thread length, 2.75 mm thread pitch).The cancellous bone block was generated by entity model (50 mm×50 mm×37 mm). According to symmetry, it was rational to simplify the pull-out model to a simpler two dimensional axisymmetric model (figure 1). There's little research on the mechanical property of cancellous bone. Earlier reports said cancellous bone had showed viscoelasticity and rate-dependence [11], but the depth of its theoretical research is still far from enough. In the current research, an elastoplastic model was employed to describe the stress-strain behavior of the cancellous bone under the axial pullout load. Material parameters associated with the stress-strain behavior of the cancellous bone, in its normal and osteoporotic conditions, are grouped in table 1 [12].Young's modulus (E) and the plastic yield stress (σp) and strain (εp) were deduced from density values based on empirical relations [13,14].

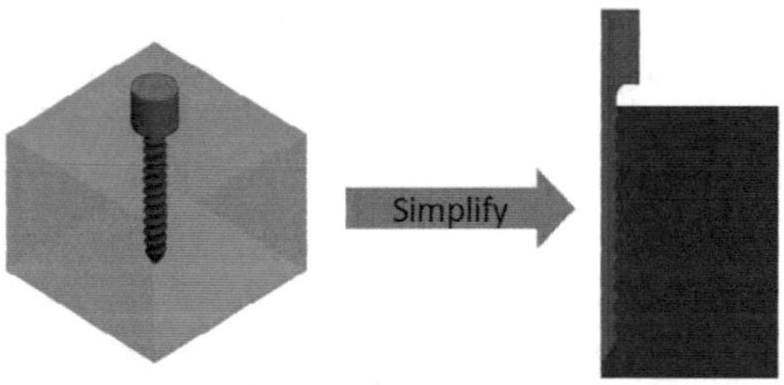

Figure 1. Model simplification of pullout.

Table 1. Material properties for normal and osteoporotic bone.

	ρ(g/cm^3)	E(MPa)	ν	σ_p	$\varepsilon_{p(\%)}$
Normal bone	0.2	100	0.2	3	4
Osteoporotic bone	0.13	75	0.2	1.5	4

2.2. Mesh generation and boundary conditions

The linear axisymmetric 4-node quadrilateral elements were opted in the mesh partition. Simplification of the finite element model and selection of the element type were based on a previous literature [15]. Mesh generation characterized by a minimal element size of 0.2 mm was implemented, leading to a total number of nodes and elements of 15,753 and 15,631, respectively (figure 2). The grid quality items that has a greater impact on the calculation were checked, such warpage, aspect ratio and jacobian. The comprehensive quality indicator value (comp.QI) was 58.01 and the poor quality elements ratio was

IOP Conf. Series: Materials Science and Engineering **629** (2019) 012005 doi:10.1088/1757-899X/629/1/012005

0.2%, largely because of the threads form. In consideration of no relative displacement between screws and bone in screw fixation, the screw-bone interfaces were defined binding constraints. To simulate the pullout condition, a prescribed axial displacement of 2 mm was imposed on the screw head via a multipoint constraint coupling condition, and a fixed boundary condition was set on edge nodes of the bone. The finite element calculation is performed by the default Abaqus standard solver after pre-processing.

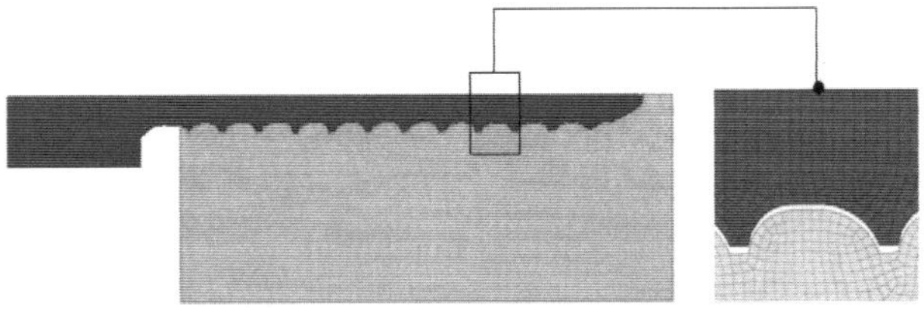

Figure 2. Finite element mesh generation.

3. Results

The stress nephograms of cancellous bone in normal and osteoporotic conditions were depicted in figure 3, showing that the bone quality had cast obvious impact on the resistance to pullout. The high-stress area of normal bone showed clearly larger than that of osteoporotic bone. The maximum shear stress resistance to pullout in normal bone was 3.95 MPa, while the value is 2.0 MPa in osteoporotic model. The load-displacement curves showed that the normal bone provided the higher pullout force of 86.09 N, and that for osteoporotic bone was 26.9 N (figure 4).

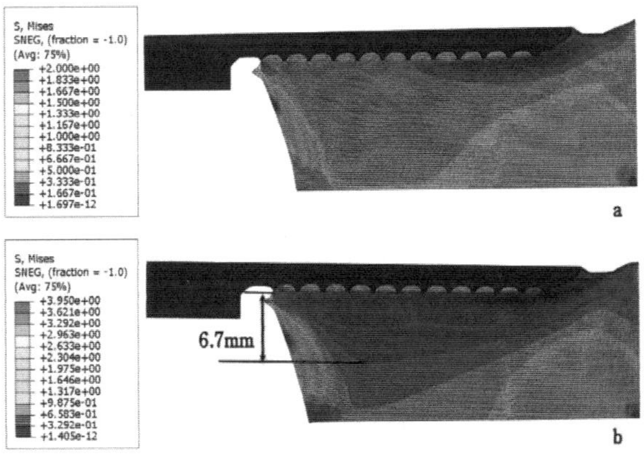

Figure 3. Stress distribution in (a) osteoporotic and (b) normal bone.

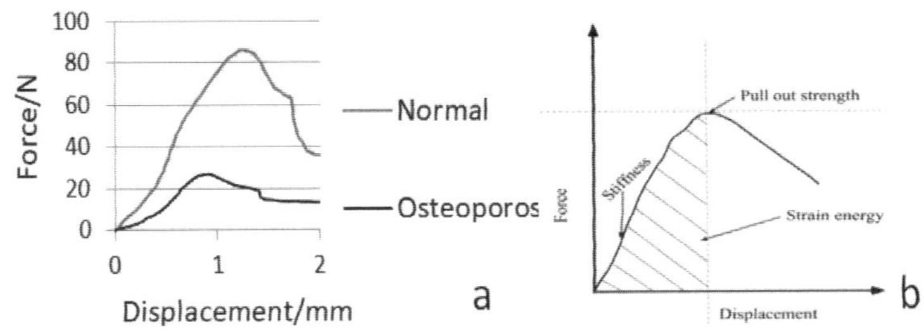

Figure 4. The load-displacement curves. (a) Result of the present study and (b) experimental result by Varghese.

4. Discussions

Cancellous bone plays a major role in the holding power of pedicle screw fixation. The magnitude and range of resistance will change with the change of bone quality. Patients needing internal fixation surgery often have the varying degree of the osteoporosis. This study used finite element technique to simulate the effect of changes in bone quality on the holding power qualitatively. To the best of the authors' knowledge, no previous study has attempted to investigate the matter by numerical calculation method.

This paper studied the resistant performance of cancellous bone to screw pullout in normal and osteoporosis conditions. The compacting structure of normal bone contributed more bone tissue to resist screw pullout. The high stress area spread to the intact screw channel, and the radial range affected by high stress was 6.7 mm. In comparison, the osteoporotic bone showed the weak ability to form solidity. The high stress area was rarely distributed in the area of the first screw thread and screw tip. The exact scope of resistance defined by the stress distribution results can provide adjustment reference for the application of technique in screw augmentation, such as the valid range of cement injection.

Pullout strength is a persuasive biomechanical parameter to demonstrate the screw fixation strength. The load-displacement curves showed that the pullout force reached the peak value first in osteoporosis model. The main reason for this was that osteoporosis decreased the deformation resistibility capacity. For our simplified two-dimensional axisymmetric model, the resultant pullout values were 86.09 N and 26.8 N for normal and osteoporotic bone respectively. As might be expected, the calculation results were far less than researches done by others. Pullout tests in which polyurethane foam with varying-density was used to represent normal and osteoporotic bone were performed by Varghese *et al* [11], the obtained mean pullout values of the foam model were 1100 N and 300 N for normal and osteoporotic bone respectively. From the above, it can be seen that the pullout strength for normal bone was more than 3 times for osteoporotic bone in the current research as well as in Varghese's tests. Besides, the computed load-displacement curves in our research had followed the same trend in comparison with the experiment results by him (figure 4). Therefore, using cement augmentation to increase bone density in osteoporotic patients can improve screw fixation, and consequently improve surgical outcome.

5. Conclusion

From the study it was found that bone quality is an important factor in pedicle screw fixation. Osteoporosis will diminish the active region of resistance to pullout around the screw and decrease the pullout strength about threefold. The results of the current study can be used for screw augmentation techniques.

References

[1] Chingchi H, Chingkong C and Jawlin W 2010 *J Orthop Res.* **23** 788-94
[2] Thompson J D, Benjamin J B and Szivek J A 1997 *Clin Orthop Relat Res.* **341** 241-9
[3] Collinge C A, Stern S and Cordes S 2000 *Clin Orthop Relat Res.* **373** 277-84
[4] Ramaswamy R, Evans S and Kosashvili Y 2010 *Injury.* **41** 179-83
[5] Dall'Oca C, Maluta T and Moscolo A 2010 *Injury.* **41** 1150-5
[6] Lee P C, Hsieh P H and Chou Y C 2010 *J Neurotraum.* **68** 954-64
[7] Kammerlander C, Gebhard F and Meier C 2011 *Injury.* **42** 1484-90
[8] Larsson S, Stadelmann V A and Arnoldi J 2012 *J Biomech.* **45** 1156-60
[9] Zehnder S, Bledsoe J G and Puryear A 2009 *Clin Biomech.* **24** 589-94
[10] Patel P S, Shepherd D E and Hukins D W 2010 *Med Eng Phys.* **32** 822-8
[11] Linde F and Hvid I 1987 *J Biomech.* **20** 83-9
[12] Nicholson P H, Müller R and Lowet G 1998 *Bone.* **23** 425-31
[13] Varghese V, Saravana K G and Krishnan V 2015 *Medical Engineering and Physics.* **40** 28-38
[14] Linde F and Sorensen H C 1993 *J Biomech.* **26** 1249-52
[15] Beckert W and Lauke B 1996 *Comp Mater Sci.* **5** 1-11

IOP Conf. Series: Materials Science and Engineering **629** (2019) 012006 doi:10.1088/1757-899X/629/1/012006

Experimental investigation of the road performance of terminal blend rubber asphalt mixture with different dry-way additives

Q B Huang[1], Z D Qian[1,2], M Zhang[1] and Y C Xue[1]

[1]Intelligent Transport System Research Center, Southeast University, 2 Sipailou, Nanjing, Jiangsu 210096, China

E-mail: qianzd@seu.edu.cn

Abstract. Using wet-way composite modification method may result in the amount of ground tire rubber (GTR) in terminal blend rubber asphalt (TB) undetectable. So, it is of great interest to find out if dry-way modification method can substitute for wet-way composite modified asphalt to address the shortcomings of TB mixture. A laboratory experiment was developed for performance evaluation of TB asphalt mixtures with dry –way use polymer additives. The results showed that: (1) 5 hours was recommended as optimum processing time, since no obvious difference existed after TB asphalt modifying for 5 h and 7 h. (2) The incorporation of anti-rutting agent PLAST.S, GTR, polyester fiber (PF) and styrene-butadiene-styrene block copolymer (SBS) improved the rutting stability, resistance against low-temperature cracking, moisture stability and fatigue life of TB asphalt mixture, especially when PLAST.S and PF were used. (3) The use of GTR and PF brought obvious increase of mixing torques, resulting in more challenges to the workability of TB asphalt mixtures, whereas mixture with PLAST.S did not have a statistically significant difference with TB asphalt mixture. Therefore, according to the results obtained, PF can be used to replace TB/SBS composited modified asphalt since the two shared the most similar improvement of the properties of TB asphalt mixtures, while PLAST.S and GTR were also considered as good choices for modifying the TB asphalt mixture in dry-way.

1. Introduction

Recently, a new kind of asphalt rubber named terminal blend rubber asphalt (TB) has been recognized as a promising alternative to the traditional one because of its construction connivance [1-3]. The production is a smooth and homogeneous asphalt processed at temperature of 210°C or higher for 16 h~24 h, ensuring the ground tire rubber (GTR) particles in asphalt matrix are totally depolymerized. Therefore, TB asphalt can be transported to an asphalt concrete mixing plant or a job site and stored in the storage tank without agitation just as the polymer modified bitumen PMB [2]. Taking advantage of compatibility and homogeneousness in asphalt matrix, TB asphalt shows many benefits at pumping, mixing, storage stability, which make it more suitable for asphalt plant to produce. What's more, it shows great advantage in the application for dense graded mixture when compared with traditional asphalt rubber (AR) [4-6]. On the other hand, due to the desulfurization of GTR in asphalt, the reduction of rubber elasticity tends to slightly weaken the high temperature rutting resistance of the asphalt pavement [7,8].

Currently, several techniques are used to make up for the reduction of elasticity of TB asphalt. One of the most widespread techniques is adding polymers to the TB asphalt during the modifying process [9-11]. Unfortunately, after introducing the polymers into TB, such as styrene-butadiene-styrene block copolymer (SBS), the viscosity of asphalt increases with an obvious amplitude which makes more challenges for pavement construction as traditional AR faces. In addition, wet-way composite modification method cost more time and money, and sometimes precipitation problem in asphalt appeared [12-14]. Besides, some researchers [14] suspected that when using wet-way composite modification method, some suppliers are probably to add either less GTR or more polymers to meet the performance requirement, which makes the exact amount of GTR undetectable. At the same time, incorporating polymer additives directly to the asphalt concrete mixer drum with plain TB asphalt (dry-way) is easy handing and economical since it requires neither the storage tank nor pumping equipment remoulded.

However, it is unclear whether the TB asphalt mixtures with dry-way polymer additives have the similar performance to TB/SBS asphalt mixture. Therefore, it's of great interest to conduct a comprehensive study on the effect of dry way polymer additives on TB asphalt mixtures.

2. Objective

The objective of this paper is to determine whether dry-way use polymer additives can substitute for TB/SBS asphalt to address the shortcomings of plain TB asphalt mixture. Processing time for plain TB asphalt was determined through conventional performance tests and fluorescence microscopy test, the engineering properties of TB asphalt mixtures were evaluated through high temperature wheel tracker test, low temperature three point beam flexure test, Marshall immersion test, modified Lottman tests, three point bending fatigue test, and mixture workability test.

3. Materials and testing methods

3.1. Materials

In this investigation, 90# base asphalt was used for modifying TB asphalt. Crushed gneiss aggregate were brought from a local quarry in Shaanxi Province in China. Three kinds of polymer additives used in dry way modification were selected, as shown in figure 1.

(a) (b) (c) (d)

Figure 1. Polymer additives (a) GTR powder; (b) coarse GTR; (c) PLAST.S (d) PF.

The GTR powder was homogenously ground at ambient temperature without particles lager than 40 mesh (69 μm). The coarse GTR was obtained from the same way, which was utilized to replace some fraction of aggregates within the size range from 0.3 mm to 4.75 mm. The selected GTRs were shown in figures 1(a) and 1(b). A commercial anti-rutting agent named PLAST.S, mainly composed of recycled polymer [15], was used to mainly improve rutting resistance of TB asphalt mixture. A commercial polyester fibber mainly comprised of polyethylene terephthalate was employed, as shown in figure 1(d).

3.2. Preparation of TB asphalt

During the process, the 190°C base asphalt was mixed with 20% fine GTR powder using a high speed stirring mixer with a speed of 4500 rpm. A kind of catalyser was introduced into the asphalt right after the GTR powder incorporation step, which aimed to accelerate the degradation speed of GTRs in asphalt and lower the process temperature. To determine the economical and effective mixing time, TB asphalt modifying 1 h, 3 h, 5 h and 7 h respectively (TB1, TB3, TB5, TB7) was evaluated through conventional performance tests. A composite-modified TB asphalt (TB/SBS) was manufactured by adding 3.0% SBS into the prepared TB5 asphalt matrix and shearing with high speed for 0.5 h at 180°C [7].

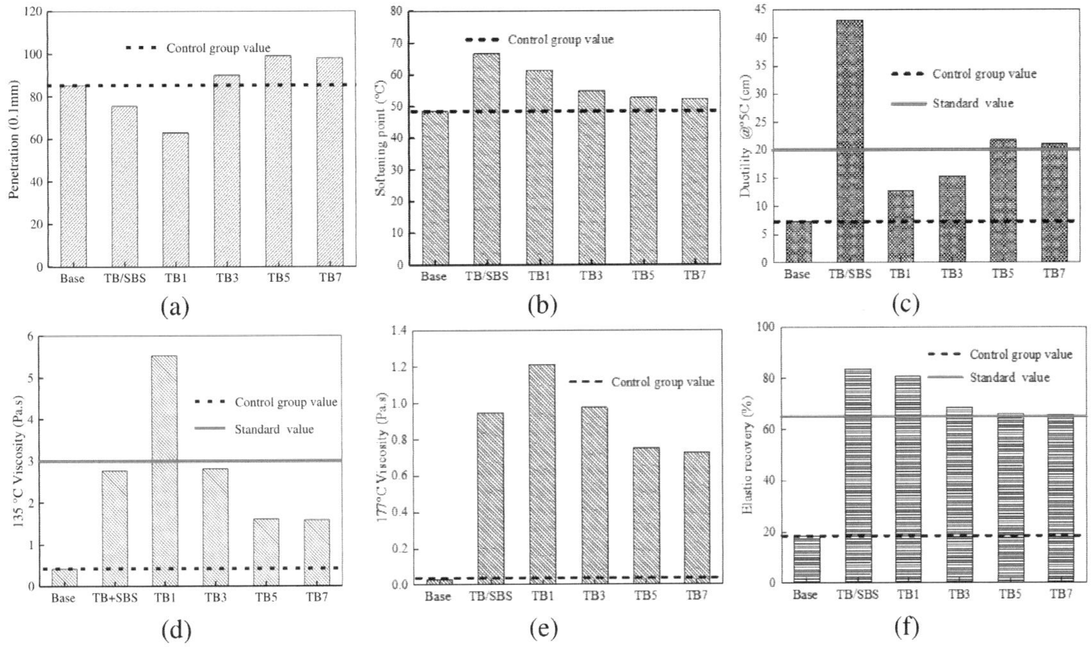

Figure 2. Convention performance test results: (a) 25°C Penetration, (b) Softening point, (c) 5°C Ductility, (d) 135°C Viscosity, (e) 177°C Viscosity, (f) Elastic recovery.

(a) (b) (c)

Figure 3. Fluorescence micrographs: (a) TB0 (magnification: X40); (b) TB5 (magnification: X100) (c) TB7 (magnification: X100).

Figure 2 shows that after a drastic change in the first 3 h modification, no obvious differences existed between the properties of TB5 and TB7. It means that all conventional engineering properties of TB asphalt achieved stability after a high speed of mixing for 5 hours. The fluorescence micrographs of TB5 and TB7 in figure 3 show the GTR particles had broken down into extremely small size and remained homogeneously in liquid phase of asphalt after modifying 5 h, while large rubber particles isolated in the TB0 matrix during the first period of modification. As a result, the TB5 was selected for further TB asphalt mixture investigation based on less time consumption in production.

It can be seen from the figure 2 that all conventional performances for TB asphalt were promoted by the addition of SBS except the kinematic viscosity at 135°C and 177°C, while still met the 3.0 Pa·s limitation. At the same time, the TB5 asphalt had considerably higher ductility and elastic recovery values as compared to the 90# base asphalt. This indicated the low temperature performance of base asphalt was improved by the fully degraded GTR polymer. In addition, the TB5 had a higher softening point value than that of 90# base asphalt, which means the high temperature performance of base asphalt was also improved. The 135°C viscosity value of TB5 was much higher than that of base asphalt, but far below the maximum viscosity limitation of 3.0 Pa·s for pump ability demand issued by CALTRANS [16]. It can be deduced that terminal blend asphalt produced in this study has a good temperature performance and shows great construction possibility.

3.3. Mixture design

Mixture design was carried out by using Marshall design method with a target air void of 4.0%. Based on the research of predecessor that TB asphalt was more suitable for dense graded mixture than the gap graded one [5], a dense asphalt concrete (DAC) gradation with the 13.2 mm nominal maximum aggregate size was selected. The synthesis aggregate gradations of all TB type mixtures were kept the same throughout the whole lab evaluation of TB asphalt mixture, as detailed in table 1. The experimental dosages of PLAST.S and PF were 1.5‰, 3.0‰ and 4.5‰ by weight of the aggregate, while the experimental dosages of GTR were 1%, 2% and 3% by weight of the aggregate.

Table 1. Synthesis aggregate gradation of mixtures.

Sieve (mm)	13.2	9.5	4.75	2.36	1.18	0.6	0.3	0.15	0.075
Passing (%)	96	76	47	30	20	16	11	9	6

The higher value of Marshall Stability is, the better high temperature rutting resistance the asphalt mixture will get. According to the test results, the recommended dosages of GTR and PF were 1% and 3.0‰, because TB asphalt mixture obtained the highest Marshall stability values at these dosages respectively, while 3.0‰ was selected for PLAST.S under the economical consideration since there is no significant improvement after the dosages increased from 3.0‰ to 4.5‰.

3.4. Pavement performance testing method

Wheel tracking test was performed at 60°C to evaluate the high temperature rutting resistance of the TB asphalt mixture by the dynamic stability (DS). The low temperature cracking resistance of TB asphalt mixtures were evaluated using a three-point flexural beam test at -10°C by the failure flexural strain and failure strain energy. Marshall Immersion test and modified Lottman test were performed to evaluate propensity of the asphalt mixtures to moisture induced damage by the residual Marshall stability (MS0) and the residual tensile strength (TSR), on the Marshall specimens. All procedures of the aforementioned tests follow the standard test methods in China test specification JTG E20-2011 [17]. The mixture workability test was performed to look into the work yield on placing asphalt mixture using an asphalt mixture workability device (AWD), which was according to the NCAT's research [18]. The torques, needed to keep a propeller rotating with a constant rate of 45 rmp in hot loose mixture weighing 4 kg, were measured. Asphalt mixtures with lower torque values were considered easier to be placed or had better workability.

4. Results and discussions

4.1. High temperature rutting resistance

Wheel tracking test was performed at 60°C to evaluate the high-temperature deformation resistance of mixtures. Figure 4 showed the TB asphalt mixture had the lowest dynamic stability (DS) and highest rut depth (RD) among the five mixtures tested, which indicated the rut resistance of TB asphalt mixture could be improved through the composite modified asphalt and dry-way polymer additives. Furthermore, mixture with TB/SBS asphalt, PLAST.S, GTR and PF were acceptable for use in any road without limitation in consideration of rutting resistance, since all of their DS met the supreme specification requirement of 3000 (cycle/mm) [19]. Mixture with PLAST.S exhibited the best rutting resistance with the minimum rut depth of 1.672 mm, and the PF could reduce the rut depth of the plain TB mixture (2.136 mm) by 14%, while coarse GTR reduce by 4%. It is clearly presented that the ranking of modification effect on mixture rut resistance was PLAST.S> TB/SBS asphalt > PF> GTR, that is, the TB asphalt mixture modified with PLAST.S through dry-way process had the best anti-rutting property, followed by TB/SBS asphalt and dry-way PF, and dry-way GTR in the least.

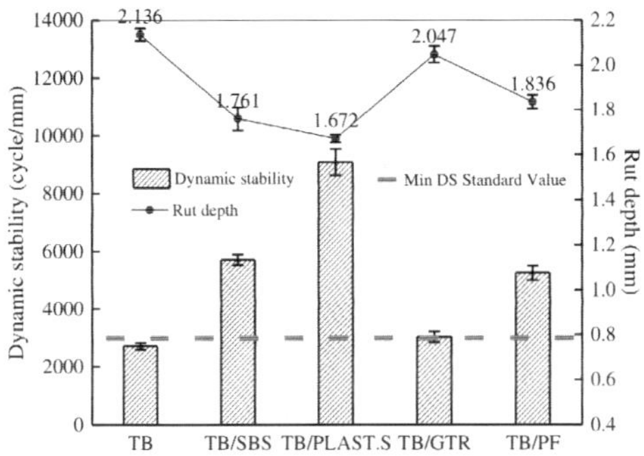

Figure 4. Wheel tracking test results.

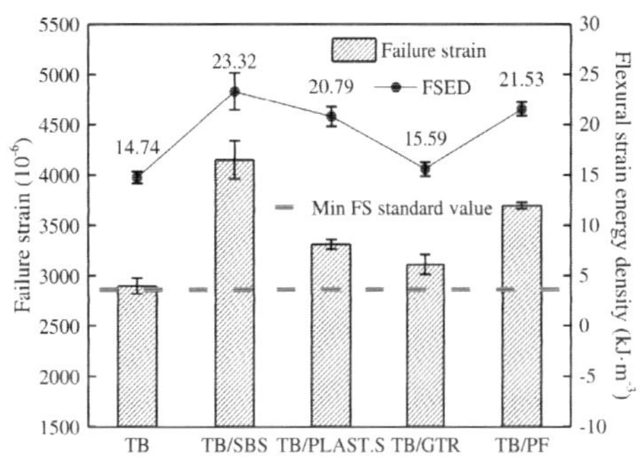

Figure 5. Three point bending beam test results.

4.2. Low temperature cracking resistance

The evaluation of low-temperature cracking resistance of the mixtures was carried out by using three point bending beam tests at -10°C. Figure 5 summarized the average values for failure strain (FS) and flexural strain energy density (FSED).

According to JTG E20-2004 [19], asphalt mixture is considered suitable for cold winter region in China when the failure strain value of three point bending beam is higher than criteria of 2800 μm, as represented by the red dash line in figure 5. It can be noted that each mixture type had great anti-low temperature cracking property since all the failure strain results passed the red dash line. Results for both of the failure strain and flexural strain energy density revealed that incorporation of either TB/SBS asphalt or dry-way additives can improve the low temperature cracking resistance of TB asphalt mixture. The PF helped to increase the FESD of the plain TB asphalt mixture from 14.74 kJ/m^3 to 21.53 kJ/m^3, and increased the low temperature flexural strain by 27%. While the anti-rutting agent PLAST.S and GTR helped to increase the low temperature flexural strain of the plain TB asphalt mixture by 14%, 7% respectively. As compared to the plain TB asphalt mixture, the mixtures with TB/SBS asphalt performed the best, followed by TB asphalt mixture blended with additives PF in dry-way in the second position, the anti-rutting agent PLAST.S in the third and GTR in the last position.

4.3. Moisture susceptibility

Marshall Immersion test and modified Lottman test were conducted to evaluate the moisture damage resistance of the asphalt mixtures. The test results were presented in figure 6.

The average values of Marshall Stability (MS) for all the mixtures in unconditioned group were around 10 kN, and the unconditioned indirect tensile strength (ITS) remained around 1.0 MPa, which were similar to previous findings [20]. It can be seen that the TB/SBS asphalt, PLAST.S and PF showed excellent benefits to improve the structure strength of TB asphalt mixture, since asphalt mixture with the modifier mentioned had much higher MS and ITS values than that of the plain one. While, the addiction of coarse GTR impaired the structure stability of the TB asphalt mixture. The reason probably could be the decrease in the contact strength of mineral skeleton caused by replacing stone-stone structure with stone-coarse GTR-stone structure.

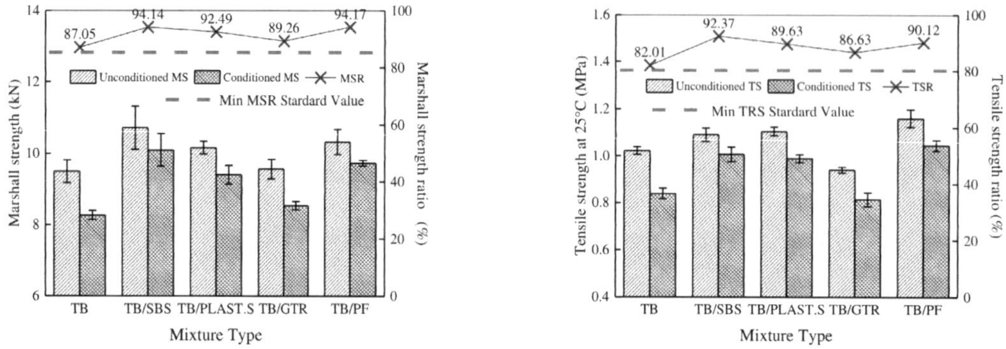

(a) (b)

Figure 6. Moisture susceptibility test results: (a) Marshall Immersion test, (b) Modified Lottman test.

On other hand, Marshall Stability (MS) and indirect tensile strength (ITS) of all asphalt mixture suffered declines after the environment simulation effect. All the asphalt mixture presented good moisture resistances since MSR and TSR values were all higher than the minimum specifications required according to JTG F40-2004 [19], namely 85% for MSR, 80% for TSR. TB/SBS composite asphalt increased the MSR and TSR values of the plain TB asphalt mixture by 7.09% and 10.36%.The PF helped to increase the MSR and TSR of the plain TB asphalt mixture from 87.05% to 94.17%, 82.01% to 90.12% respectively. While the anti-rutting agent PLAST.S and GTR helped to increase 5.44% and 2.21% for MSR of the plain TB asphalt mixture, 7.62% and 4.62% for TSR respectively. As a result, composite-modified TB/SBS asphalt was the most effective way to improve the moisture damage resistance of TB asphalt mixture, followed by PF, PLAST.S and GTR in order.

4.4. Fatigue cracking resistance

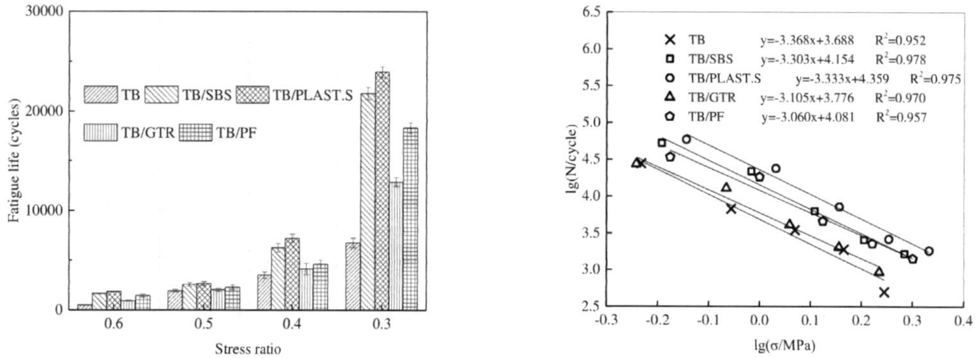

(a) (b)

Figure 7. Three point bending fatigue test results: (a) Fatigue life, (b) Linear regression formula.

The evaluation of fatigue cracking resistance of the mixture was performed by employing three point bending fatigue test in stress controlled mode. The curves of fatigue life and fatigue stress for each mixture, generated by test values of fatigue life (Nf) at each stress ratio, were shown in figure 7. The regression equations of fatigue curve complied with a linear relationship in denary logarithm, in which K was the slope and b was the intercept. It can be seen that TB asphalt mixture had the largest K and smallest b among the five mixtures tested, which indicated the mixture had the highest fatigue sensitivity and would result in the shortest fatigue life (in cycles) at the same applied load respectively. Conversely, the fatigue stress required to promote a crack for TB asphalt mixture was lowest for equal fatigue lives [21]. Taking the stress ratio of 0.3 as an example, mixture with PLAST.S exhibited the longest fatigue life, and the TB/SBS asphalt increased fatigue life (Nf) of the plain TB mixture by 222.1%, while PF coarse GTR increased Nf by 170.6% and 90.3% respectively. This suggested that the incorporation of composite-modified TB/SBS asphalt and dry-way additives PLAST.S, GTR and PF improved the fatigue life of TB asphalt mixture. And the rank of modification effect was PLAST.S> TB/SBS asphalt> PF> GTR, that is, the TB asphalt mixture with PLAST.S had the best anti-fatigue cracking property, followed by TB/SBS asphalt and PF, and with GTR backmost.

4.5. Workability performance
For workability evaluation, asphalt workability device was employed to test the mixing torque of each loose mixture. Figure 8 depicted the average values of mixing torque obtained in the test. It can be noted that each loose asphalt mixture showed lower mixing torque when the test temperature was increased from 135°C to 177°C. On the other hand, an obvious increase of mixing torque in TB asphalt mixture was observed after the incorporation of TB/SBS asphalt and dry-way additives except the PLAST.S, which slightly reduced the mixing torque of the TB asphalt mixture. It clearly proved that the TB asphalt mixtures with PLAST.S had better workability than that with TB/SBS asphalt or with other dry-way additives. This was probably because the TB/SBS asphalt had higher viscosity than that of TB asphalt, and at the same time, the intact GTR and PF increased the viscosity of TB asphalt for absorbing large amount of asphalt, while the PLAST.S melted completely with a viscosity far below the TB asphalt at the mixing temperatures, and consequently, lowest torque.

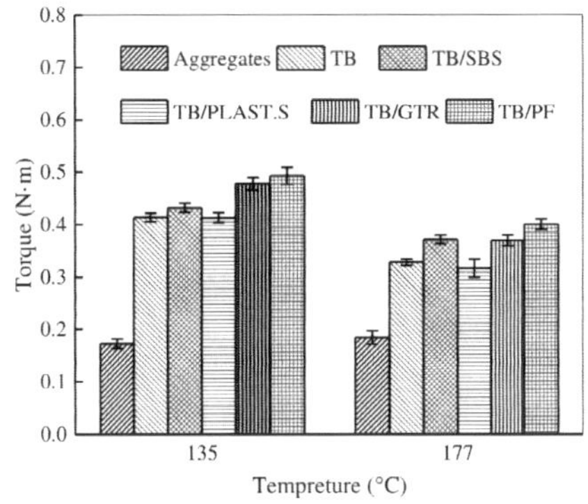

Figure 8. Asphalt mixture workability test results.

5. Conclusion
Dry-way additives directly mixed with heated aggregate before the incorporating TB asphalt was an easier way to produce asphalt concrete when comparing with wet-way production, since wet-way production faced more challenges in TB composite asphalt processing, pumping and asphalt mixture mixing. Based on the comprehensive laboratory testing results, the following conclusions can be summarized:

- Based on conventional performance results and morphologies in fluorescence micrographs, the optimum laboratory processing time at 190°C with catalyst for terminal blend rubber asphalt was determined to be 5 hours.
- Mixture with PLAST.S exhibited greater dynamic stability (DS) and longer fatigue life but higher sensitiveness to low-temperature cracking and moisture damage than that of mixture with TB/SBS composite asphalt, while still increased 14% low temperature flexural strain and 7.62% TSR as compared to the plain TB asphalt mixture. No statistical difference existed in workability between the mixture with PLAST.S and the plain one.
- The positive effect of GTR incorporated was far junior to TB/SBS composite asphalt, while still slightly increased the resistance against rutting deformation (4%), low-temperature flexural strain (7%), TSR (4.62%) and fatigue life (90.3%) of the plain TB asphalt mixture. Besides,

mixture with GTR had the highest mixing torque, indicating a worst workability.

- PF got the most similar effect to TB/SBS modified asphalt on enhancing road performances of mixture, since the incorporation of PF increased 14% the resistance against rutting deformation, 27% low-temperature flexural strain, 8.11% TSR and 170.6% fatigue life of the original mixture, which were close to 17%, 43%, 10% and 13% of improvements respectively provided by TB/SBS composite asphalt. Although mixture with PF faced more challenges in mixing.

It seems more practicable to use PF as surrogates to TB/SBS composite asphalt to overcome the defect of plain terminal blend rubber asphalt mixture in high temperature stability, and still provide other satisfactory performances. The comprehensive analysis of lab tests still shows that terminal blend rubber asphalt mixtures performed good durability for structural aspect.

Acknowledgments

The authors are grateful to the financial support of KRDPC (National Key R & D Program of China, Nos. 2017YFC0840201), and NSFC (National Natural Science Foundation of China, Nos. 51878167).

References

[1] Lo Presti D, Airey G and Partal P 2012 Manufacturing terminal and field bitumen-tyre rubber blends: The importance of processing conditions *SIIV-5th International Congress-Sustainability of Road Infrastructures* **53** 485-94

[2] Lo Presti D 2013 Recycled tyre rubber modified bitumens for road asphalt mixtures: a literature review *Constr Build Mater* **49** 863-881

[3] California Department of Transportation 2003 Materials engineering and testing services *Asphalt Rubber Usage Guide* (5900 Folsom Blvd, Sacramento, CA 95819 USA)

[4] Federal Highway Administration 2014 The use of recycled tire rubber to modify asphalt binder and mixtures (FHWA-HIF-14-015, FHWA, U.S. Department of Transportation)

[5] Han L, Zheng M and Wang C 2016 Current status and development of terminal blend tyre rubber modified asphalt *Constr Build Mater* **128** 339-409

[6] Hicks R G, Cheng D and Duffy T 2010 Evaluation of terminal blend rubberized asphalt in paving application (California pavement preservation center report CP2C-2010-102TM)

[7] Shatnawi S 2011 Comparisons of rubberized asphalt binders: Asphalt rubber and terminal blend (White Paper Rubber Pavements Assoc.)

[8] Lin P, Huang W and Tang N 2017 Performance characteristics of terminal blend rubberized asphalt with SBS and polyphosphoric acid *Constr Build Mater* **141** 171-82

[9] Charles J G, Richard R D, Jerry A B *et al* 2000 A comprehensive laboratory and field study of high-cure crumb-rubber modified asphalt materials (FHWA/TX-01/1460-1) (Technical report by Texas Transportation Institute)

[10] Yan K, Xu H and You L 2015 Rheological properties of asphalts modified by waste tire rubber and reclaimed low density polyethylene *Constr Build Mater* **83** 143-9

[11] Xu O, Xiao F, Han S *et al* 2016 High temperature rheological properties of crumb rubber modified asphalt binders with various modifiers *Constr Build Mater* **112** 49-58

[12] Attia M and Abdelrahman M 2009 Enhancing the performance of crumb rubber modified binders through varying the interaction conditions *Int J Pavement Eng* **10** 423-34

[13] Fang C, Wu C, Hu J *et al* 2014 Pavement properties of asphalt modified with packaging-waste polyethylene *J Vinyl Addit Techn* **20** 31-5

[14] Navarro F J, Partal P, Martínez-Boza F *et al* 2004 Thermo-rheological behaviour and storage stability of ground tire rubber-modified bitumens *Fuel* **83** 2041-9

[15] Bahia H, Coenen A *et al* 2011 White paper on the effects of digesting crumb rubber in modified binders (MB) (The University of Wisconsin-Madison Asphalt Research Group)

[16] Niu D, Han S, Li X *et al* 2016 Mechanical property of asphalt mortar with additive *J Traffic Transp Eng* **16** 8-16

[17] California Department of Transportation 2003 Materials engineering and testing services

IOP Conf. Series: Materials Science and Engineering **629** (2019) 012006 doi:10.1088/1757-899X/629/1/012006

Asphalt Rubber Usage Guide (5900 Folsom Blvd, Sacramento, CA 95819 USA)

[18] Research Institute of Highway Ministry of Transport 2011 Standard Test Method of Bitumen and Bituminous Mixtures for Highway Engineering (JTG E20-2011, RIOH, CHN. Ministry of Transport)

[19] Gudimettla G J, Cooley L and Brown E 2003 Workability of hot mix asphalt (Report #03-03) (Auburn, AL: National Center for Asphalt Technology (NCAT))

[20] Research Institute of Highway Ministry of Transport 2004 Technical Specification for Construction of Highway Asphalt Pavements (JTG F40-2004, RIOH, CHN. Ministry of Transport)

[21] Chiu C and Lu L 2007 A laboratory study on stone matrix asphalt using ground tire rubber *CONSTR BUILD MATER* **21** 1027-33

IOP Conf. Series: Materials Science and Engineering **629** (2019) 012007 doi:10.1088/1757-899X/629/1/012007

Formation of weld defects in cold metal transfer arc welded 7075-T6 plates and its effect on joint performance

G Ipekoğlu[1] and G Çam[1,2]

[1]Iskenderun Technical University, Faculty of Engineering and Natural Sciences, Department of Mechanical Engineering, 31200 Iskenderun, Hatay, Turkey

E-mail: gurel.cam@iste.edu.tr

Abstract. This study aims at investigating the formation of weld defects (i.e., pores) and determining the effect of pores on mechanical performance of the AA7075-T6 plate joints produced using cold metal transfer (CMT) gas metal arc welding (GMAW) technique. For this purpose, AA7075-T6 Al-alloy plates with a thickness of 2 mm were joined using CMT welding technique. The microstructural and mechanical properties of the welded plates were investigated by detailed optical microscopy investigations, micro-hardness measurements and tensile tests. A correlation between the joint performance and the formation of porosity in the fusion zone (FZ) was also attempted to show the effect of the presence of large pores on the mechanical behavior of the joint.

1. Introduction

Al-alloys, are widely used in transportation industry particularly in applications where lightweight is required owing to their excellent specific strength and fracture properties, high-formability, good corrosion resistance [1]. Among Al-alloys, age hardened AA7075 alloy (i.e., T6) has a superior strength which is comparable with that of steel [2]. The weldability is one of the main challenges for the use of a material in commercial application. Conventional fusion welding of Al-alloys, especially in age hardened condition, have several difficulties such as porosity formation, solidification cracking and significant strength loss in the joint [2-7]. The main reason for these problems is the high heat input occurred during these welding processes. One way of avoiding these problems is to use solid state welding techniques such as friction stir welding (FSW), which are inherently low heat input joining techniques. Thus, numerous studies have been conducted over the last three decades on FSW of Al-alloys [2,3,8-19] as well as other structural materials, such as pure Pb [20], Cu-alloys [21], Mg-alloys [22], and higher melting temperature materials (i.e., steels) [23,24]. Another approach to avoid weld defects and to produce sound joints in age hardened Al-alloys is to use lower heat input fusion welding processes. For instance, Cold Metal Transfer (CMT) welding process, which is based on the Gas Metal Arc Welding (GMAW) – also known as MIG, developed by Fronius of Austria in 2004, has an advantage for welding of Al-alloys [25,26]. CMT technology has different variants, such as conventional CMT, CMT pulse (CMT-P), CMT advanced (CMT-ADV) and CMT pulse advanced (CMT-PADV) and experiments were performed on both single layer deposits and multilayer deposits [27].

There are very limited reports on CMT welding of AA7075 alloy [28]. Thus, there is certainly a need for further research on CMT welding of this alloy, which was the motivation for this research. In this study, AA7075-T6 Al-alloy plates with a thickness of 2 mm have been welded using CMT welding technique. Detailed optical microscopy investigations, microhardness measurements and tensile tests have been conducted to investigate the microstructural and mechanical properties of the welded plates and to correlate the joint performance with microstructural aspects, such as the formation of porosity in the fusion zone (FZ).

2. Experimental procedure

AA75075-T6 plates with a thickness of 2 mm were used in this study. In order to conduct the robotic CMT welding technique, specimens with dimensions of 15 x 25 x 2 mm were extracted from as received base plate. The weld surfaces to be welded were only mechanically cleaned using metal brush prior to the joining process. No further cleaning was conducted. ER5356 with a diameter of 1.2 mm was sued as filler wire. The weld parameters used during welding are listed in the table 1. Argon shielding gas was switched on 0.1 s before and was switched off 0.5 s after the arc was struck.

Table 1. The weld parameters used during welding.

Material	Current (A)	Voltage (V)	Weld speed (mm/min)	Wire feed rate (mm/min)	Shielding gas rate (l/min)
AA7075-T6	120	15.2	900	1100	15

A metallography specimen and four tensile test specimens were extracted from each joint produced in order to evaluate microstructural and hardness variations within the weld region and to determine tensile properties, i.e. weld performance values. Metallography specimens were ground and polished and then etched for 120 s using Keller's reagent for microstructural investigations and microhardness measurements. A detailed optical microscopy was conducted along the cross-sections of the joints produced in order to evaluate the microstructural alterations taking place within the weld region (i.e., FZ and heat affected zone - HAZ) and the formation of porosity in the FZ. Vickers micro-hardness measurements ($HV_{0.2}$) were conducted across the weld cross-section on a line in the mid-thickness of the welded joint at numerous locations using a load of 200 g in order to determine the hardness profile of the joint produced. The tensile tests were conducted on the standard transverse tensile specimens extracted from the as-received base plate and the welded joint using a loading rate of 0.1 mm/min. Furthermore, a detailed optical microscopy was also conducted on the specimens after the tensile test in order to correlate the microstructural aspects to the mechanical properties of the joints and to determine the fracture mode of the joint under tensile loading.

3. Results and discussion

3.1. Microstructural aspects

A macrograph taken from the welded joint produced in this study is given in figure 1. As seen from this figure, the welded joint did not exhibit any cracks. On the other hand, there is relatively extensive porosity formation in the FZ. This is in contrast to the results reported by Elrefaey [28] who observed very small amount of relatively small pores within the FZ of the CMT welded 7075-T6 joints produced using very similar weld parameters. However, they employed cleaning of the surfaces with acetone prior to welding whereas only mechanically cleaning using metal brush was used in the current study. Thus, the reason of the porosity formation in the FZ of the joint is attributed to the insufficient cleaning procedure applied prior to the joining process to remove the oxide layer and other contaminants in the current study. Another reason for extensive pore formation might be the high cooling rates involved after the welding in the current study.

The microstructure of AA7075-T6 base plates used in this study are shown in figure 2(a). As seen from these micrographs, base plate exhibits a microstructure consisting of alpha grains containing undesirable inhomogeneously distributed iron- and silicon-rich particles in the form of coarse constituent particles which are readily visible in optical microscopy. As reported in earlier publications [13,16,17], the particles randomly oriented in alpha matrix are Al_7Cu_2Fe, $Al1_2Fe_3Si$, and Mg_2Si particles. The microstructure of AA7075 alloy also contains very fine grained $MgZn_2$ precipitates

IOP Conf. Series: Materials Science and Engineering **629** (2019) 012007 doi:10.1088/1757-899X/629/1/012007

homogeneously distributed within the alpha grains, which result in strengthening in T6 temper condition of this alloy. However, these strengthening precipitates ($MgZn_2$) are extremely fine, so that they are not visible under optical microscope, and even in scanning electron microscopy.

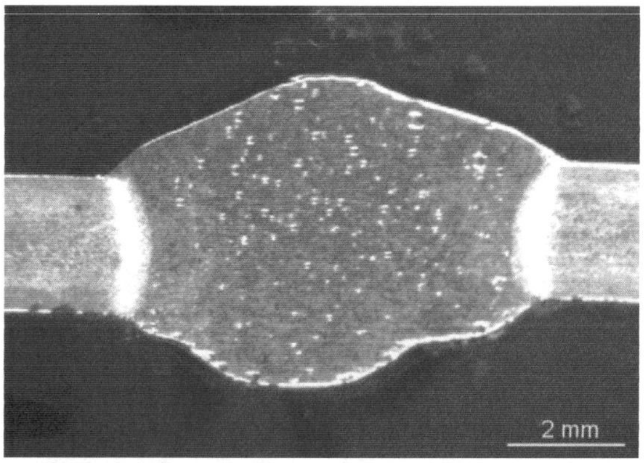

Figure 1. Macrograph taken from the CMT Welded joint.

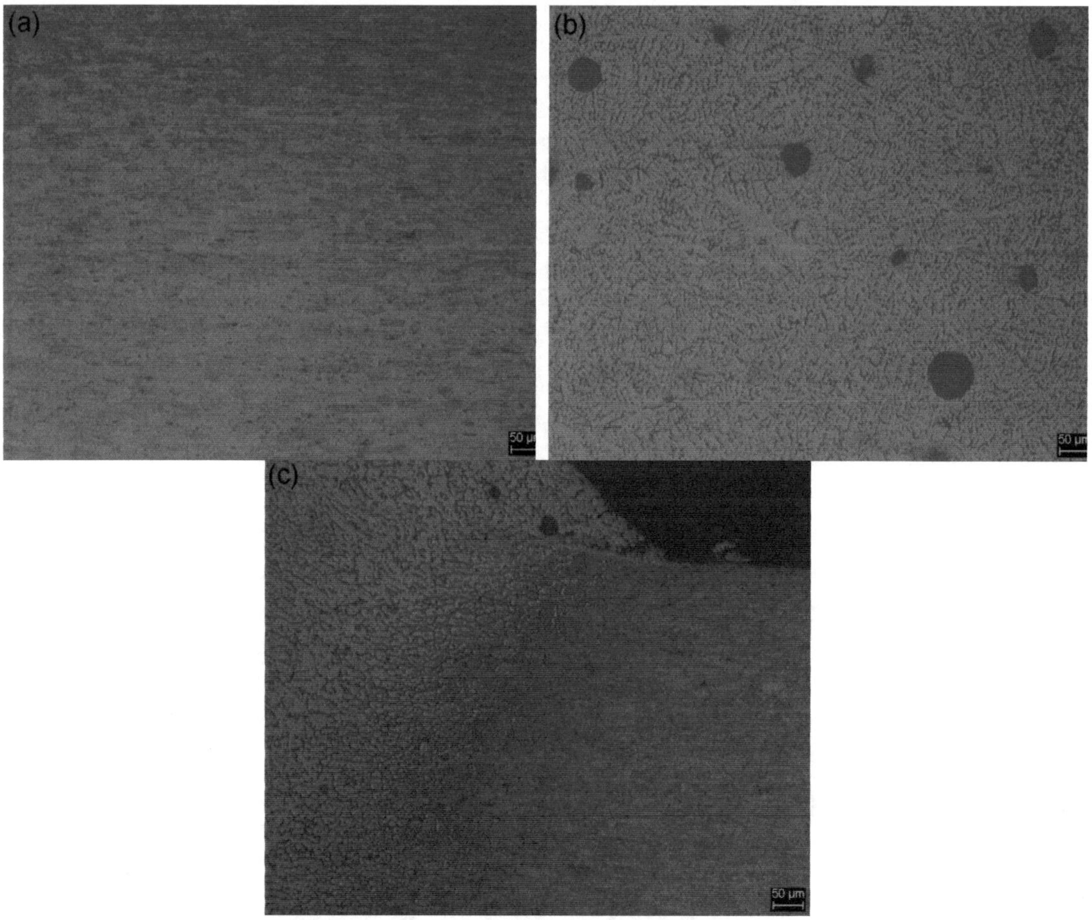

Figure 2. Optical micrographs of 2 mm thick base plate and different weld zones of the joint produced on this plate: (a) base plate, (b) weld zone, (c) transaction zones.

As clearly seen from figure 2(b), FZ exhibited a dendritic structure with some particles due to the solidification of the fusion zone taking place after the welding. Furthermore, significant amount of large pores were clearly observed in the FZ as mentioned earlier. On the other hand, a refinement of alpha grains and coarsening/dissolution of strengthening particles took place in the HAZ regions as shown in figure 2(c).

3.2. Mechanical properties

Figure 3 illustrates the hardness profile of the joint produced. As clearly seen from figure, the hardness loss took place in the weld region, the minimum hardness, i.e., about 65 HV, lying in the FZ. This is not surprising since a hardness loss usually takes place in the weld region of fusion welded age-hardened Al-alloys [4-7]. The reason of the hardness loss in the FZ and HAZ is the evolution of dendritic structure and the coarsening of strengthening precipitates due to overaging, respectively. The base plate used in this study is in the artificially aged temper condition (i.e., T6) and the strength in this temper condition originates mainly from very fine homogeneously distributed strengthening precipitates within the alpha grains. When this alloy is exposed to heat after aging, such as welding, dissolution of precipitates and evolution of a dendritic structure and coarsening of the strengthening precipitates takes place in FZ and HAZ regions, respectively as mentioned above, thus resulting in a loss in hardness in both regions. Similar results were also reported by Elrefaey [28] for CMT arc welded AA7075-T6 plates. The hardness level observed in the FZ in current study is slightly lower than that reported by [28].

The hardness profile of this fusion welded joint is somewhat different than those usually observed in fusion welded or friction stir welded AA7075-T6 joints in which the hardness minimum lies within the overaged HAZ regions on both sides of the weld nugget since hardness loss in the weld nugget is partly recovered by the grain refinement taking place there, giving rise to a W-shaped hardness profile [16]. Although there is a hardness loss in the overaged HAZ regions on both sides of the FZ of this joint as well the hardness loss is not as significant as those in the HAZ regions of conventional fusion welded or friction stir welded AA7075-T6 joints [16]. On the other hand, the hardness profile obtained is very similar to those reported for lower heat input fusion welded (i.e., electron beam or laser beam welded) age-hardened Al-alloys [4-7].

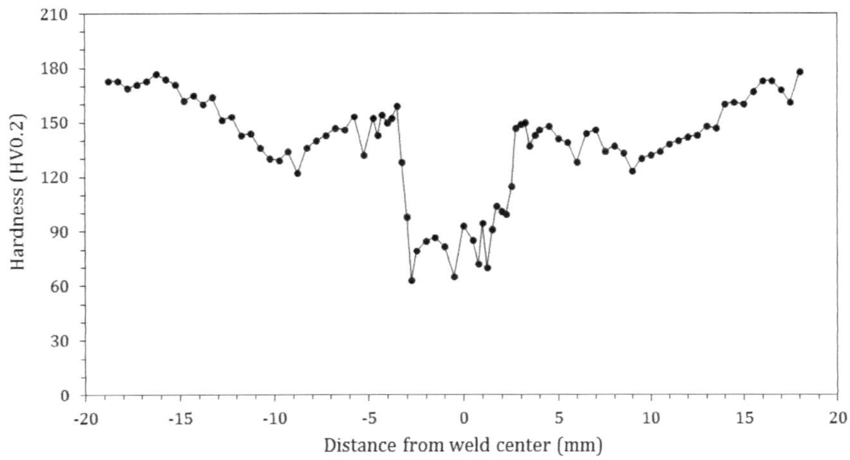

Figure 3. Hardness profile of the joint produced.

The tensile test results are summarized in table 2. Three tensile tests were conducted for the base plate and the joint and the average of these three test results were calculated, and the average values calculated are given in table in bold and parenthesis. These average values were used in the calculations of the joint performance value in terms of tensile strength. The tensile strength performance value was determined as explained below:

Tensile strength performance (%) = (Tensile strength of welded joint / tensile strength of base plate) x 100

Table 2. Tensile test results.

Specimen	0.2% Proof stress (MPa)	Tensile strength (MPa)	Elongation (%)	Tensile strength performance (%)
Base Metal	526; 550; 541 **(539)**	589; 605; 597 **(597)**	14.2; 14.7; 14.9 **(14,6)**	---
Welded Joint	---	304; 307; 315; 323 **(312)**	0.002; 0.004; 0.005; 0.002 **(0.003)**	52

As seen from table 2, the tensile strength performance value was found to be about 52%. This value was somewhat lower than that reported by [28], i.e, 60%. These results are in agreement with the fact that there is a significantly higher amount of pores in the FZ of the joint produced in this study those reported by [28]. Moreover, the pores present are much larger in size than those reported by [28]. The reason of the formation of more and larger pores in the current study is apparently the inadequate surface cleaning prior to welding as discussed earlier in microstructural aspects section. This result is also in good agreement with the hardness profile of the joint, figure 3. As seen from figure 3, the hardness minimum in the FZ of the joint and the hardness loss in this region is significant, i.e., about 110 HV (the hardness value being about 65 HV in the FZ compared to about 175 HV in the base plate). Moreover, the elongation value exhibited by the joint is extremely low, being 0.03%. This is partly due to the strength undermatching in the FZ. Since the strength is much lower within the FZ than that of the base plate the elongation takes place only within the weld region section of the transverse tensile test specimens and the base plate sections do not yield (it is in the elastic region throughout the test) and thus do not contribute to the total elongation. In the case of strength undermatching joints, the stress concentration and, thus, the fracture take place in the lower strength weld metal region (confined plasticity), leading to an increase of constraint within the weld region and, thus, significantly lower ductility levels. This confined plasticity is also quite common in fusion welded or diffusion bonded joints with a strength undermatching weld region [2-7,29-32]. Moreover, another reason of this extremely low elongation is the presence of large pores within the FZ, as seen in figure 2(b). As He *et al* [33] pointed out as well, the local deformation in the FZ was concentrated around the porosities in the case of the existence of pores. Thus, this accelerates the failure of the specimen and reduces the global ductility remarkably.

The detailed metallographic study conducted on the fractured transverse tensile specimen of the joint clearly indicated that the failure took place within the FZ next to the transition region between FZ and HAZ regions. This is not surprising due to the strength undermatching of the joint. On the other hand, this detailed fractography clearly revealed that the presence of large pores influenced the failure mode by dictating the crack growth path. As clearly seen from figure 4, the failure of the specimen took place along the transition region between the FZ and HAZ within the FZ by the jumping of the crack from pore to pore, i.e., bridging of pores. Thus, this led to a kind of brittle failure, consequently to an extremely low elongation, although the FZ is very low strength (i.e., soft) indeed.

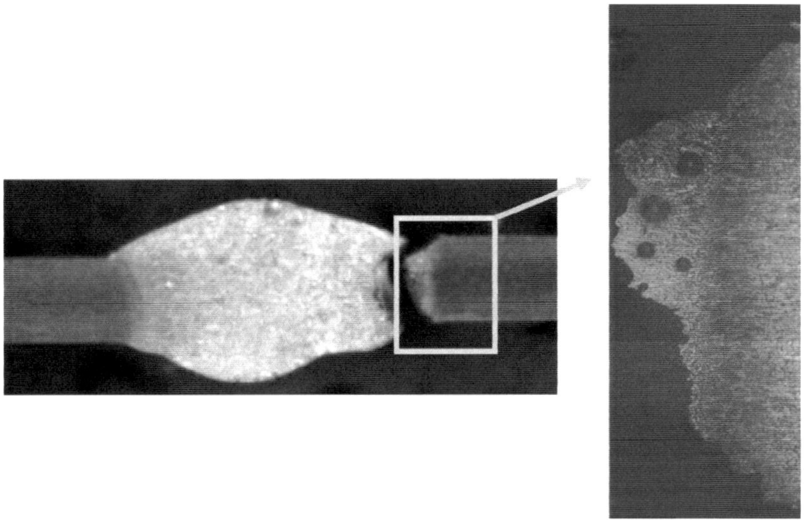

Figure 4. Macrograph of the transverse tensile test specimen showing the fracture location and higher magnification micrograph taken from the fracture area marked with a rectangular illustrating the fracture path. Note the crack propagation within the FZ next to the HAZ and the crack growth taking place by bridging from one pore to another.

4. Conclusions

In this study, AA7075-T6 Al-alloy plates with a thickness of 2 mm, have been welded using CMT arc welding technique. The following conclusions have been withdrawn from this experimental work:

- A dendritic structure containing a significantly high volume percent of large pores was observed in the FZ. The reason for the formation large pores is the inadequate surface cleaning prior to welding.
- The hardness minimum was exhibited by the FZ where a dendritic structure is present. A hardness loss was also observed in the HAZ regions on both sides of the FZ. However, the hardness decreases in these regions are much less significant than that taking place in the FZ. This can be attributed to the lower heat inputs involved during CMT welding leading to a finer grained structure and reduced coarsening of strengthening particles in the HAZ regions.
- The tensile strength performance of the joint was found to be about 52%, which is in good agreement with the hardness profile obtained from the joint, in which the hardness minimum lies in the FZ.
- The elongation value exhibited by the joint was extremely low. This is not surprising since lower ductility levels are usually obtained from the transverse tensile specimens extracted from the joints with strength undermatching weld zone due to confined plasticity. Another reason for this low ductility is the presence of large pores in the FZ, which results in concentrated local deformation within the FZ and thus accelerates the failure of the specimen and reduces the global ductility remarkably.
- The presence of large pores also influenced the failure mode by dictating the crack growth path. The crack propagated by jumping from one pore to other within the FZ, thus rendering brittle failure.

References

[1] Anderson T 2010 *Welding Aluminium - Questions and Answers: A practical guide for troubleshooting aluminum welding-related problems 2nd ed* (American Welding Society, Doral, FL, USA) pp 1-9
[2] Çam G and İpekoğlu G 2017 *Int. J. Adv. Manuf. Technol.* **91** 1851-66

[3] Kashaev N, Ventzke V and Çam G 2018 *J. Manuf. Proc.* **36** 571-600
[4] Pakdil M, Çam G, Koçak M and Erim S 2011 *Mater. Sci. Eng. A* **528** 7350-6
[5] Çam G and Koçak M 2007 *J. Mater. Sci.* **42** 7154-61
[6] Çam G *et al* 1999 *Prakt. Metallogr.* **36** 59-89
[7] Çam G *et al* 1999 *Sci. Technol. Weld. Join.* 4 317-23
[8] Çam G and Mıstıkoğlu S 2014 *J. Mater. Eng. Perform.* **23** 1936-53
[9] Çam G, Güçlüer S, Çakan A and Serindağ H T 2009 *Mat.-wiss. u. Werkstofftech.* **40** 638-42
[10] Çam G 2005 *Mühendis ve Makine (Engineer and Machinery)* **46** 30-9 (in Turkish)
[11] Von Strombeck *et al* 2001 *Proc. of the TMS 2001 Annual Meeting Aluminum, Automotive and Joining* (New Orleans, February 12-14, 2001), eds: S K Das *et al* pp 249-64
[12] dos Santos J F *et al* 2000 *Weld World* **44** 42-64
[13] İpekoğlu G, Kıral B G, Erim S and Çam G 2012 *Mater. Tehnol.* **46** 627-32
[14] İpekoğlu G, Erim S, Kıral B G and Çam G 2013 *Kovove Mater.* **51** 155-63
[15] İpekoğlu G, Erim S and Çam G 2014 *Metall. Mater. Trans. A* **45** 864-77
[16] İpekoğlu G, Erim S and Çam G 2014 *Int. J. Adv. Manuf. Technol.* **70** 201-13
[17] İpekoğlu G and Çam G 2014 *Metall. Mater. Trans. A* **45** 3074-87
[18] Çam G, İpekoğlu G and Serindağ H T 2014 *Sci. Technol. Weld. Join.* **19** 715-20
[19] Bozkurt Y, Salman S and Çam G 2013 *Sci. Technol. Weld. Join.* **18** 337-45
[20] Günen *et al* 2018 *Indian J. Eng. Mater. Sci.* **25** 26-32
[21] Küçükömeroğlu T *et al* 2016 *J. Mater. Eng. Perform.* **25** 320-6
[22] Çam G 2011 *Int. Mater. Rev.* **56** 1-48
[23] Çam G, İpekoğlu G, Küçükömeroğlu T and Aktarer S M 2017 *J. Achv. Mater. Manuf. Eng.* **80** 65-85
[24] İpekoğlu G *et al* 2019 *Mater Res Express* **6** 046537
[25] Selvi S, Vishvaksenan A and Rajasekar E 2018 *Defence Technology* 14 28-44
[26] Gungor B, Kaluc E, Taban E and Şık A 2014 *Mater. Des. (1980-2015)* **54** 207-11
[27] Cong B, Ding J and Williams S 2015 *Int. J. Adv. Manuf. Technol.* **76** 1593-606
[28] Elrefaey 2015 *Sci. Technol. Weld. Join.* **20** 280-5
[29] Koçak M, Pakdil M and Çam G 2002 *Sci. Technol. Weld. Join.* **7** 187-96
[30] Çam G, Koçak M, Dobi D, Heikinheimo L and Siren M 1997 *Sci. Technol. Weld. Join.* **2** 95-101
[31] İpekoğlu G, Akçam Ö and Çam G 2018 *The Paton Welding Journal* **12** 77-82
[32] Çam G 2005 *Mühendis ve Makina (Engineer and Machinery)* **46** 30-9 (in Turkish)
[33] He E *et al* 2018 *Int J Adv Manuf Technol.* **95** 659-71

IOP Conf. Series: Materials Science and Engineering **629** (2019) 012008 doi:10.1088/1757-899X/629/1/012008

Techniques to accelerate thermo-mechanical simulations in large-scale FE models with nonlinear plasticity and cyclic input

D Benasciutti[1,3], F De Bona[2], L Moro[2] and J Srnec Novak[2]

[1]Department of Engineering, University of Ferrara, via Saragat 1, 44122 Ferrara, Italy
[2]Politechnic Department of Engineering and Architecture (DPIA), University of Udine, via delle Scienze 208, 33100 Udine, Italy

E-mail: denis.benasciutti@unife.it

Abstract. A procedure is proposed to reduce the computation time of thermo-mechanical simulations with large nonlinear finite element (FE) models that involve cyclic plasticity. The procedure is helpful when it is practically unfeasible to simulate the huge amount of cycles needed to bring the material model to its fully stabilised state (an unfavourable situation that often occurs when small plastic strains are present), as required before assessing the structural durability. A "reference" test case, with combined kinematic and isotropic nonlinear model calibrated on actual material properties, is compared to accelerated models as well as pure kinematic models. Guidelines on how to set up the accelerated model are finally discussed.

1. Introduction

Structural durability of complex structures undergoing cyclic thermo-mechanical loadings may exploit results from elasto-plastic finite element (FE) analyses. On the one hand, the fatigue life assessment requires that the simulated cyclic material behavior reaches its complete stabilized condition (which occurs about at half the number of cycles to failure). On the other hand, attaining such a stabilized condition in large-scale nonlinear FE models may require that a huge number of cycles are simulated, which may become computationally unfeasible.

In a recent study dealing with elasto-plastic analysis of a copper mold for steelmaking plant [1], it was estimated that the three-dimensional FE model would had required thousands of cycles (around 60000) to simulate the material behavior until complete stabilization, due to the unfavorable combination of small plastic strains and low stabilization speed of the mold alloy. In turn, thousands of cycles would have last hundreds of simulation days, i.e. beyond any practical limit.

The use of accelerated material models, as proposed in [2], was considered a mean to keep the computation time within more acceptable limits. In accelerated models, the stabilization speed is increased up to a "fictitious" value (see figure 1), which permits the number of cycles for simulating the material behavior from the onset of plasticization up to the stabilized condition to be reduced considerably (and so the computation time, too). In some cases, however, the stabilization speed cannot be increased arbitrarily and unreasonably (figure 1) if one needs to prevent numerical converge issues or "unrealistic" results in simulations [2].

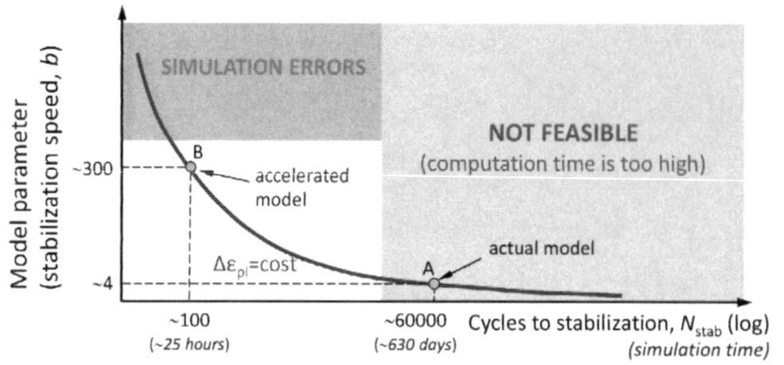

Figure 1. Actual and accelerated model. Simulation time is decreased if the stabilization speed b is increased from its actual value (A) to a "fictitious" value (B).

Checking the correctness of acceleration techniques is possible only by comparison with a "reference" case that permits the cyclic material behavior to be simulated until complete stabilization. In this work, the reference case is an axisymmetric FE model of a hollow copper mold subjected to cyclic thermal loadings.

2. Plasticity model: Theoretical background

The cyclic material behaviour is modelled through a combined kinematic and isotropic model [3]:

$$\sqrt{\frac{3}{2}(\boldsymbol{S}-\boldsymbol{X}):(\boldsymbol{S}-\boldsymbol{X})}-R-\sigma_0=0 \tag{1}$$

where $\boldsymbol{S}$ is the deviatoric stress tensor, $\boldsymbol{X}$ is the deviatoric back-stress tensor, R is the drag stress and σ_0 is the initial yield stress. In the combined model, the yield surface both translates (controlled by $\boldsymbol{X}$) and expands (controlled by R) as function of plastic strain increment $\mathrm{d}\varepsilon_{\mathrm{pl}}$ and accumulated plastic strain $\varepsilon_{\mathrm{pl,acc}}$. The kinematic part is based on the Chaboche's nonlinear model [3]:

$$\boldsymbol{X}=\sum_{i=1}^{n}\boldsymbol{X}_i \quad ; \quad \mathrm{d}\boldsymbol{X}_i=\frac{2}{3}C_i\mathrm{d}\varepsilon_{\mathrm{pl}}-\gamma_i\boldsymbol{X}_i\mathrm{d}\varepsilon_{\mathrm{pl\,acc}} \tag{2}$$

where C is the hardening modulus and γ is the recovery parameter controlling the decay of C as plastic strain accumulates. For i=1, equation (2) yields the Armstrong-Frederick model with only two pairs (C_1, γ_1). For γ=0, equation (2) gives the Prager model (linear kinematic), in which $\mathrm{d}\boldsymbol{X}=(2/3)C_{\mathrm{lin}}\mathrm{d}\varepsilon_{\mathrm{pl}}$.

The isotropic model, which controls the material hardening/softening, is based on the incremental relationship $\mathrm{d}R=b(R_\infty-R)\mathrm{d}\varepsilon_{\mathrm{pl,acc}}$, in which R_∞ is the saturated stress and b is the stabilization speed for hardening ($R_\infty>0$) or softening ($R_\infty<0$). Integrating the previous expression gives [3]:

$$R=R_\infty\left[1-\exp\left(-b\varepsilon_{\mathrm{pl\,acc}}\right)\right] \tag{3}$$

The material stabilizes when R reaches R_∞, which (according to [3]) occurs approximately when the exponent $b\varepsilon_{\mathrm{pl,acc}}\cong5$. In strain-controlled fully reversed cycles characterized by a plastic strain range $\Delta\varepsilon_{\mathrm{pl}}$, the total plastic strain accumulated after N cycles is approximately $2N\Delta\varepsilon_{\mathrm{pl}}$ and the stabilized condition becomes $2bN_{\mathrm{stab}}\Delta\varepsilon_{\mathrm{pl}}\cong5$ (where N_{stab} is the number of cycles to stabilization). This relationships shows that N_{stab} is inversely proportional to both b and $\Delta\varepsilon_{\mathrm{pl}}$ (see figure 1) and explains why N_{stab} may become really large in those situations in which plastic strain accumulates rather slowly, due to a material that has a small b and is subjected to cycles where plastic strain is not predominant.

Increasing b is but one of many techniques proposed in literature. If both creep and thermal fatigue are present, some authors [4,5] suggest that only a limited number of cycles can be simulated. Although not well defined, this procedure could be justified by considering that visco-elasticity tends to reduce the time to stabilization. If creep rupture constitutes the damage criterion in design, an extrapolation technique aimed to speed up the simulation is proposed in [6]. In situations with no creep, other authors [7,8] adopt a kinematic model with stabilized material properties from the beginning of simulation (no isotropic model simulates the cyclic material hardening/softening).

3. Numerical case study

3.1. Component description

As benchmark, numerical simulations consider a mold for continuous casting of steel, which is a hollow tube where the initial solidification of steel takes place (see figure 2(a)). The actual shape of mold cross section (round, square, rectangular) depends on the shape of the final product. On its inner surface, the component is subjected to a massive thermal flux q that varies along its length, from a maximum q_{max} (corresponding to the position of liquid steel) to a constant value. The overall thermal flux distribution also changes over time, from its full value (plant is in service) to zero (plant switched off). On its outer surface, the mold is water-cooled. The component is simply supported to the surrounding structure and is free to expand. Molds are made of copper alloys, as they have a favorable combination of high thermal conductivity and good mechanical strength. The mold analyzed in this study has length 1000 mm, inner diameter 200 mm and thickness 16 mm, and it is made of CuAg0.1 alloy, whose cyclic plasticity properties (estimated from experiments) are summarized in [9].

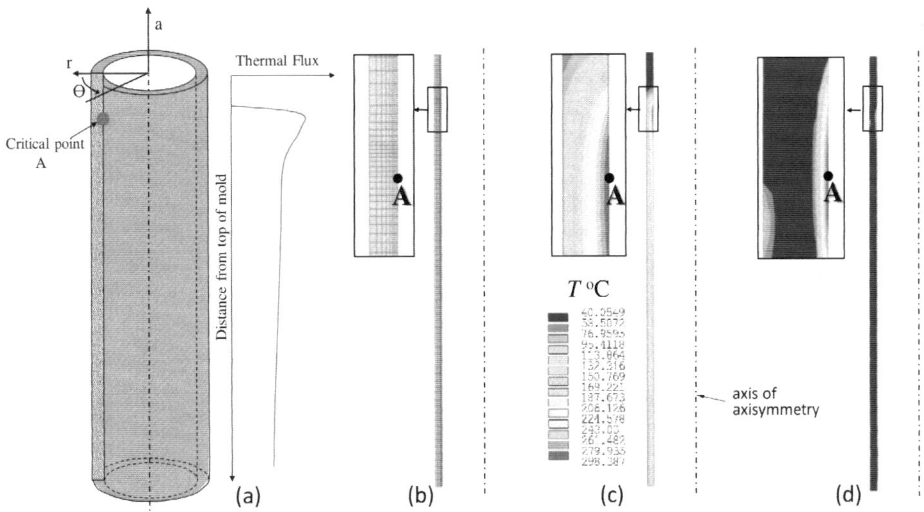

Figure 2. (a) component; (b) axisymmetric finite element model and mesh zoom; (c) temperature distribution; (d) equivalent plastic strain distribution.

3.2. Finite element model

Thanks to the axial symmetry of geometry and boundary conditions, the mold can be represented by a two-dimensional (2D) axisymmetric FE model (see figure 2(b)), which moreover permits the computation time to be greatly reduced even when simulating hundreds of cycles. The model is constituted by 760 axisymmetric 8-node isoparametric elements, for a total of 2487 nodes. Mesh is refined in the region of maximum thermal flux (point A), where the highest temperatures and stress gradients are expected to occur. The mold thermo-mechanical response is simulated by a first thermal analysis that yields the temperature distribution, which is then input in following mechanical analysis.

3.3. Thermal analysis results

In thermal analysis, a thermal flux of 3.2 MW/m^2 proposed in [10] is applied on the inner surface, a convection on the outer surface (with bulk temperature 40°C and convection coefficient 48000 W/m^2K).

The thermal flux from [10] was incremented by about 50% so that the maximum temperature reached the value 300°C for which material properties are available in [9]. The model also includes temperature dependent material properties. The change of heat flux is simulated by a sequence of steady-state analyses, are explained in [1].

Figure 2(c) displays the calculated temperature distribution. The largest temperature gradients are located close to the region of maximum thermal flux, where also the maximum temperature 298°C occurs at point A (this point will be monitored in structural analysis).

3.4. Mechanical analysis results

The input is the temperature field calculated in previous thermal analysis. No mechanical constraints are applied to the model, as the actual mold is free to expand. A nonlinear combined kinematic and isotropic material model is used and brought up to complete stabilization ("reference" case). Material parameter values are taken from [9] (the CuAg01 exhibits cyclic softening).

Results from the "reference" case were compared to those from other models, namely:

- accelerated (combined kinematic-isotropic) models, with an increased stabilization speed b_a. In order to make the sensitivity analysis the widest possible, eight values – covering a wider range than that suggested in [2] – were scrutinized (b_a=0.01b, 10b, 20b, 30b, 100b, 200b, 300b, 421b);
- only kinematic model (linear) according to Prager;
- only kinematic model (nonlinear), with the static parameters calibrated on either first or stabilized cycle;

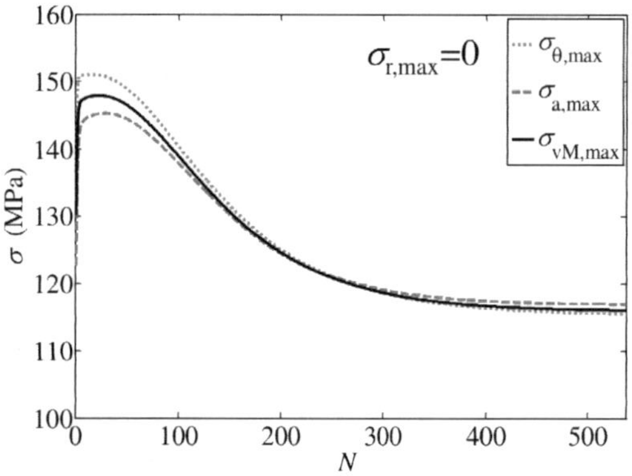

Figure 3. Stress components vs. loading cycles.

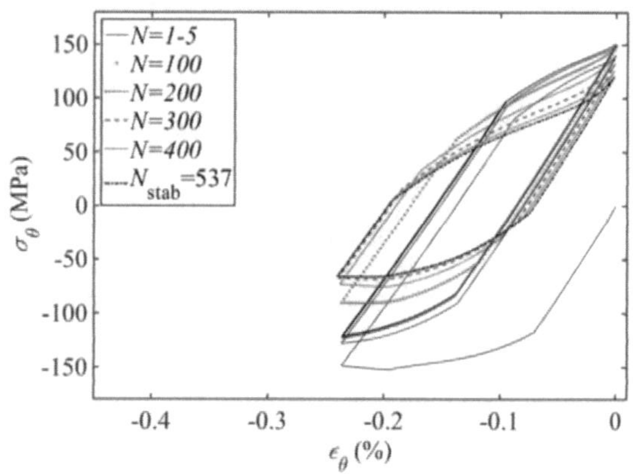

Figure 4. Evolution of hoop stress-strain at cycles (for clarity, only some cycles are plotted).

For accelerated model, the value 0.01b represents a lower bound for which the contribution of isotropic part to the combined model becomes negligible (i.e. material does not show softening). By contrast, value 421b constitutes an upper bound for which cyclic softening occurs so rapidly that – as it will be discussed later on – unrealistic results ("distorted" stress-strain cycles) are obtained (for even higher b, simulation do not converge at all). For Prager and nonlinear kinematic models, the difference in material properties between initial state and stabilized state are considered by calibrating the elastic modulus and yield stress either on the first cycle (E_1, σ_0) or on the stabilized cycle (E_s, σ_{0*}). Note that the first quarter of first cycle corresponds to the monotonic uniaxial stress-strain curve.

The "reference" combined model takes about 537 cycles to make the material stabilize (adopting the stabilization criterion described in [11]), as also confirmed by the cyclic evolution of stress components (radial, hoop and axial stress), see figure 3. Figure 4 depicts the evolution of hoop stress-hoop strain cycles at certain intervals (for more clarity, not all cycles are shown). The change of shape thorough cycling is evident.

Figure 5 shows how, for various accelerated models, the maximum von Mises stress in each cycle changes over the applied cycles N (which are normalized to the cycles N_{stab} required to complete stabilization, in order to better emphasize the various trends in the very first cycles). All models attain almost the same asymptotic stress value, except the model with b_a=0.01b in which the isotropic model, is somehow "deactivated". It follows that the material to exhibit any softening (the hardening shown is only due to the kinematic model and ends in the first 8 cycles).

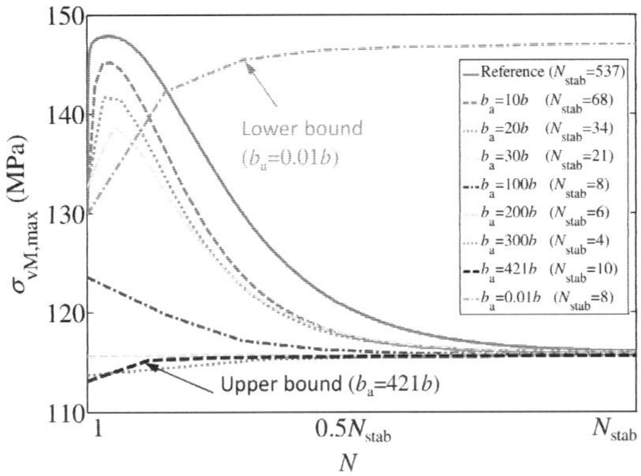

Figure 5. Evolution of max. von Mises stress over cycles for different accelerated models.

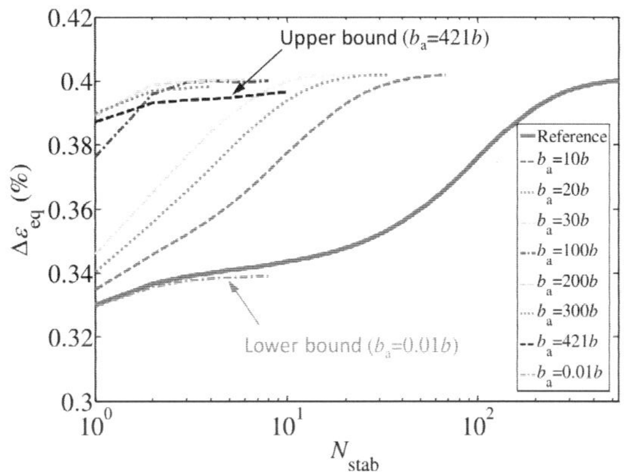

Figure 6. Evolution of equivalent strain range over cycles for different accelerated models.

Of more interest is to monitor the evolution of the equivalent strain range $\Delta\varepsilon_{eq}$ as defined in [12] (see figure 6), as this parameter is used to assess the low-cycle fatigue life. For the "reference" model (not accelerated), the strain range $\Delta\varepsilon_{eq}$ increases throughout the whole loading cycles, with a maximum rise of about 18%. This trend confirms that, if cyclic loading is stopped before material had reached complete stabilization and $\Delta\varepsilon_{eq}$ attained its asymptotic value, the fatigue life may be overestimated (in a Manson-Coffin curve, the life is indeed a nonlinear inverse function of strain range as $\Delta\varepsilon_{eq}^{-c}$, where c is a material parameter). Models with a stabilization speed only moderately increased (b_a=10b÷30b) display a trend similar to the "reference" case, whereas others with much higher values (b_a=100b÷300b) follow a different behavior, in which stabilization is almost completed in the early cycles. In all models examined, though, the equivalent strain range never stays constant during cycles, which confirms how important is that, in simulations, the material reaches its stabilized state in order to avoid errors in following fatigue life assessment.

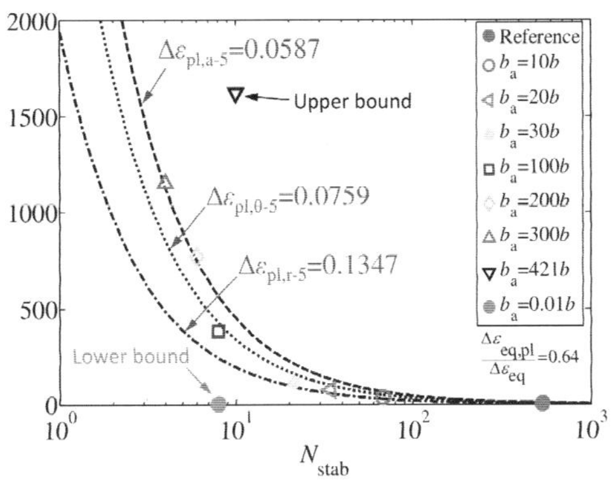

Figure 7. Design diagram for T=300°C.

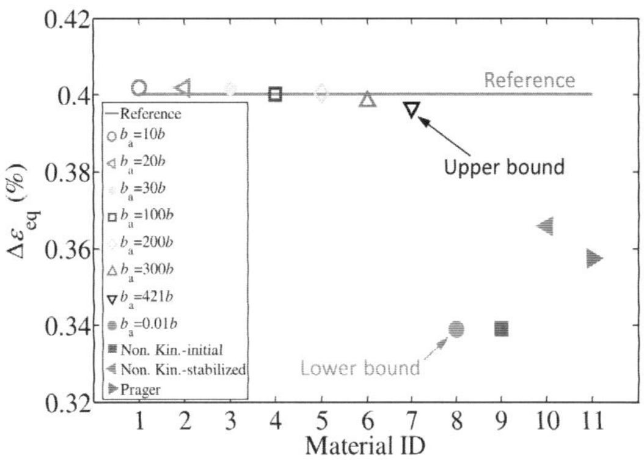

Figure 8. Comparison of $\Delta\varepsilon_{eq}$ from all models.

Figure 7 represents a sort of “design diagram” that applies the concepts sketched in figure 1. It correlates parameters b, N and $\Delta\varepsilon_{pl}$ through the relationship $2bN_{stab}\Delta\varepsilon_{pl}\cong 5$. The plastic strain range is calculated at the 5th simulated cycle. The three curves correspond to the three plastic strain range components ($\Delta\varepsilon_{pl,a}$=0.0554% axial, $\Delta\varepsilon_{pl,\theta}$=0.0734% hoop, $\Delta\varepsilon_{pl,r}$=0.1289% radial). Except for the two bound cases (b_a=0.01 and b_a=421b), the other points lie close to the analytical curves, whose validity is thus confirmed.

Table 1. Number of cycles to stabilization and $\Delta\varepsilon_{eq}$ for point A (T=300°C).

Material ID	1	2	3	4	5	6	7	8	9	10	11
	Comb. model	Accelerated models							NKI[a]	NKS[b]	Prager model
		10b	20b	30b	100b	200b	300b	421b	0.01b		
N_{stab}	537	68	34	21	8	6	4	10	8	8	10	5
$\Delta\varepsilon_{eq}$ (%)	0.400	0.402	0.402	0.402	0.400	0.401	0.398	0.397	0.339	0.339	0.366	0.357
Δe (%)		0.41	0.40	0.36	-0.02	0.5	-0.48	-0.91	-15.30	-15.26	-8.60	-10.71

[a] NKI – nonlinear kinematic model with initial parameters
[b] NKS - nonlinear kinematic model with stabilized parameters

Figure 8 compares the equivalent strain range $\Delta\varepsilon_{eq}$ in the stabilized state, from all the material models examined (pure kinematic models–both linear and nonlinear–are included, too). The values of $\Delta\varepsilon_{eq}$ provide an indirect measure of structure durability, as they are inversely proportional to the low-cycle fatigue life in a Manson-Coffin diagram. Values from figure 8 are also listed in Table 1.

Accelerated models yield comparable $\Delta\varepsilon_{eq}$ (with a maximum relative deviation of 0.5% from the “reference” case), which in turn confirms that a moderate increase of the stabilization speed (below the upper bound b_a=421b) would leave the estimated fatigue life unaltered if compared to the life estimated by the “reference” model. The upper bound value, of course, cannot be determined beforehand and needs be established for the specific engineering application under study.

Of more interest is to observe how pure kinematic models always underestimate the $\Delta\varepsilon_{eq}$ of the “reference” case, with a maximum deviation up to 15%. The lowest $\Delta\varepsilon_{eq}$ comes from the nonlinear kinematic model considering “initial” state of material (E_1, σ_0). Only a small improvement yet occurs if the nonlinear kinematic model is calibrated based on static parameters obtained from the stabilized state, or it is replaced by the linear Prager model calibrated on the initial state.

Either way, it seems apparent that neglecting the isotropic part in the combined model does not seem very appropriate. One the one hand, using a linear kinematic model greatly simplifies the model calibration, as only the monotonic tensile curve is required without any cyclic experimental data. On the other hand, figure 8 confirms the conclusions in [2], which caution engineers about the use of pure kinematic models that – regardless of being calibrated on the initial or stabilized material state – do completely neglect the material hardening/softening and the corresponding change in elastic-plastic properties over cyclic loading.

4. Guidelines for practical applications

A test case is now briefly discussed to show how accelerated models can be applied in practice. For the FE model in figure 2, the thermal flux from [10] is incremented by only 30%, so that the maximum temperature is close to the value 250°C for which material parameters (and, in particular, the stabilization speed b) are available in [9].

The relationship $2bN_{stab}\Delta\varepsilon_{pl}\cong 5$ governs the set-up of the accelerated model. To be calibrated, that relationship needs the value of plastic strain range $\Delta\varepsilon_{pl}$ in one structure point. This value can be approximated by $\Delta\varepsilon_{pl\text{-}5}$ after five cycles, which normally demands for a relatively short simulation time even with large-scale nonlinear FE models. The b parameter is known from experimental data.

The actual b value normally determines a rather high number of stabilization cycles N_{stab} (and so is the corresponding simulation time). Therefore, after establishing a target value $N^*<N_{stab}$ that yields a more affordable simulation time, the "fictitious" $b_a \cong 5/(2N^* \Delta\varepsilon_{pl\text{-}5})$ for the accelerated model can be derived accordingly (for the example under study, $N^*=40$ gives $b_a=200$), see figure 9. Running 40 cycles with b_a gives an equivalent strain range $\Delta\varepsilon_{eq}=0.3144\%$, which only deviates by 2.27% from the value obtained by simulating 1033 cycles with the actual b. This result then validates the procedure.

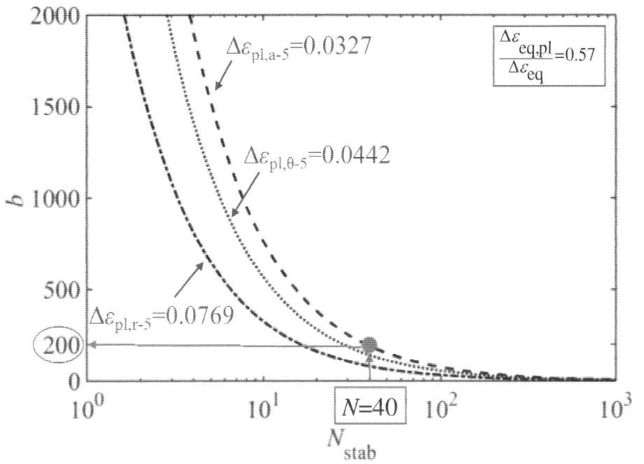

Figure 9. Design diagram for T=250°C [11].

5. Conclusions

Selecting the most appropriate cyclic plasticity model is a crucial step in thermo-mechanical finite element analysis. It may happen that the material model that fits more closely the experimental cyclic plasticity behavior is also the one that, in finite element simulations, requires a too large number of cycles to reach its stabilized state. This would result in a very long computation time, especially with large-scale three-dimensional FE models.

This work implemented several plasticity models (combined kinematic and isotropic, accelerated, Prager and nonlinear kinematic with "initial" and "stabilized" parameters) and compared them in terms of the equivalent strain range. As a basis for comparison, a "reference" test case was considered, in which the plasticity model is fitted on actual experimental data and the simulation carried out until complete material stabilization. The comparison showed that pure kinematic models, even if calibrated on either the initial or stabilized material state, deviated from the "reference" equivalent strain range, thus overestimating the actual structure fatigue life. Pure kinematic models, neglecting the material hardening/softening, seem inadequate and their use not recommended. By contrast, accelerated kinematic-isotropic models were always closed to the "reference" case, if the stabilization speed is increased up to two orders of magnitude (which allows the computation time to be shortened significantly). Guidelines were finally provided on how, in practical numerical cases, the "increased" stabilization speed in accelerated models can properly be set up.

References

[1] Srnec Novak J, Benasciutti D, De Bona F, Stanojević A and Huter P 2015 *Procedia Engineering* **133** 688-97
[2] Chaboche J L and Cailletaud G 1986 *Comput. Struct.* **23** 23-31
[3] Chaboche J L 2008 *Int. J. Plasticity* **24** 1642-93
[4] Arya V K, Melis M E and Halford G R 1990 *Finite Element Elastic-Plastic-Creep and Cyclic Life Analysis of a Cowl Lip* (NASA Technical Memorandum 102342)
[5] Amiable S, Chapuliot S, Constantinescu A and Fissolo A 2006 *Fatigue Fract. Eng. Mater. Struct.* **29** 209-17
[6] Kontermann C, Scholz A and Oechsner M 2014 *Mater. High Temp.* **31** 334-42

[7] Li B, Reis L and de Freitas M 2006 *Int. J. Fatigue* **28** 451-8
[8] Campagnolo A, Berto F and Marangon C 2016 *Theor. Appl. Fract. Mec.* **81** 76-88
[9] Benasciutti D, Srnec Novak J, Moro L, De Bona F and Stanojević A 2018 *Fatigue Fract. Eng. Mater. Struct.* **41** 1364-77
[10] Galdiz P, Palacios J and Thomas B G 2014 *Proc. European Cont. Casting Conf.* (Graz, Austria, June 23–26, 2014)
[11] Srnec Novak J, De Bona F, Benasciutti D and Moro L 2018 *MATEC Web of Conf.* **165** 19010
[12] Manson S S 1966 *Thermal Stress and Low-cycle Fatigue* (New York: McGraw-Hill)

Mechanical properties of reactive powder concrete: A review

R J Shi[1], Z Xiang[1], S G Liu[1], D H Wang[2,3] and Y Z Ju[2]

[1]Shenzhen Power Supply Co., Ltd, Shenzhen, 518001, China
[2]School of Civil Engineering and Architecture, Northeast Electric Power University, Jilin, 132012, China

E-mail: hitwdh@126.com

Abstract. To promote engineering application of reactive powder concrete (RPC), based on the domestic and foreign research achievements, the preparation, mix proportion, mechanical properties of RPC were discussed. The results showed that different admixtures have different influence on mechanical properties of reactive powder concrete. High temperature curing can excitant material's activity and improve the compressive strength of RPC. The axial tensile strength of RPC was usually in the range of 4~12 MPa, and the splitting tensile strength and flexural strength of RPC were close and higher than that of axial tensile strength. The tensile properties of RPC can be greatly improved by increasing the fiber volume fraction.

1. Introduction

Reactive Powder Concrete (RPC) was a new type of ultra-high performance cement matrix composite, also known as ultra high performance fiber reinforced concrete (UHPFRC). Compared with normal strength concrete, RPC has higher strength, toughness and durability, and has a broad application prospect [1]. In the past 20 years, many researchers and engineers have extensively studied the preparation and properties of reactive powder concrete. Based on the domestic and foreign research results, the preparation technology and mechanical properties of RPC were summarized, and the problems in the development of RPC and the future development direction were analyzed.

2. Materials and mix proportions of RPC

Pierre *et al* [2] introduced the manufacture principle and production process of RPC for the first time, and put forward the mix proportions of RPC200 and RPC800. The RPC200 of compressive strength 170~230 MPa was prepared by using quartz sand as aggregate and heat curing at 90°C. The RPC of 490~650 MPa was successfully prepared by 50 MPa compression molding and autoclaved curing at 250~400°C. By adding steel aggregate with particle size less than 0.8 mm, combined with compression molding and autoclaved curing, the RPC800 of compressive strength 650~810 MPa was prepared successfully.

The basic mechanical properties of fly ash and granulated blast furnace slag RPC under different curing conditions were studied by Yazcl *et al* [3,4]. The results showed that autoclaved pressure and steam curing can improve the hydration process and compressive strength of RPC, but autoclaved pressure and steam curing decreased flexural strength and toughness of RPC. Moreover, increasing the content of slag and fly ash can reduce shrinkage and deformation of RPC and improve compressive strength and toughness. The development of ultra-high performance concrete with high efficiency cement and mineral admixture was studied by Yu *et al* [5]. The influence of fly ash, blast furnace slag and limestone powder instead of cement on the performance of UHPC was analyzed. The effect of the three materials on the hydration reaction of UHPC in the first five days was similar. Compared with the fly ash or limestone concrete of 28 and 91 days curing, the mechanical properties of ultra-high performance concrete of blast furnace slag were better than those of the concrete. When the water-binder ratio increased from 0.165 to 0.18, compressive strength and flexural strength of UHPC increased. When the water-binder ratio increased from 0.18 to 0.2, the mechanical properties of UHPC decreased. The effects of sand gradation ratio, cement content and silica fume content on the fluidity and mechanical properties of RPC were studied by Ahmad [6]. The compressive strength and elastic modulus of RPC were significantly affected by water-binder ratio, and the fracture modulus was influenced by cement content. With the increase of cement content, the compressive strength of RPC increased, and the fracture modulus and elastic modulus decreased. The fracture modulus of RPC was three times higher than that of normal strength concrete, and the elastic modulus of RPC was 20% higher than that of other high strength concrete. A regression equation was fitted according to the content of main admixtures and the mechanical properties of RPC to optimize the mix proportion.

He *et al* [7] used P.O 62.5 cement, took standard sand as aggregate, removed mold after standard curing 24 h, and adopt three curing systems (20°C water curing 28 days, 90°C hot water curing 48 h and hot water curing 200°C dry heat curing 8 h). When the mix proportions was cement: sand: silica fume: quartz powder: DFS water reducer = 1: 1: 1: 0.35: 0.32.5, the ratio of water-binder was 0.2, and the volume fraction of steel fiber was 3% (the diameter was 0.175 mm, the length was 13 mm), the compressive strength of RPC was 178.8 MPa and 298.6 MPa, respectively.

Liu [8,9] used natural fine aggregates and admixtures to replace quartz powder and silica fume in RPC, respectively. Under the condition of hot water or autoclaved curing, RPC with compressive strength > 200 MPa, flexural strength > 50 MPa and pull-compression ratio > 0.27 was successfully prepared. The cement, silica fume, ultrafine fly ash and ultrafine slag powder quaternary composite system were used to give full play to the superposition and composition complementary effect among all kinds of mixed materials. A optimum mix proportion of 420 kg/m^3 cement, 105 kg/m^3 silica fume, 262 kg/m^3 ultrafine fly ash, 262 kg/m^3 superfine slag powder, 1258 kg/m^3 natural sand, 21 kg/m^3 water reducer, 147 kg/m^3 water were obtained. Standard curing 24 h after test specimen forming, mold removal, and hot water curing at 90°C for 48 h. The compressive strength of the cement sand specimen was 192.7 MPa, and the compressive strength of the cement sand specimen was 208.5 MPa if the mould was cured in 1.7 MPa autoclaves for 8 h.

Zhang *et al* [10] studied the development of high aluminum RPC. By adding high aluminum material, under the general curing mode, the early strength RPC material with 28 days compressive strength up to 184 MPa and elastic modulus up to 66 GPa was obtained. The material also had the advantages of low water absorption (0.94%) and good carbonation resistance.

Zhu *et al* [11] studied the mechanical properties and durability of RPC contained rice husk ash, and compared the effects of quartz sand and natural sand on the properties of RPC. Test results showed that flexural strength, compressive strength and fluidity of RPC changed little when natural sand was substituted for quartz sand. The optimum water-binder ratio of RPC mixed with rice husk ash was 0.20~0.22. With the increase of the substitution rate of rice husk ash, the shrinkage of rice husk ash decreased and the change of rice husk ash decreased with the increase of age, and the resistance to chloride ion permeation decreased. It was recommended to select RPC that appropriate replacement rate of rice husk ash according to different performance requirements.

3. Mechanical properties of RPC

3.1. RPC compression performance

3.1.1. RPC compressive strength. The compressive strength tests of 21 groups of RPC cubes were carried out by Ju *et al* [12] the effects of curing conditions and steel fiber content on the compressive strength of RPC were studied. The size of specimen was 100 mm× 100 mm×100 mm, volume fraction of steel fiber was 0.6%, 0.8%, 1.3% and 5.0%, respectively. Curing conditions include natural curing, standard curing and high temperature steam curing. Test results showed that the compressive strength of RPC can be improved slightly by increasing the steel fiber content. When the volume content of steel fiber was less than 2.0%, the enhancement effect of high temperature curing on RPC strength was significant, and when the content of steel fiber was more than 3.5%, the effect of high temperature curing on improving the compressive strength of RPC decreased. The compressive strength of standard curing and high temperature curing RPC was higher than that of natural curing RPC, and the increase was 11.5% and 34.2% respectively. But a high amount of steel fiber significantly increases the costs of RPC.

Kazemi and Lubell [13] studied the effect of specimen size and fiber content on mechanical properties of UHPFRC. The dimensions of the two cube test specimens were 50 mm×50 mm×50 mm and 100 mm×100 mm×100 mm, respectively. The diameters of the three kinds of cylinder specimens were 50 mm, 70 mm and 100 mm, respectively. The volume fractions of 5 kinds of steel fibers were 0%, 2%, 3%, 4% and 5%, respectively. The compressive strength test results of ultra-high performance fiber reinforced concrete decreased with the increase of the size of the specimen. When the side length of the cube was equal to the diameter of the cylinder, the compressive strength of the cube was greater than that of the cylinder. The ratio of compressive strength of cube specimen with 50 mm length to that of cube specimen with side length 100 mm, cylinder specimen with diameter 100 mm and cylinder test specimen with diameter 50 mm were 1.09 and 1.14, respectively. The compressive strength of super high performance fiber reinforced concrete increases significantly with the increase of fiber volume fraction. Compared with ultra-high performance concrete without steel fiber, when the volume fraction of steel fiber was increased to 5%, the compressive strength of the cube specimens with 50 mm and 100 mm increased by 25% and 26%, respectively. And the compressive strength of cylinder specimens with diameter 100mm was increased by 15%.

Graybeal and Davis [14] studied the compressive strength of RPC by using three cylinder specimens of three sizes and three cubes of different sizes. The RPC compressive strength ranged from 80 MPa to 200 MPa. The diameters of the three kinds of cylinder specimens were 51 mm, 76 mm and 102 mm, respectively. The side lengths of the three cubes were 51 mm, 70.7 mm and 100 mm. respectively. The conversion coefficient of compressive strength of different sizes was shown in table 1.

Table 1. Coefficients for conversion of compressive strength results [14].

Conversion value / Experiment value	Cube with a side length of 100 mm	Cube with a side length of 100 mm	Cube with a side length of 100 mm	Cylinder with diameter of 76 mm	Cylinder with diameter of 76 mm	Cylinder with diameter of 76 mm
Cylinder with diameter of 76 mm	1.0 (R^2=0.9672)	0.94 (R^2=0.9857)	0.96 (R^2=0.9541)	1.01 (R^2=0.9853)	-	1.08 (R^2=0.9645)
Cylinder with diameter of 102 mm	1.0 (R^2=0.9791)	0.93 (R^2=0.9694)	0.96 (R^2=0.9472)	-	0.99 (R^2=0.9839)	1.07 (R^2=0.9360)

Yu [15] carried out the conventional triaxial compression test for RPC under different confining pressures. The results showed that when the confining pressure was less than or equal to 60 MPa, the conventional triaxial compression failure mode of RPC was mainly split failure, and when the confining pressure was 65 MPa, the failure showed the characteristics of squeeze flow. Under different confining pressures, the stress-strain curves of RPC specimens were similar in shape, and all of them undergo four stages: compaction, elasticity, stress softening and steady load reduction. The triaxial compressive strength was about 3~4 times of the uniaxial compressive strength, and the maximum compressive strength appears when the ratio of the second principal stress to the third principal stress was 0.25. The peak strain in the triaxial direction was 5~10 times of the peak strain in uniaxial compression. With the increase of the second principal stress, the peak strain in the direction of the principal pressure increased gradually. But a mathematical expression for calculating compressive strength under different confining pressures wasn't proposed.

3.1.2. Modulus of elasticity. Yazici *et al* [3] prepared RPC with mineral admixture (granulated blast furnace slag and fly ash) instead of cement, and analyzed its mechanical properties. Test results showed that the elastic modulus of mineral admixture RPC was 46~57 GPa. When the cement substitution level reached 30%, the elastic modulus of RPC decreased slightly, this may be due to the decrease of the compressive strength of RPC. When the cement substitution level reached 40%, the elastic modulus decreased by 18%. Graybeal *et al* [16] studied the RPC cylinder uniaxial compression. The relationship between elastic modulus of RPC and the compressive strength of the cylinder was obtained by fitting, it was $E = 3840\sqrt{f_c}$.

Ju *et al* [17] measured the compressive strength and elastic modulus of five kinds of steel fibers with different content of RPC. The compressive strength of cube specimens with the length of 100 mm was 155.20~178.04 MPa, and the elastic modulus was 45.6~51.3 GPa, respectively. The volume content of steel fiber had no obvious effect on the elastic modulus of RPC. The uniaxial compressive mechanical properties of RPC were performed by Wu *et al* [18]. Test results showed that the elastic modulus of RPC200 increased with the increase of compressive strength. According to the experimental results, the relationship between the elastic modulus of RPC and the compressive strength of the prism was obtained by fitting, it was $E = 840\sqrt{f_c} + 34900$. Lv *et al* [19] collected 70.7 mm×70.7 mm×210 mm, 100 mm×100 mm×300 mm prisms and measured their elastic modulus. The relation between the elastic modulus and prism compressive strength was obtained with fitting tests data, and it was $E = 3027\sqrt{f_c} + 9533$, as shown in figure 1.

3.1.3. Compressive strain at peak load. The experimental of the compressive properties of RPC200 without fiber added was studied by Wu *et al* [19]. Test results showed that the compressive strain of RPC at peak load was higher than that of normal strength concrete and high strength concrete, and it was 0.00286~0.00384. Wang [20] measured the compressive strain of basalt polypropylene hybrid fiber RPC at peak load was 0.00285~0.00335. The relationship between RPC compressive strain and axial compressive strength of hybrid fiber was obtained by fitting, it was $\varepsilon_0 = (1093.24 + 23.19f_c) \times 10^{-6}$ (see in figure 2). The uniaxial compression test of RPC with 0.1% volume fraction of steel fiber was carried out by Jin *et al* [21]. The relation between the peak strain of RPC and the compression resistance of prism was $\varepsilon_0 = (1030 + 23.10f_c) \times 10^{-6}$.

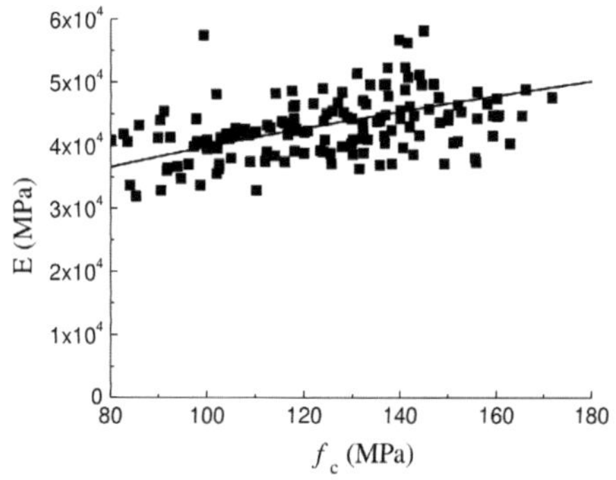

Figure 1. Relationship between RPC elastic modules and prism compressive strength [19].

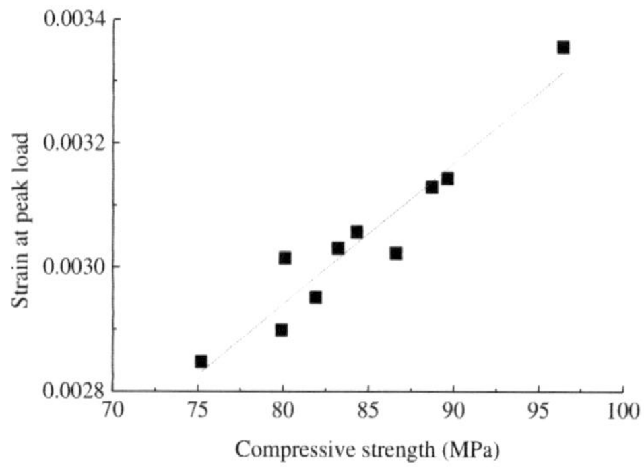

Figure 2. Relationship between compressive strain at peek load and axial compressive strength [20].

3.2. Tensile mechanical properties of RPC

Nguyen *et al* [22] studied the relationship between direct tensile stress and strain of ultra-high performance fiber reinforced concrete with different sizes and geometric shapes. Test results showed that with the increase of length, cross-section area and volume, the crack resistance strength of UHPFRC decreased slightly, the strain capacity and absorption ability decreased obviously, and the crack spacing increased greatly. With the increase of the thickness of the specimen, the strength of the specimen increases slightly, the strain capacity and the absorptive capacity increased obviously, and the crack spacing decreased obviously. The axial tensile tests of three groups of RPC specimens with different volume content of steel fiber were carried out by Yuan [23]. The diameter of steel fiber was 0.22 mm, the length of steel fiber was 12~15 mm, the volume content was 0, 1% and 2%, respectively. The average cube compressive strength of the three groups was 98.7 MPa, 140.9 MPa and 159.1 MPa, respectively, and the corresponding tensile strength was 4.75 MPa, 5.94 MPa and 6.77 MPa, respectively. The experimental results showed that the steel fibers in the cracks were pull-out and broken, and the pull-out fibers were distributed in a random direction and appeared to be interlaced.

The effect of steel-polypropylene hybrid fiber on the splitting tensile strength of RPC was studied by Smarzewski *et al* [24]. Test results showed that the splitting tensile strength of RPC increases with the increase of steel fiber volume content. Among them, the concrete with 1% steel fiber content has the highest splitting tensile strength, which was 55% higher than that of non-fiber concrete. Du *et al* [25] studied the tensile strength of RPC with different content of steel fiber. Table 2 showed splitting tensile strength, axial tensile strength and flexural strength of RPC. It can be seen from the test results that the splitting tensile strength and the flexural strength of the steel fiber RPC were close, which were significantly higher than the axial tensile strength. The axial tensile strength, splitting tensile strength and flexural strength of RPC increased with the increase of volume content of steel fiber, and the increase of splitting tensile strength was larger.

Table 2. Test results of RPC tensile strength [25]

Steel fiber volume fraction (%)	0	1	2	3
Axial tensile strength (MPa)	4.07	6.16	8.13	11.15
Splitting tensile strength (MPa)	6.59	15.47	19.96	26.00
Flexural strength (MPa)	7.35	14.95	19.58	30.95

4. Conclusion

The above studies showed that a great deal of research results have been made on the basic mechanical properties of RPC. The basic mechanical parameters of RPC have been measured and by referring to the test method of normal strength concrete mechanical performance, including the compressive strength, the flexural strength, splitting strength and elastic modulus and so on. The RPC stress-strain curves were derived and fitted. These results can basically meet the requirements of determining the basic mechanical parameters and constitutive relations of RPC. However, there were still some problems to be solved.

- Due to there was no uniform specification and standard for the design of RPC mix proportions, the mechanical properties of RPC may differ greatly under different mix proportions. It was necessary to study the relationship between the mechanical properties and the mix proportions of RPC, and to develop an optimal design method of RPC mix proportions, which was easy to calculate, which can realize the design of RPC mixture properties based on performance.
- The tensile strength of RPC was affected by the direction of fiber distribution in the matrix. The development of a method to design fiber distribution in matrix was an effective way to further improve the tensile properties and reduce the brittleness of RPC.
- At present, the high strength and high performance of RPC mainly depend on high temperature curing, which was difficult for cast-in-place construction, so its application in practical structural engineering was greatly restricted. It was necessary to further study the mix proportions and construction technology of RPC materials, develop a RPC material that can meet the requirements of field construction, and further broaden its application scope.

Acknowledgments
The authors gratefully acknowledge Shenzhen Power supply Bureau Co., Ltd. (project No.: 090000KK52170163) and the National Natural Science Foundation of China (Project No. 51878128

and No. 51378095).

References

[1] Ju Y Z, Wang D H and Zhang C 2011 Advance of research and application on reactive powder concrete *Journal of Northeast Dianli University* **31** 9-15

[2] Richard P and Cheyrezy M 1995 Composition of reactive powder concretes *Cement Concrete Res* **25** 1501-11

[3] Yazıcı H, Yardımcı M Y, Aydın S and Karabulut A Ş 2009 Mechanical properties of reactive powder concrete containing mineral admixtures under different curing regimes *Constr Build Mater* **23** 1223-31

[4] Yazıcı H, Aydın S, Yiğlter H S, Yardımcı M Y and Alptuna G 2010 Improvement on SIFCON performance by fiber orientation and high volume mineral admixtures *J Mater Civil Eng* **22** 1093-101

[5] Yu R, Spiesz P and Brouwers H J H 2014 Development of an eco-friendly ultra-high performance concrete (UHPC) with efficient cement and mineral admixtures uses *Cement Concrete Comp* **55** 383-94

[6] Ahmad S, Zubair A and Maslehuddin M 2015 Effect of key mixture parameters on flow and mechanical properties of reactive powder concrete *Constr Build Mater* **99** 73-81

[7] He F and Huang Z Y 2000 Study on the preparation technology of 200 ~ 300 MPa reactive powder concrete (RPC) *China Concrete and Cement Products* **4** 3-7

[8] Liu S F, Sun W, Lin W and Lai J Z 2003 Peraration and durability of a high performance concrete wiyh natural ultra-fine particles *Journal of the Chinese Ceramic Society* **31** 1080-5

[9] Liu S F, Sun W, Zhan Y S, Lai J Z and Lin W 2003 Study on preparation technique for green reactive powder concrete *Architecture Technology* **34** 38-9

[10] Zhang J Z, Ye W Z and Ji M J 2012 Study on properties of high aluminium active powder concrete *The 3rd Int. Symp. on High-Performance Concrete between the Two Sides of the Taiwan Strait (Wuhan, China)* 15 58-72

[11] Zhu P, Li Z Y, Qu W J, Mao W J and Mao W T 2015 Experiment on mix proportion of reactive powder concrete with rice husk ash *Journal of Architecture and Civil Engineering* **32** 58-65

[12] Ju Y Z, Shao A L and Wang D H 2017 Study on factors affecting the compressive strength of reactive powder concrete *Concrete* **2017** 130-2

[13] Kazemi S and Lubell A S 2012 Influence of specimen size and fiber content on mechanical properties of ultra-high-performance fiber-reinforced concrete *ACI Mater J* **109** 675-84

[14] Graybeal B A and Davis M 2008 Cylinder or cube: strength testing of 80 to 200 MPa (11.6 to 29 ksi) ultra-high-performance fiber-reinforced concrete *ACI Mater J* **105** 603-9

[15] Yu Z R, Zhao H Z, An M Z and Liu Y Q 2017 Mechanical properties of reactive powder concrete under triaxial compression *Journal of the China Rail Way Society* **39** 117-22

[16] Graybeal B A 2007 Compressive behavior of ultra-high-performance fiber-reinforced concrete *ACI Mater J* **104** 146-52

[17] Ju Y Z, Wang D H and Kang M X 2013 Mechanical properties of RPC with different steel fiber contents *Journal of Basic Science and Engineering* **21** 299-306

[18] Wu H Y and Lin Z Y 2004 Experimental research on mechanical properties of RPC *Journal of China University of Architecture* **2004** 7-11

[19] Lv X Y, Wang Y, Fu C J and Zheng W Z 2014 Basic mechanical property indexes of reactive powder concrete *Journal of Harbin Institute of Technology* **46** 1-9

[20] Wang D H, Ju Y Z, Shen H and Xu L B 2019 Mechanical properties of high performance concrete reinforced with basalt fiber and polypropylene fiber *Constr Build Mater* **197** 464-73

[21] Jin Z L, Li Y X and Fu Q 2014 Compressive strength test of reactive powder concrete with different admixture contents *Journal of Henan University of Science and Technology (Natural Science Edition)* **35** 55-62+110-111

[22] Nguyen D L, Ryu G S, Koh K T and Kim D J 2014 Size and geometry dependent tensile behavior of ultra-high-performance fiber-reinforced concrete *Compos Part B-Eng* **58** 279-92
[23] Yuan H Y, An M Z, Jia F F and Yu Z R 2011 Experimental research on uniaxial tensile performance *Engineering Mechanics* **28(s1)** 141-4
[24] Smarzewski P and Barnat-Hunek D 2017 Property Assessment of hybrid fiber-reinforced ultra-high-performance concrete *Int J Civ Eng* **16** 593-606
[25] Du R Y, Chen B C and Shen X J 2016 Tensile strength of UHPC by different testing methods *Material Guide* **30(s2)** 483-6

Investigation of mechanical and microstructural properties of friction stir welded dual phase (DP) steel

T Küçükömeroğlu[1,4], S M Aktarer[2] and G Çam[3]

[1]Department of Mechanical Engineering, Karadeniz Technical University, 61100 Trabzon, Turkey
[2]Department of Automotive Technology, Recep Tayyip Erdogan University, 53080 Rize, Turkey
[3]Department of Mech. Engineering, Iskenderun Technical University, 31200 İskenderun/Hatay, Turkey

E-mail: tkomer@ktu.edu.tr

Abstract. The application of dual phase (DP) steels has been increasing significantly in the automotive industry because of their high strength as well as good ductility, thus, cold formability. These steels are generally joined using conventional welding methods such as resistance spot welding and laser welding in the production of automotive parts. In recent years, several studies have been conducted to investigate the possibility of joining the advanced high-strength steels such as DP steels using the solid-state friction stir welding (FSW) method due to its advantages over conventional fusion joining methods such as metallurgical benefits, energy efficiency, and environmental friendliness. The aim of this study is to investigate the microstructure, hardness and tensile properties of friction stir welded DP 600 steel plates. Thus, 1.5 mm thick DP 600 steel plates were friction stir butt-welded by a tungsten carbide stirring tool consisting of a concave shoulder having a diameter of 14 mm and a conical pin (angle=30°) with a diameter and length of 5 mm and 1.25 mm, respectively. In the weld trials conducted, the tool was tilted 2° and the down-force of the tool was kept constant at 6 kN. The tool rotation and traverse speeds used in FSW trials were 1600 rpm and 170 mm.min-1, respectively. The microstructure of friction stir welded zone comprised of main martensite, bainite, and refined ferrite. The average hardness of the stir zone has increased to about 400 HV. The tensile specimens failed in the base plate away from the weld zone and tensile strength as high as that of the base plate was obtained from the welded specimens, i.e., about 640 MPa. However, the elongation of the welded plates was significantly reduced, i.e. about 55% of that of the base.

1. Introduction

Today, the automotive industry is increasingly focusing on reducing the dependence on fossil fuels, increasing security and increasing fuel economy. The use of advanced high strength steel (AHSS) in automotive parts has been a success in meeting the requirements of the lightweight and safe vehicle. Dual-phase (DP) steels which are one kind of the advanced high strength steels employed in the automotive industry consist of soft ferrite matrix contributing to ductility and a hard martensite phase distributed throughout the ferrite [1-3]. Dual-phase steels are the most commonly used in the automotive industry among advanced high-strength steels due to their high strength and good formability [4]. In many automotive applications such as door panels, roofs, and A and B columns, joining of dual phase steels is needed [5]. However, during the welding of the dual phase steels using conventional fusion welding methods, undesirable hardness reduction occurs in the heat affected zone [6].

The friction stir welding (FSW), which is a new solid-state welding method originally developed for joining difficult-to-weld Al-alloys [7-12], are widely used to join several structural materials, such as pure Pb [13], Cu and Cu-alloys [14-16], Mg-alloys [10,17], similar and dissimilar Al-alloys [18-25] and even high melting temperature materials such as Ti-alloys and steels [9-11,26]. FSW results in lower residual stresses and distortion of the plates and the grain growth is kept to the lowest level in steels as in other structural materials due to the low heat input involved [10,11,25,27]. In recent years, many researchers have been working on joining of high strength steels via friction stir welding owing to its metallurgical advantages [28-30]. Most of the studies reported on advanced high strength steels have been conducted on friction stir spot welding [31-36]. The number of studies conducted on the friction stir butt welding of advanced high strength steels is very scarce [37-39]. Thus, there is still a need for a systematic study of the microstructure and mechanical properties of dual-phase steels joined with the friction stir welding. Therefore, the purpose of this study is to investigate the microstructure, microhardness and mechanical properties of DP 600 steel plates joined by FSW.

2. Experimental procedure

Hot rolled DP600 steel with a chemical composition of 0.060% C, 1.476% Mn, 0.330% Si, 0.045% Al, 0.040% Cu, 0.667% Cr, 0.034% Ni, 0.003% Mo, 0.062% P, 0.004% S and balance Fe was used in this study. DP600 steel sheet plates with dimensions of 200 mm × 50 mm × 1.5 mm were joined by FSW. The shoulder diameter, pin diameter and pin length of the tungsten carbide tool used for FSW are 14 mm, 5 mm, and 1.3 mm, respectively. The cone angle of the pin is 30 degrees. The tool rotation speed used was 1600 rpm and the traverse speed was 170 mm/min. The tool force was kept constant at 6 kN and tool tilt angle 2°. Optical microscopy (OM) and scanning electron microscope (SEM) was used for the investigation of the microstructures evolved in the FSW region. Tensile test specimens and metallographic examination samples were extracted from the joint by electro-discharge machine (EDM) method in a direction perpendicular to the welding direction as shown schematically in figure 1.

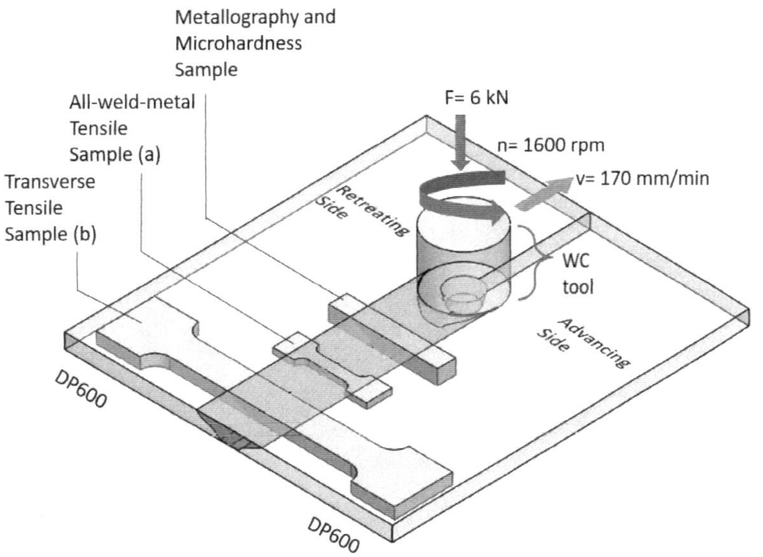

Figure 1. Schematic representation of the FSW trial and the extraction of the test samples (both transverse tensile and all-weld-metal tensile specimens) from the joint.

Transverse dog-bone-shaped tensile and all-weld-metal dog-bone-shaped tensile specimens with gage sections of 1.4 mm × 3 mm × 8 mm and 1.4 mm × 8 mm × 35 mm, respectively, were subjected to tensile testing. The tensile tests were carried out using an Instron-3382 universal test machine with a video type extensometer at a strain rate of $5\times10^{-3}s^{-1}$. The metallography sample was etched for 20 s with 2% Nital (2 ml. HNO_3 + 98 ml. C_2H_6O) reagent after standard grinding and polishing. Vickers microhardness measurements were performed using 200 g load for 10s. In order to measure the temperatures that were experienced in various regions of the joint during the FSW, the thermocouples were located at the center of the stir zone and heat affected zone (figure 2(a)).

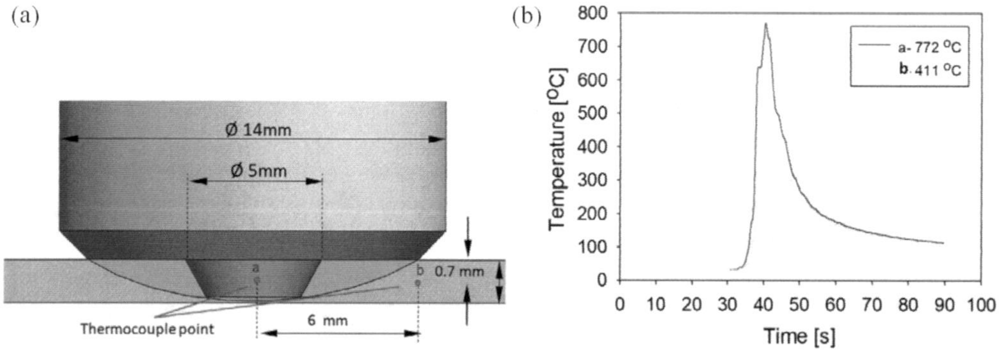

Figure 2. (a) The positions where the thermocouples were placed to measure the temperatures experienced in the FSW region (in both the SZ and HAZ regions) and (b) the corresponding temperature curves obtained from the SZ and HAZ regions.

3. Results and discussion

3.1. Microstructural aspects

The positions of the thermocouples placed in the stir zone (SZ) and the heat affected zone (HAZ) are shown in figure 2(a) and the temperature curves recorded from these regions during FSW are given in figure 2(b). As seen from these curves the peak temperature reached in the SZ during FSW was about 772°C.

As clearly seen from figure 3(a), no defects such as slot formation or microvoids were observed in the weld region. The microstructure of the DP 600 steel base plate consists of ferrite grains containing martensite islands (figure 3(b)). This microstructure has changed significantly after FSW. The microstructure of the stir zone was observed to be consisting of lath martensite (figure 3(c)) and bainite (figure 3(d)). The heat affected zone mainly consisted of tempered martensite (figure 3(e)). The average ferrite grain size of the DP 600 steel base plate was 6 μm and the martensite fraction was 25%. The microstructure evolved in the center of the stir zone after the FSW consisted of mainly lath martensite. Similarly, Ohashi *et al* [39] also reported that the microstructure formed in the SZ of DP 590 steel consisted of entirely lath martensite after the friction stir spot welding. The stir zone is exposed to both severe plastic deformation and friction heat due to the rotating tool during FSW. The peak temperature value of the stir zone center and HAZ were measured to be 772°C and 411°C, respectively. Therefore, it is said that the austenite can be transformed into martensite or bainite depending on the cooling rate during FSW. On the other hand, the formation of the bainite structure in the stir zone neighboring the HAZ region may be due to the decrease in the peak temperature experienced in this region and the plastic deformation taking place. Xie *et al* [40] also reported the formation of similar bainitic microstructure which they attributed to the low heat input because of insufficient plastic material flow and friction. Martensite islands in the HAZ began to decompose into ferrite and cementite as this region experienced a temperature of 411°C during FSW, and subsequently resulted in a structure consisting of tempered martensite in this region (figure 3(e)).

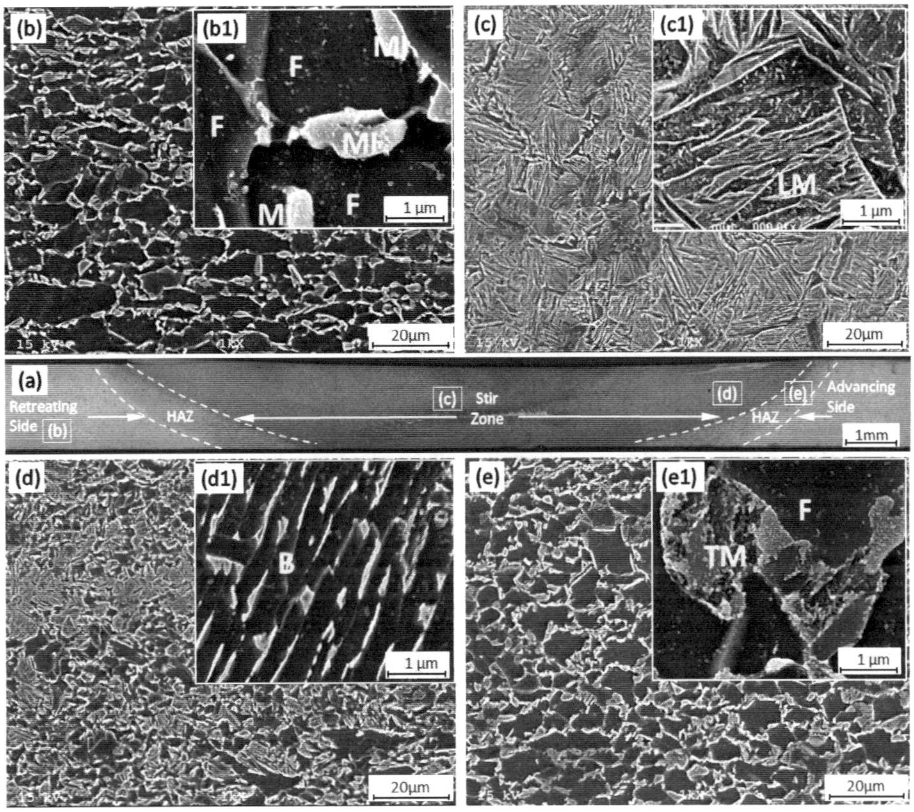

Figure 3. (a) Optical macrograph showing the cross-section of DP600 steel joint, (b)-(b1) SEM image of DP 600 steel base material microstructure, (c)-(c1) SEM image of the microstructure evolved in the SZ, (d)-(d1) SEM image of the microstructure of the SZ close to the HAZ, (e)-(e1) SEM image of the HAZ microstructure.

3.2. Microhardness and tensile properties

Optical macrograph of the weld region and a microhardness map with a color code illustrating the hardness distribution within the weld region are given in figures 4(a) and 4(b), respectively. A region corresponding to the geometry of the stirring tool in welded plate experienced significant plastic deformation and heat during FSW resulting in a drastic change in the microstructure as discussed above. This significant alteration in the microstructure in the stir zone led to a dramatic increase in the microhardness of this region. The hardness of the DP 600 steel base plate is about 220 HV. The region around the stirring tool tip subjected to severe plastic deformation, exhibiting a high hardness value (i.e. 400-450 HV), seen as yellow areas in figure 4(b). The presence of these local areas with a high hardness value within the weld region was also observed by other researchers [41,42] and they attributed this to the complex material flow and variation of plastic deformation from region to region, during FSW. The average hardness value of the stir zone was found to be about 400 HV. It was also observed that the microhardness in the stir zone decreased near the HAZ region. Thus, the average microhardness value in the SZ next to the HAZ was about 300 HV. The hardness of HAZ region, the microstructure of which composed of tempered martensite, displayed the lowest microhardness value, i.e., 205 HV. Although the hardness value of the HAZ was lower than that of the base plate, it was only 5% lower than that of the base material. Similar results were also reported by others [43,44]. Thus, it can be said that the hardness loss in the HAZ after FSW is less significant compared to the hardness loss experienced in fusion welding of this steel [43,44].

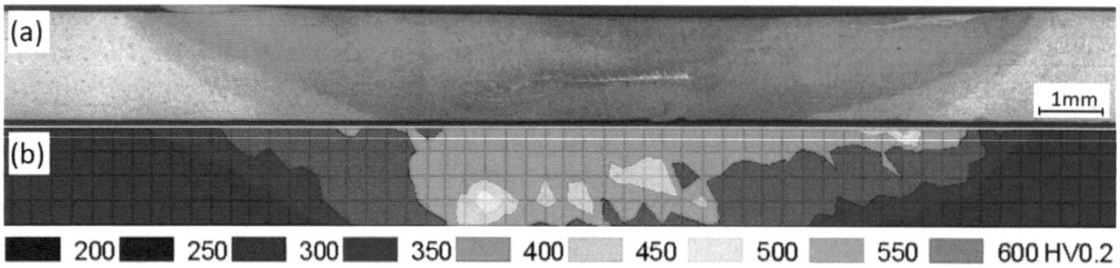

Figure 4. (a) A cross-section image of the FSWed sample, and (b) a contour map showing the microhardness distribution in the welded joint across the weld region.

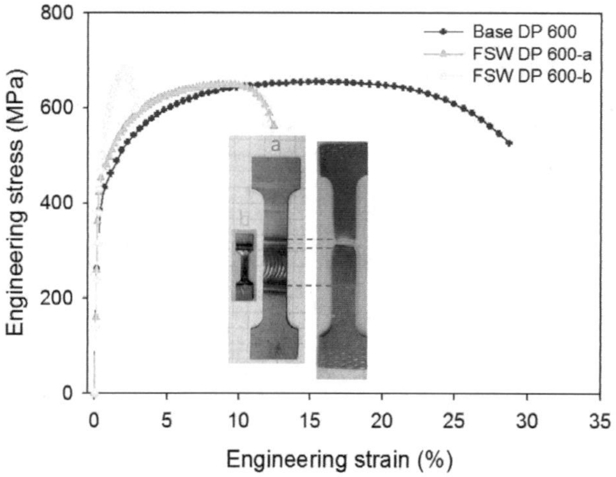

Figure 5. Engineering stress-strain curves obtained from base material specimen and friction stir welded samples (a: transverse tensile specimen and b: all-weld-metal tensile specimens).

The stress-strain curves of the base material and FSWed samples (both transverse tensile and all-weld-metal tensile specimens) are given in figure 5 and the strength and ductility values obtained from these curves are summarized in table 1. The base material exhibited a typical stress-strain curve of dual phase steel with high elongation and a large deformation hardening region. The yield strength and tensile strength values of the DP 600 steel base plate were 410 MPa and 645 MPa, respectively. Transverse tensile specimen (specimen a) failed in the transition region between the HAZ and base material, as seen in figure 5. The stress-strain curve displayed by the transverse tensile specimen is similar to that of the base material. This is not surprising since about two-thirds of the total gauge length of the transverse tensile specimen is the base material. The yield strength and tensile strength values obtained from the transverse tensile sample were 414 MPa and 640 MPa, respectively. These values are very similar to those of the base material since the failure took place away from the weld region between HAZ and BM due to strength overmatching nature of the joint as seen from the hardness distribution map of the joint (figure 4). This is not surprising since the joints with strength overmatching usually fail in the BM next to the weld region. Similar results were also reported for laser beam welded steel joints with strength overmatching [45-48] or for diffusion bonded joints with strength overmatching [49,50]. The elongation value exhibited by transverse tensile specimen extracted from the joint, however, showed a significantly lower total elongation to failure, i.e., 13%, than that of the base plate specimen, i.e., 28% (the reduction in elongation being 55%). This is due to the fact that the FSWed region (which is one-third of the total gauge length) having a yield stress of 590 MPa does not contribute to the total elongation during the tensile test, this, in turn, leads to a reduction in the total elongation. Moreover, a 5% hardness drop in the HAZ compared to BM apparently led to early necking in the HAZ region next to the BM, and thus failure took place there. On the other hand, the stress-strain curve obtained from the all-weld-metal tensile specimen (sample b) exhibited a yield strength value of 590 MPa and a tensile strength of about 680 MPa. This result is in good agreement with the hardness distribution within the weld region. The reason for these relatively high strength values is apparently due to the evolution of hard microstructural constituents such as martensite and bainite within the weld region after FSW and to a lesser extent due to grain refinement.

Table 1. Tensile test results obtained.

Specimen	Yield Strength (MPa)	Tensile Strength (MPa)	Uniform Elongation (%)	Elongation to Failure (%)
DP 600 Base Plate	410 ± 6	645 ± 8	17 ± 1	28 ± 1
DP600-DP600-a (transverse tensile)	412± 7	640 ± 10	9 ± 2	13± 2
DP600-DP600-b (all-weld-metal)	590 ± 8	683 ± 12	3 ± 1	4 ± 1

4. Conclusions

DP 600 steel plates were joined by friction stir welding, and the microstructural evolutions, microhardness and tensile properties of the joint were investigated in detail. The main conclusions of this study can be summarized as follows:

- The DP 600 steel plates were successfully butt-joined by the FSW without any defects.
- The FSW led to a refined microstructure consisting of ferrite, bainite and lath martensite in the weld region due to dynamic recrystallization.
- Evolution of hard microstructural constituents in the weld region after FSW as well as grain refinement resulted in a significant increase in the hardness value within the weld region.
- The transverse tensile specimen failed within the transition region between the HAZ and the base material away from the stir zone (SZ). This is not a surprising due to the strength overmatching.
- All-weld-metal tensile specimen displayed very high yield and tensile strength values due to the evolution of hard phases within the weld region after FSW such as martensite and bainite. This is in good agreement with the hardness distribution across the weld region. On the other hand, this specimen exhibited a very low elongation.

Acknowledgments
This study was financed by the Scientific Research Projects Council of Karadeniz Technical University, under Grant No: FBA-2016-5509. Authors would like to thank Dr. Cemil Günhan ERHUY and Ermetal Automotive and Goods (ERMETAL) Inc., Bursa, Turkey for their support in supplying the initial material.

References

[1] Rashid M S 1981 Dual phase steels *Ann. Rev. Mater. Sci.* **11** 245-67
[2] Kim N J and Thomas G 1981 Effects of morphology on the mechanical behavior of a dual phase Fe/2Si/0.1 C steel. *Metall. Trans. A* **12** 483-9
[3] Coldren A P and Tither G 1978 Development of a Mn-Si-Cr-Mo as-rolled dual-phase steel *J. Met.* **30** 6-9
[4] Nishimoto A, Hosoya Y and Nakaoka K 1981 A new type of dual-phase steel sheet for automobile outer body panels *Trans. Iron Steel Inst. Japan* **21** 778-82
[5] Antunes W D, Sergio M and Lima F De 2016 Experimental development of dual phase steel laser-

arc hybrid welding and its comparison to laser and gas *Metal Arc Welding* **21** 379-86
[6] Ghosh M, Gupta R K and Husain M M 2014 Friction stir welding of stainless steel to al alloy: Effect of thermal condition on weld nugget microstructure *Metall. Mater. Trans. A Phys. Metall. Mater. Sci.* **45** 854-63
[7] Kashaev N, Ventzke V and Çam G 2018 Prospects of laser beam welding and friction stir welding processes for aluminum airframe structural applications *J. Manuf. Process.* **36** 571-600
[8] Çam G and Ipekoglu G 2017 Recent developments in joining of aluminum alloys *Int. J. Adv. Manuf. Technol.* **91** 1851-66
[9] Cam G and Mistikoglu S 2014 Recent developments in friction stir welding of al-Alloys *J. Mater. Eng. Perform.* **23** 1936-53
[10] Cam G 2011 Friction stir welded structural materials: Beyond Al-alloys *Int. Mater. Rev.* **56** 1-48
[11] Çam G, Ipekoglu G, Küçükömeroglu T and Aktarer S M 2017 Applicability of friction stir welding to steels *J. Achiev. Mater. Manuf. Eng.* **80** 65-85
[12] Çam G 2005 Friction stir welding: A novel welding technique developed for Al-alloys *Mühendis ve Makine (Engineer Mach.)* **46** 30-9
[13] Günen A, Kanca E, Demir M, Demir M, Çavdar F, Mistikoğlu S and Çam G 2018 Microstructural and mechanical properties of friction stir welded pure lead *Indian J. Eng. Mater. Sci.* **25** 26-32
[14] Küçükömeroglu T, Sentürk E, Kara L, Ipekoglu G and Çam G 2016 Microstructural and mechanical properties of friction stir welded nickel-aluminum bronze (NAB) alloy *J. Mater. Eng. Perform.* **25** 320-6
[15] Çam G, Mistikoglu S and Pakdil M 2009 Microstructural and mechanical characterization of friction stir butt joint welded 63% Cu-37% Zn Brass Plate *Weld. J.* **88** 225-32
[16] Çam G, Serindag H T, Mistikoglu S and Yavuz H 2008 The effect of weld parameters on friction stir welding of brass plates *Mat.-wiss. u. Werkstofftech* **39** 394-9
[17] Xie G M, Ma Z Y and Geng L 2008 Effect of microstructural evolution on mechanical properties of friction stir welded ZK60 alloy *Mater. Sci. Eng. A* **486** 49-55
[18] Bozkurt Y, Salman S and Çam G 2013 Effect of welding parameters on lap shear tensile properties of dissimilar friction stir spot welded AA 5754-H22/2024-T3 joints *Sci. Technol. Weld. Join.* **18** 337-45
[19] Çam G, İpekoğlu G and Serindağ H T 2014 Effects of use of higher strength interlayer and external cooling on properties of friction stir welded AA6061-T6 joints *Sci. Technol. Weld. Join.* **19** 715-20
[20] Çam G, Güçlüer S, Çakan A and Serindag H T 2009 Mechanical properties of friction stir butt-welded Al-5086 H32 plate *Mat.-wiss. U. Werkstofftech* **40** 638-42
[21] İpekoğlu G, Erim S and Çam G 2014 Effects of temper condition and post weld heat treatment on the microstructure and mechanical properties of friction stir butt-welded AA7075 Al alloy plates *Int. J. Adv. Manuf. Technol.* **70** 201-13
[22] İpekoğlu G, Erim S and Çam G 2014 Investigation into the influence of post-weld heat treatment on the friction stir welded AA6061 Al-alloy plates with different temper conditions *Metall. Mater. Trans. A* **45** 864-77
[23] İpekoğlu G, Gören Kıral B, Erim S and Çam G 2012 Investigation of the effect of temper condition on friction stir weldability of AA7075 Al-alloy plates *Mater. Teh.* **46** 627-32
[24] İpekoğlu G, Erim S, Gören Kıral B and Çam G 2013 Investigation into the effect of temper condition on friction stir weldability of AA6061 Al-alloy plates *Kov. Mater* **51** 155-63
[25] Küçükömeroğlu T, Aktarer S M, İpekoğlu G and Cam G 2018 Mechanical properties of friction stir welded St 37 and St 44 steel joints *Mater. Test.* **60** 1163-70
[26] Mishra R S and Ma Z Y 2005 Friction stir welding and processing *Mater. Sci. Eng. R Reports* **50** 1-78
[27] Küçükömeroğlu T, Aktarer S M, İpekoğlu G and Çam G 2018 Microstructure and mechanical properties of friction-stir welded St52 steel joints *Int. J. Miner. Metall. Mater.* **25** 1457-64

[28] Nene S S, Liu K, Frank M, Mishra R S, Brennan R E, Cho K C, Li Z and Raabe D 2017 Enhanced strength and ductility in a friction stir processing engineered dual phase high entropy alloy *Sci. Rep.* **7** 16167

[29] Tehrani-Moghadam H G, Jafarian H R, Salehi M T and Eivani A R 2018 Evolution of microstructure and mechanical properties of Fe-24Ni-0.3C TRIP steel during friction stir processing *Mater. Sci. Eng. A* **718** 335-44

[30] Razmpoosh M H, Zarei-Hanzaki A and Imandoust A 2015 Effect of the Zener–Hollomon parameter on the microstructure evolution of dual phase TWIP steel subjected to friction stir processing *Mater. Sci. Eng. A* **638** 15-9

[31] Xie G M, Cui H B, Luo Z A, Yu W, Ma J and Wang G D 2016 Effect of Rotation Rate on Microstructure and Mechanical Properties of Friction Stir Spot Welded DP780 Steel *J Mater Sci Technol.* **32** 326–32

[32] Feng Z, Santella M L, David S A, Steel R J, Packer S M, Pan T-Y, Kuo M and Bhatnagar R S 2005 Friction stir spot welding of advanced high-strength steels - A feasibility study *SAE Tech. Pap.* **1** 1-7

[33] Saunders N, Miles M, Hartman T, Hovanski Y, Hong S-T and Steel R 2014 Joint strength in high speed friction stir spot welded DP 980 steel *Int. J. Precis. Eng. Manuf.* **15** 841-8

[34] Santella M, Hovanski Y, Frederick A, Grant G and Dahl M 2010 Friction stir spot welding of DP780 carbon steel **15** 271-8

[35] Khan M I, Kuntz M L, Su P, Gerlich A, North T and Zhou Y 2007 Resistance and friction stir spot welding of DP600: A comparative study *Sci. Technol. Weld. Join.* **12** 175-82

[36] Das H, Mondal M, Hong S-T, Lim Y and Lee K-J 2018 Comparison of microstructural and mechanical properties of friction stir spot welded ultra-high strength dual phase and complex phase steels *Mater. Charact.* **139** 428-36

[37] Miles M P, Pew J, Nelson T W and Li M 2006 Comparison of formability of friction stir welded and laser welded dual phase 590 steel sheets *Sci. Technol. Weld. Join.* **11** 384-8

[38] Kucukomeroglu T and Aktarer S M 2017 Microstructure, microhardness and tensile properties of FSWed DP 800 steel *J. Achiev. Mater. Manuf. Eng.* **81** 56-60

[39] Ohashi R, Fujimoto M, Mironov S, Sato Y S and Kokawa H 2009 Effect of contamination on microstructure in friction stir spot welded DP590 steel *Sci. Technol. Weld. Join.* **14** 221-7

[40] Xie G M, Cui H B, Luo Z A, Yu W, Ma J and Wang G D 2015 Effect of rotation rate on microstructure and mechanical properties of friction stir spot welded DP780 steel *J. Mater. Sci. Technol.* **32** 326-32

[41] Cho J-H, Boyce D E and Dawson P R 2005 Modeling strain hardening and texture evolution in friction stir welding of stainless steel *Mater. Sci. Eng. A* **398** 146-63

[42] Tutunchilar S, Haghpanahi M, Givi M K B, Asadi P and Bahemmat P 2012 Simulation of material flow in friction stir processing of a cast Al–Si alloy *Mater. Des.* **40** 415-26

[43] Pazooki A M A, Hermans M J M, Richardson I M, Pazooki A M A, Hermans M J M and Richardson I M 2016 Finite element simulation and experimental investigation of thermal tensioning during welding of DP600 steel Finite element simulation and experimental investigation of thermal tensioning during welding of DP600 steel *Sci. Technol. Weld. Join.* **22** 7-21

[44] Xu W, Westerbaan D, Nayak S S, Chen D L, Goodwin F and Zhou Y 2013 Tensile and fatigue properties of fiber laser welded high strength low alloy and DP980 dual-phase steel joints *Mater. Des.* **43** 373-83

[45] dos Santos J, Çam G, Torster F, Insfran A, Riekehr S, Ventzke V and Koçak M 2000 Properties of power beam welded steels, al-and ti-alloys: Significance of strength mismatch *Weld. World* **44** 42-64

[46] Çam G, Erim S, Yeni Ç and Koçak M 1999 Determination of mechanical and fracture properties of laser beam welded steel joints *Weld. J.* **79** 193-201

[47] Çam G, Yeni Ç, Erim S, Ventzke V and Koçak M 1998 Investigation into properties of laser

welded similar and dissimilar steel joints *Sci. Technol. Weld. Join.* **3** 177-89

[48] Çam G, Koçak M and dos Santos J 1999 Developments in laser welding of metallic materials and characterization of the joints *Weld. World* **43** 13-26

[49] Çam G, Koçak M, Dobi D, Heikinheimo L and Sirén M 1997 Fracture behaviour of diffusion bonded bimaterial Ti–Al joints *Sci. Technol. Weld. Join.* **2** 95-101

[50] Koçak M, Pakdil M and Çam G 2002 Fracture behaviour of diffusion bonded titanium alloys with strength mismatch *Sci. Technol. Weld. Join.* **7** 187-96

Microstructure and wear behaviour of high alloyed hot-work tool steels 1.2343 and 1.2367 under thermo-mechanical loading

I Y Malik[1,2], U Lorenz[1], A Chugreev[1] and B-A Behrens[1]

Institute of Forming Technology and Machines (IFUM) - Leibniz University Hannover, An der Universität 2, 30823 Garbsen, Germany

E-mail: malik@ifum.uni-hannover.de

Abstract. Tools and their maintenance costs in hot forging processes account for a considerable proportion of the total components' costs. Forging tools undergo extreme heating and subsequent cooling during the forging process and between the forging cycles, respectively. This cyclic heating and cooling of the tool surfaces leads to local changes in the tool microstructure which result in hardening or softening of the material in different regions of the tool and consequently influence the tool strength. Temperature in the tool areas experiencing high thermo-mechanical loadings can exceed the austenitic temperature. Hence, a strong cooling, for example by spraying or lubrication, can lead to formation of a martensitic layer in the boundary zone of the tool. Due to its higher hardness, martensitic layer has greater resistance to wear as compared to the basic or tempered materials. In the scope of this paper, the austenitisation behaviours of two high alloyed hot-work tool steels, 1.2343 and 1.2367, have been characterized by means of dilatometer tests to obtain time-temperature-austenitisation (TTA) diagrams for specimen under thermo-mechanical loads. Moreover, continuous-cooling-transformation (CCT) diagrams were recorded. Metallographic investigations were carried out to gather a detailed understanding of the microstructure behaviour and its resulting hardness. With the results of this works, it is aimed to gather a detailed and accurate insight into the arising hardening and softening effects. This would eventually lead to an optimisation of the numerical modelling for tool wear prediction.

1. Introduction

Hot-working tool steels are generally employed as tool material in hot forging processes. The tools in these processes are exposed to high mechanical, thermal and tribological stresses which lead to various types of damage [1]. The mechanical and thermal crack formation, plastic deformation and tool surface wear are, in particular, the primary reasons of failure in hot forging tools [2]. The tool surface wear is mainly caused due to the loss of material hardness in the areas close to the surface. This hardness loss is because of thermal stress induced as a result of the large heat transfer from the workpiece to the tool surface during the contact [3]. Owing to the high pressure in the tool-workpiece contact area, in combination with high tribological stresses, the surface temperature of forging tools increases considerably [4]. The alternating forging operations and lubricant application lead to a cyclic heating and cooling of the tool surface during the process. This in return results in material annealing and to a softening of the material in the surface layer. However, there are certain die regions, like convex radiuses, where the surface temperatures exceed the material austenitisation temperature [5]. In combination with high cooling rates, martensitic structures with a high hardness are generated in these regions. Both, softening as well as hardening of the tool material have a great influence on the wear resistance of the forging tools [6]. Thus, the knowledge about the service life and reliability as well as the associated efficiency of the forging tools is of great importance [7].

Due to the above mentioned loadings, the forging tools undergo rapid temperature changes during forming processes as well as during heat treatment operations [8]. The temperature-time profile experienced by the tools during hot forging processes significantly influences their behaviour. As the material diffusion is a time-dependent variable, the heating or cooling rate has a considerable influence on the crystalline state of the metallic material. Different representations which take into account the time dependency of the transformations are considered during heating and cooling of the considered steel tools. During the heating, time-temperature-austenitisation diagram (TTA) is chosen which describes the microstructural transformations as well as the initial microstructure including grain growth. TTA diagrams are prepared in order to show achievable austenitisation states with as little grain growth as possible with several heating rates (θ_h). On the contrary, during cooling the behaviour of the microstructure is represented by means of continuous-cooling-transformation (CCT) and time-temperature-transformation (TTT) diagrams. For a CCT diagram, the specimen is continuously cooled after austenitisation to room temperature at various cooling rates (θ_c). Whereas, for the TTT diagram, the specimen is quenched to a desired testing temperature and then held until all transformations have taken place [9]. These diagrams are either provided by steel manufacturers or listed in heat treatment databanks [10, 11]. Nonetheless, the experimental process boundary conditions have a great influence on these diagrams. Thus, Behrens et al. in an earlier work have suggested using transformation diagrams for the process conditions being considered [12]. This paper describes the austenitisation behaviour (TTA) under influence of thermo-mechanical loads as well as the continuous cooling transformation (CCT) behaviour at various cooling rates for two high alloyed hot-work tool steels.

2. Material and specimen production method

In the scope of this work, two high alloyed hot-work tool steels 1.2343 (X37CrMoV5-1) and 1.2367 (X38CrMoV5-3) have been analysed. The chemical composition for both of these steels was measured by optical emission spectroscopy. The content of main alloying elements for both the steels is listed in table 1. Both consist of high chromium content. Other alloying elements are manganese, molybdenum, silicon and vanadium.

Table 1. Chemical composition of the tool steels under consideration (wt. %).

	C	Si	Mn	Cr	Mo	V
1.2343	0.41	0.97	0.32	4.77	1.30	0.46
1.2367	0.42	0.28	0.37	4.90	2.95	0.52

In order to obtain a homogenous cooling inside the specimen, hollow cylindrical specimen with length 10 mm, inner diameter 3.2 mm and outer diameter of 4 mm, i.e. 0.4 mm wall thickness (S_0) are used for the investigation performed in this work (figure. 1, a). The production of these thin-walled hollow specimens poses a major challenge. The manufacturing process of the hollow specimen must ensure that the initial structure is not influenced during specimen production and that the specimen faces are plane-parallel to each other. For this purpose, holes of diameter 3.2 mm were drilled into plates with a thickness of 10 mm and subsequently the plates are heat treated to a hardness of 45 HRC (450 HV0.2) by vacuum hardening. The hollow specimens were then separated from the hardened plates using the erosion by wire method. Due to the precise positioning of the wire, very small cutting widths as well as sharp-edged holes can be achieved with this method, which are also characterized by a very low heat-affected zone.

3. Experimental investigation

In order to carry out the experiments, the quenching and forming dilatometer DIL 805 A/D+T (manufacturer: TA Instruments) available at the IFUM was used. During the tests, a temperature measurement of the specimen is necessary to evaluate the thermal loads that actually occur. For this purpose, type S thermocouples from Bähr Thermoanalyse GmbH with a wire thickness of 0.1 mm were welded onto the outer surface of the hollow specimens to measure and regulate the temperature. The connection of specimen and thermocouple was made by spot welding. During spot welding, the joint is flooded with gas to prevent oxidation or corrosion of the joint. With this method, a measuring range from 0 °C to 1700 °C with a measuring accuracy of 0.5 °C at the specimen surface can be measured. The change in length (ΔL) of the specimen due to heating and phase transformation is measured continuously via the push rods. Figure 1 shows the quenching (b) as well as deformation test (c) setup inside the dilatometer.

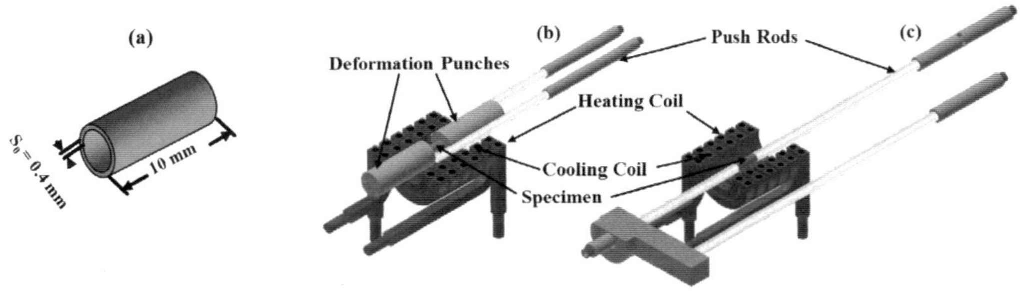

Figure 1. Schematic diagram of the hollow specimen (a); deformation module (b) and quenching module (c) of the quenching and forming dilatometer.

3.1. Time-temperature-austenitisation (TTA) diagram

The deformation and forming module of the dilatometer was used for characterizing the austenisation behaviour of the tool steels. As a result of the phase transformation, the specimen volume changes. For steel materials, austenitisation is characterised by the transformation of the lattice from body-centred cubic (bcc) into the surface-centred cubic (fcc) microstructure. With a constant temperature increase, an approximately linear increase in the change in length is observed until the transformation range is reached. This is characterised by a decrease in volume of the sample length despite a further increase in temperature and indicates the bcc to fcc lattice transformation. The temperature at the beginning of the phase transformation corresponds to the austenite start temperature (Ac_1), the temperature at complete austenitisation is called the austenite end temperature (Ac_3). An intermediate state exists between Ac_1 and Ac_3 whereas the Ac_3 indicates that the entire microstructure has converted to austenite. The Ac_1 or Ac_3 temperatures can be determined from the change in length - temperature curve (figure 2). The required data were recorded on the dilatometer during the tests. The hollow specimens were inductively heated to 1100 °C in the dilatometer at different heating rates (1, 10, 100, 1000, 2000 and 2500 °Cs^{-1}). With the help of these tests involving different heating rates, the austenitisation process could be characterized in detail for different areas in the surface zone of the hot work tool steel to be investigated. To avoid oxidation, the tests were carried out under vacuum.

To characterize the austenitisation behaviour, the Ac_1 and Ac_3 temperatures at different constant heating rates were plotted to TTA diagram. Figure 3 shows the comparison of the Ac_1 and Ac_3 temperatures with respect to the heating rates for the materials under consideration. The two steels show similar austenitisation behaviour. An increase in Ac_1 and Ac_3 temperatures of up to 1000 °Cs^{-1} can be recorded. At heating rates greater than 1000 °Cs^{-1}, a drop in the Ac_1 and Ac_3 temperatures can be observed. The minimum Ac_1 temperatures are 803 °C for the 1.2367 and 814 °C for the 1.2343.

During a forging process, in addition to high thermal stresses, high mechanical tool stresses also occur in the die. These have a significant influence on the austenitization behaviour and can lead to a reduction of the Ac_1 temperature in hot work steels. To take this effect into account, the austenitization behaviour of both materials was additionally investigated by applying a mechanical load. In order to avoid plastic deformations in the subsequent austenitization tests under influence of mechanical load, the yield stress was determined at the minimum Ac_1 temperature. For this purpose, compression tests with cylinder samples (diameter = 5 mm and height = 10 mm) were used to record the yield curves of the two materials (figure 4) and the corresponding load cases at yield stress (k_{f0}) were determined (table 2). A strain rate of 0.1 s^{-1} was considered for this purpose.

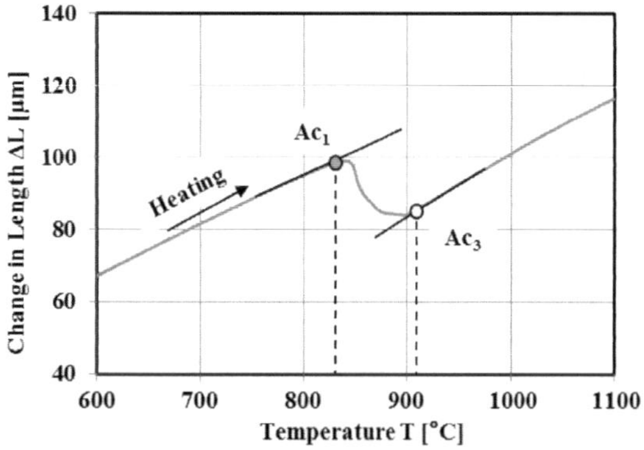

Figure 2. Determination of Ac_1 and Ac_3 temperatures by means of change in length (ΔL) – temperature (T) diagram.

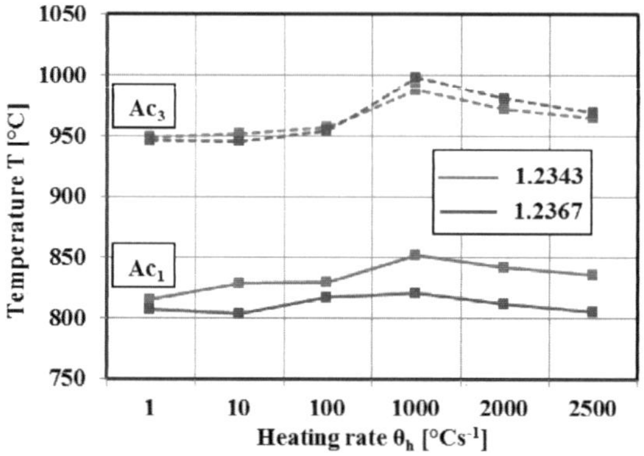

Figure 3. Comparison of Ac_1 and Ac_3 temperatures with respect to heating rates (θ_h) for tool steel 1.2343 and 1.2367.

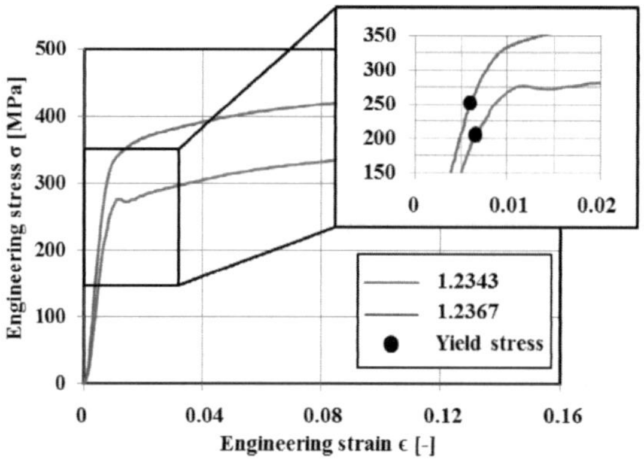

Figure 4. Evaluation of the yield stress from engineering stress-strain diagram for the tool steels 1.2343 and 1.2367.

Based on the calculated yield stress, the respective load cases to be tested (30 %, 50 % and 80 % of the k_{f0} stress) were defined for both materials (table 2). The force to be applied in the load superimposed tests was calculated corresponding to the specimen surface area (4.52 mm^2). The calculated mechanical loads were constantly applied to the hollow specimens during heating.

Table 2. Yield stress and evaluated load cases.

	k_{f0}	$F(k_{f0})$	$F(k_{f0}^{30\%})$	$F(k_{f0}^{50\%})$	$F(k_{f0}^{80\%})$
	MPa	N	N	N	N
1.2343	210	949	284	474	759
1.2367	250	1130	339	565	904

The results of the tests with superimposed loads are shown in figure 5 for both materials in relation to the different loads along with the initial results without influence of mechanical loads. It can be observed that the Ac_1 temperature decreases with increasing load. The dependence on the heating rate corresponds to the previous tests without load superposition. The highest Ac_1 temperatures occur at 1000 °Cs^{-1} as before. At a heating rate of 1000 °Cs^{-1}, the Ac_1 temperature of material 1.2343 is reduced by 54 °C, 93 °C and 104 °C with respect to a load of 30 %, 50 % and 80 % of k_{f0} mechanical load as compared to the unloaded state (figure 5, left). For the material 1.2367, a reduction of the Ac_1 temperature by 14 °C, 26 °C and 44 °C can be observed at the heating rate of 1000 °Cs^{-1} for the material 1.2367, respectively for the load cases 30 %, 50 % and 80 % (figure 5, right). This shows that the mechanical load has a significant influence on the Ac_1 temperature and thus on the transformation behaviour of the tool steels. A lower Ac_1 temperature increases the probability of new hardening effects occurring, especially in areas subject to high mechanical stress.

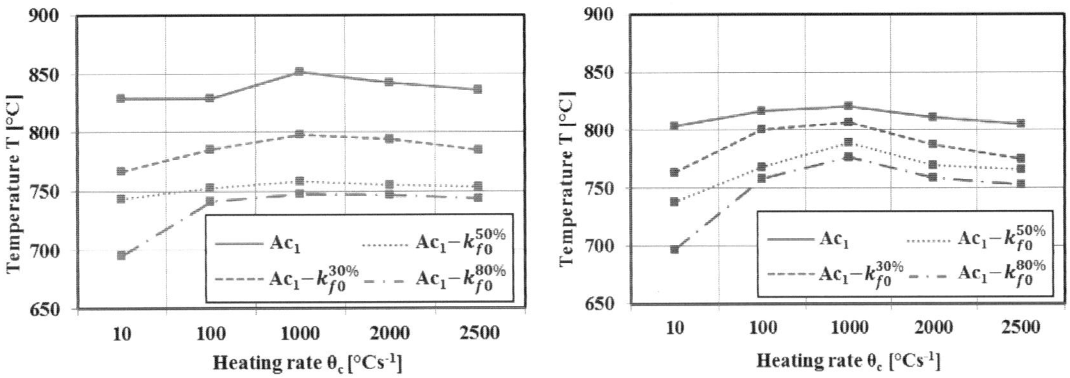

Figure 5. Effects of mechanical loads on the Ac_1 temperature for 1.2343 (left) and 1.2367 (right).

3.2. Continous-cooling-transformation (CCT) diagram

The continuous-cooling-transformation diagrams of the investigated materials are required to predict the mechanical properties that arise from different hardening conditions. The CCT tests were carried out using the quenching module of the deformation dilatometer and the results of experiments were evaluated according on the specifications of based on corresponding standards SEP 1680 and SEP 1681 [13, 14]. A constant heating rate of 1000 °Cs^{-1} was used to heat the hollow cylinder specimen till maximum temperature (T_{max} = 1100 °C) in vacuum conditions followed by a 10 seconds holding time at the maximum temperature. Subsequently, the heated specimens were quenched by using nine different cooling rates (0.1, 0.3, 0.5, 1, 5, 10, 25, 50 and 100 °Cs^{-1}). Nitrogen was used as a cooling medium for these tests. Following the experiments, the specimens were prepared for hardness measurements. For this purpose, the quenched hollow specimens were embedded in an epoxy resin. Micro-hardness measurements (HV0.2) were employed to evaluate the hardness of the embedded specimens using a 200 g test load with the Leica VMHT MOT hardness tester available at the IFUM. The temperature data obtained from experimental results were plotted against logarithmic time for all of the considered cooling rates mentioned above to obtain the CCT diagram (figure 6) of both the tool steels 1.2343 and 1.2367. The corresponding Vickers hardness values are also listed in the figure.

Based on experimental findings, it has been observed that both the steels have relatively low critical cooling rates ($\theta_{c,crit}$) of about 1 °Cs^{-1} due to their high carbon content. Cooling the steels at cooling rates higher than the critical cooling rate resulted in a significant increase in hardness of the specimen, namely hardness values greater than 800 HV0.2. Whereas, the lower cooling rates (0.1, 0.3 and 0.5 °Cs^{-1}) led to increase in bainite content in the microstructure. The formation of pearlitic (P), bainitic (B) and martensitic (M) and carbidic (C) phase fractions from the initial austenitic (A) phase was observed (figure 6, left) for the tool steel 1.2343. On the contrary, pearlitic phase was not observed for the tool steel 1.2367 (figure 6, right). Moreover, the high carbide percentage existing due to increased chromium content in both the steels prevents formation of bainite and instead majority of the austenite grains are transformed into martensite at cooling rates higher than the critical cooling rates. The minimum martensitic start (M_s) and end (M_f) temperatures were observed as 280 °C and 125 °C respectively for both the tool steels.

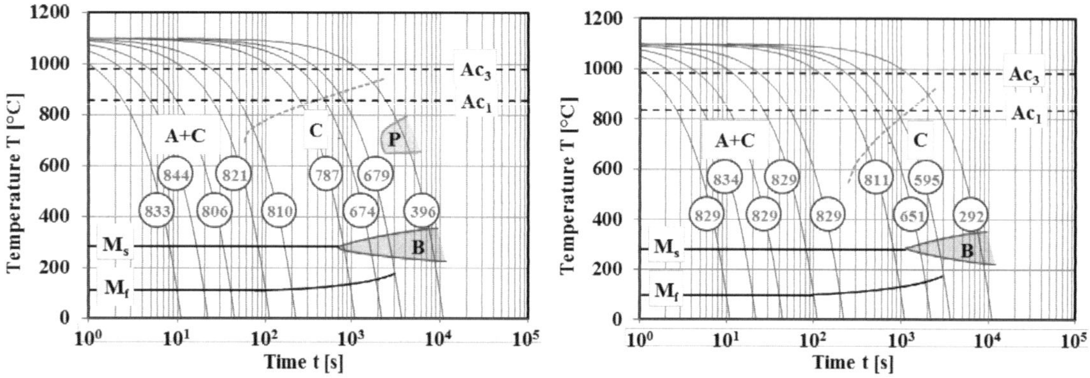

Figure 6. Continuous-time-temperature diagram for both the tool steels, 1.2343 (left) and 1.2367 (right).

Metallographic tests were performed on the specimens from experimental CCT experiments and a purely martensitic microstructure was observed for all specimens cooled with critical cooling rates or higher. The microstructure images of specimen for tests performed at the critical cooling rate are exemplarily shown in figure 7.

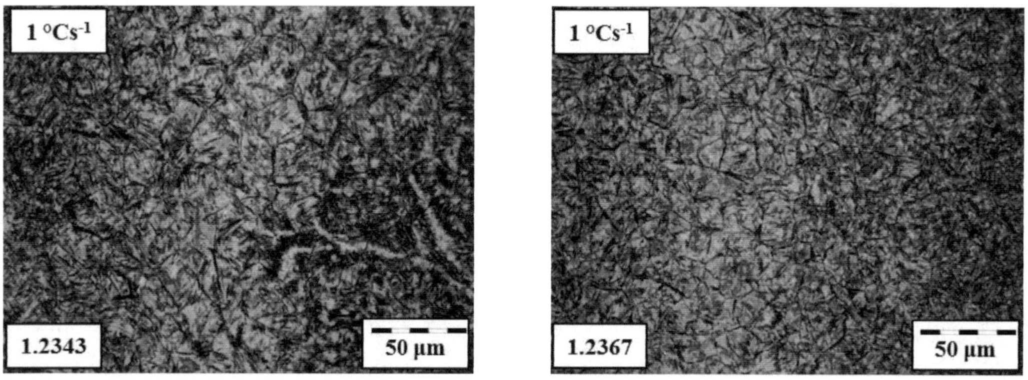

Figure 7. Specimen microstructure for tests at critical cooling rate ($\theta_{c,crit}$), 1.2343 (left) and 1.2367 (right).

4. Conclusion

Tools in hot forging experience cyclic high heating and subsequent cooling which can lead to local changes in the tool microstructure and hence a hardening or softening of the material in various tool regions consequently influencing the tool strength. Besides, high thermo-mechanical loadings acting on the tools during such processes result in temperatures exceeding the austenitic temperatures locally. In the scope of this paper, the austenitisation behaviours of two hot-work tool steels, 1.2343 and 1.2367, have been characterized by means of time-temperature-austenitization (TTA) diagrams for specimens under thermo-mechanical loads. It was observed that the application of mechanical loads, along with thermal loads, leads to lowering of the Ac_1 temperature. Moreover, continuous cooling transformation (CCT) diagrams were recorded to examine the phase transformation and hardness behaviour in the tested specimens. A fully martensitic transformation was observed for both the materials at all the cooling rates above 1 °Cs^{-1}. In future work, these findings will be used to design the temperature profiles for cyclic thermo-mechanical tests and subsequently tool hardness profiles in relation with the number of process cycles, mechanical loads and the process temperature will be evaluated in order to develop an improved wear methodology.

Acknowledgements
Results presented here are from the research project "New hardening and tempering effects II" (grant number IGF 19647N). This project has been funded by the German Federal Ministry of Economics and Energy via the German Federation of Industrial Cooperative Research Associations "Otto von Guericke" (AiF) in the program to encourage the industrial Community research by a resolution of the German Bundestag and the Steel Forming Research Society (FSV).

References
[1] Kannappan A 1969 Wear in Forging Dies – A Review of World Experiences *Metal Forming* **36** 335-43
[2] Summerville E, Venkatesan K and Subramanian C 1995 Wear processes in hot forging press tools, *Materials & Design* **16** 289-94
[3] Bartz J W and Barnert L 2004 Handbook of Tribology and Lubrication *Tribology and Lubrication in Massive Forming* (Expert-Verlag Renningen) **13** Chapter 6 140-86
[4] Saiki H, Marumo Y, Minami A and Sonoi T 2001 Effect of the surface structure on the resistance to plastic deformation of a hot forging tool *J. Mat. Proc. Tech.* **113** 22-7
[5] Behrens B-A, Bouguecha A, Vucetic M and Chugreev A 2016 Advanced Wear Simulation for Bulk Metal Forming Processes *MATEC Web Conf.* **80** 04003
[6] Dean T A and Silva T M 1979 Die temperatures during production drop forging *ASME J. Eng. Ind.* **101** 385-90
[7] Jeong D J, Kim D J, Kim J H, Kim B M and Dean T A 2001 Effects of surface treatments and lubricants for warm forging die life *J. Mat. Proc. Tech.* **113** 544-50
[8] Behrens B-A, Bouguecha A, Bonk C and Chugreev A 2017 Numerical and experimental investigations of the anisotropic transformation strains during martensitic transformation in a low alloy Cr-Mo steel 42CrMo4 *Procedia Engineering* **207** 1815-20
[9] Behrens B-A, Bouguecha A, Bonk C and Chugreev A 2017 Experimental investigations on the transformation-induced plasticity in a high tensile steel under varying thermo-mechanical loading *Computer Methods in Materials Science* **17** 36-43
[10] Wever F, Rose A, and Peter W 1961 *Atlas zur Wärmebehandlung der Stähle* **1&2** (Düsseldorf: Verlag Stahleisen).
[11] Orlich J, Rose A, and Wiest P 1973 *Atlas zur Wärmebehandlung der Stähle* **3** (Düsseldorf: Verlag Stahleisen).
[12] Behrens B-A, Chugreev A and Kock C 2018 Experimental-numerical approach to efficient TTT-generation for simulation of phase transformations in thermomechanical forming processes *IOP Conf. Ser. Mater. Sci. Eng.* **461** 012040
[13] SEP 1680 1990 STAHL-EISEN-Prüfblätter (SEP) 3rd Ed.
[14] SEP 1681 1998 STAHL-EISEN-Prüfblätter (SEP) 2nd Ed.

Modelling and analysis on the mechanism of electromagnetic driven weft insertion

Z P Zhou[1], C Zhang[2,3], C J Zhang[2] and X Z Cui[2]

[1]Wuchang Shouyi University, Wuhan, Hubei, 430064, China
[2]School of Mechanical Engineering and Automation, Wuhan Textile University, Wuhan, Hubei, 430073, China

E-mail: czhang1@wtu.edu.cn

Abstract. In view of the difficulty in improving the weaving efficiency of the projectile weaving machine based on the traditional technology principle, This research presents the electromagnetic-drive gripper weft insertion mechanism, change the existing mechanical pick principle of weft insertion, its goal is to achieve "zero transmission" friction-free high-speed weft insertion, omit complex mechanical transmission mechanism in picking shaft, and breakthrough the bottleneck of the weft insertion rate, wide fabric problem in principle. Based on the theoretical study of electromagnetic projection and electromagnetic levitation, the structural model and motion model of electromagnetic pick and electromagnetic weft are established by comparing the electromagnetic and permanent magnetic model schemes. In order to meet the requirement of electromagnetic levitation, the linear hall sensor is selected and its driving circuit is designed to obtain the real-time height value of the chip shuttle. Through the analysis of the coil response, the advantages and disadvantages of DC-DC regulation and PWM wave regulation are compared, and the high response hardware circuit of DC-DC regulation is designed. MATLAB simulation ensures the feasibility of PID regulation. The feasibility of the scheme was verified by Maxwell electromagnetic picking simulation and electromagnetic filling insertion simulation. The first phase of picking experiment and static filling insertion experiment were carried out. In order to ensure the stability, the double-row soft iron chip shuttle was used, and good experimental results were obtained, which proved the feasibility of electromagnetic suspension weft insertion scheme.

1. Introduction

The gripper loom is a kind of textile machinery used in the production of high-quality fabrics, it can be used in various yarn weaving, jacquard and industrial wide fabrics [1]. The shuttles are projected by store mechanical energy and torsion shaft to meet the weaving requirements of the shuttles. Due to the limitation of mechanism principle, there exits high impact force, noise, high friction energy consumption, the speed and shuttle efficiency of the machine is low. According to literature [2], the maximum mechanical efficiency is 18.3%. In order to further improve the efficiency of the shuttle loom, it is necessary to optimize its mechanism and improve its material properties.

Due to the low efficiency of the traditional mechanism, electromagnetic projection technology has made great progress in theory and engineering [3]. At the same time, in order to reduce the matter of high friction, vibration and material loss in the process of weft insertion of traditional gripper loom, the new idea of electromagnetic levitation was combined. In this paper, an electromagnetic driving insertion model combined with electromagnetic projection and electromagnetic levitation is presented. The

IOP Conf. Series: Materials Science and Engineering **629** (2019) 012012 doi:10.1088/1757-899X/629/1/012012

magnetic coil is used to project the shuttle, and the levitation process is realized by the magnetic levitation array system. This method is a new method to replace torsion shaft and guide in weft insertion process.

2. The principle of electromagnetic driven weft insertion

In the control system of the electromagnetic driver, the shuttle is driven by the speed-response coil and the suspension coil is directly driven, in which the speed-response coil is controlled by the speed-response controller and the suspension coil is controlled by the electromagnetic suspension controller.

The following figure shows the three different states of the shuttle.

2.1. The principle and structure of electromagnetic projecting

The basic principle of electromagnetic projection is that the electromagnetic force drives an electrically charged conductor or a ferromagnetic object in a magnetic field. The electromagnetic projection mechanism is mainly composed of coil and shuttle, and the shuttle body is composed of the shed skeleton and embedded soft iron. Electromagnetic emission model is composed of power supply, switch, coil and shuttle. When the projectile is in front of the coil, the electromagnetic force attracts the projectile to move toward the center of the coil and accelerates the projectile motion. The force slows the coil as it passes through the center of the coil. The continuous acceleration of the coil can be guaranteed by controlling the on-off sequence of the coil [4].

The mechanism uses the electromagnetic force of the coil to accelerate the shuttle. After the shuttle moves to the other side of the coil, the electromagnetic force slow it down reversely. After the bobbin is loaded onto the yarn, the electromagnetic force will re-accelerate the shuttle and repeat the filling process, so that the shuttle does not need to be reversed itself.

The formula derivation is shown in figure 1 as follows:

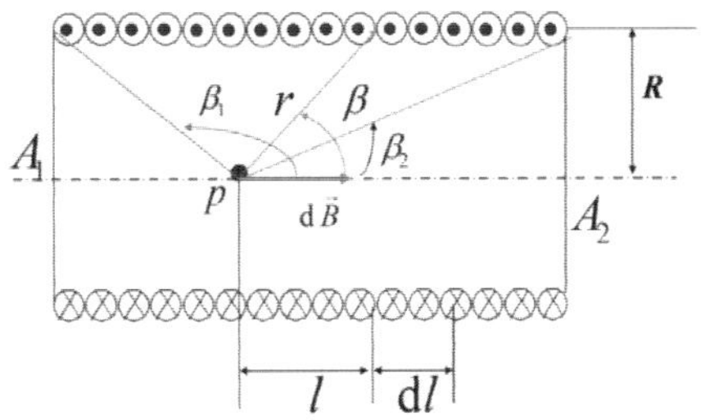

Figure 1. Coil magnetic field anatomy.

According to electromagnetism theory [5], the magnetic flux density B of the axial component generated in the coil is:

$$B = \frac{nI\mu_0}{2}(cos\beta_1 - cos\beta_2) \tag{1}$$

In the equation: n represents the number of turns per unit length; N represents the total turns of the coil; L represents the length of the coil; I represents the coil current; μ_0 represents the permeability of air; β_1 and the β_2 represent the angles between the position of the coil on the inner axis and the two ends of the coil.

The magnetic flux Φ on every turn of the coil in equation (1) is as follows [6]:

$$\Phi = BS = \frac{\pi R^2 nI\mu_0}{2}(cos\beta_1 - cos\beta_2) \tag{2}$$

The magnetomotive force generated by the energizing coil is:

$$F_m = N_i = \Phi R_m \; F_m = N_i = \Phi R_m \tag{3}$$

In this equation: R_m is magnetoresistance, $R_m = l_m/(\mu A)$; μ *is* the magnetic permeability of the shuttle material; l_m is the length of magnetic circuit of the shuttle; A is the cross-sectional area of the magnetic circuit. For the coil, the magnetic flux Φ is constant, Φ and Ni show linear relationship, the magnetic energy stored in the coil W_m as follows:

$$W_m = \int_0^{\Phi} N_i d\Phi = \frac{1}{2}\Phi^2 R_m \tag{4}$$

Equation (4) shows that magnetic energy W_m stored in magnetic field is a function of flux Φ and reluctance R_m. $W_m = f(\emptyset, R_m)$. Based on the principle of virtual work, the force exerted on the bobbin in the coil can be derived, that is, the differential of the x direction of the moving axis in equation (5) can be obtained to get the thrust generated by the drive coil on the bobbin [7]:

$$F_p = \frac{dW_m}{dx} = \frac{d\left(\frac{1}{2}\Phi^2 R_m\right)}{dx} = \frac{1}{2}NI\frac{d\Phi}{dx} \tag{5}$$

In equation (5), $d\emptyset/dx$ express the change rate of magnetic flux with displacement.

2.2. *The principle of electromagnetic levitation*

Since the speed of the shuttle is with high speed, it is necessary to design a new reed sitting to reduce the friction between the shuttle and the sitting. The suspension controller is a magnetic levitation device composed of closed - loop electromagnetic structure. When the shuttles are projected onto an electromagnet array system, the suspended shuttles can overcome gravity with the force of the magnetic field and remain suspended in the centre of the reed sitting.

Due to the influence of gravity, the flying shuttle will do parabolic movement, the width of the new shuttle weft is 8 meters. Without the sitting, the vertical difference in altitude during the projectile flight is about 20 mm at the speed of 30 m/s. Due to the difference in height, the new loom will carry out double weft insertion, which indicates that it needs an effective guide. The basic structure of electromagnetic levitation is shown in figure 2.

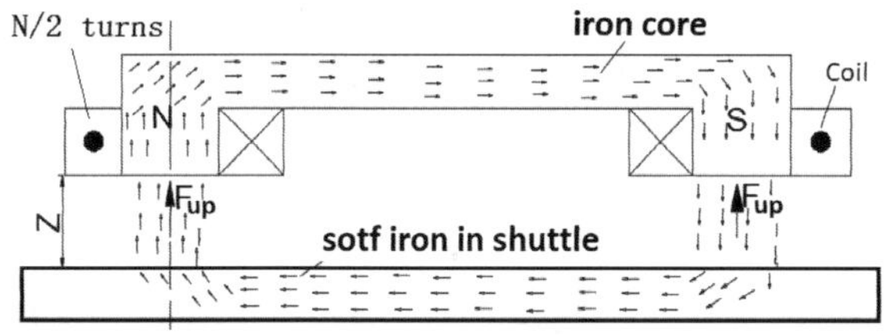

Figure 2. Soft iron stress diagram of the shuttle.

Among the picture, soft iron and iron core are ferric materials to improve magnetic permeability of magnetic circuit; The suspension height between the gripper and the coil is *Z*; Coil current size is *I*, coil turns number is *N/2* turns per side, coil inner wall area is *S*, field strength generated by coil is *B*, magnetic conductivity is μ.

From the above formula:

$$\emptyset = \frac{F_m}{R_m} = \frac{NI}{\frac{2Z}{\mu_0 S} + \frac{l_{Fe}}{\mu_{Fe} S}} \tag{6}$$

From Maxwell's suction formula:

$$F_d = \frac{B^2 S}{2\mu_0} \tag{7}$$

Formula:

$$F_d = \frac{S}{2\mu_0} \times (\frac{NI}{\frac{2Z}{\mu_0} + \frac{l_{Fe}}{\mu_{Fe}}})^2 \tag{8}$$

In equation (13), F_d is the magnitude of the total suction; S is the effective area of magnetic circuit of magnetic pole; while, μ_0 is magnetic conductivity in vacuum; *N* is coil turns; *I* is coil current; *Z* is air gap height; and l_{Fe} is the equivalent total length of coil core and soft iron in gripper.

2.3. Dynamic analysis
When the soft iron takes the gripper along flying in the suspension electromagnet array, the dynamic stress analysis is shown in figure 3 as follows:

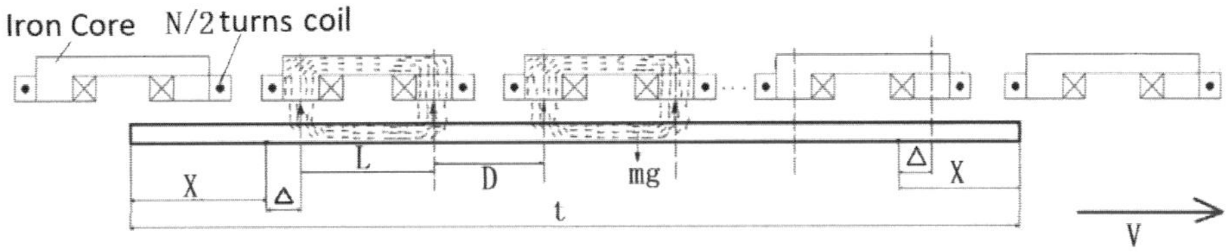

Figure 3. Structure of filling insertion.

In figure 4, when the gripper is flying to from left to the right, the two red mark points on the gripper are the coil-end detecting point and the coil-start detecting point respectively. Suppose the number of driving sources to work simultaneously is *n*, then

$$t = 2 \times X + (n + 1) \times L + n \times D \tag{9}$$

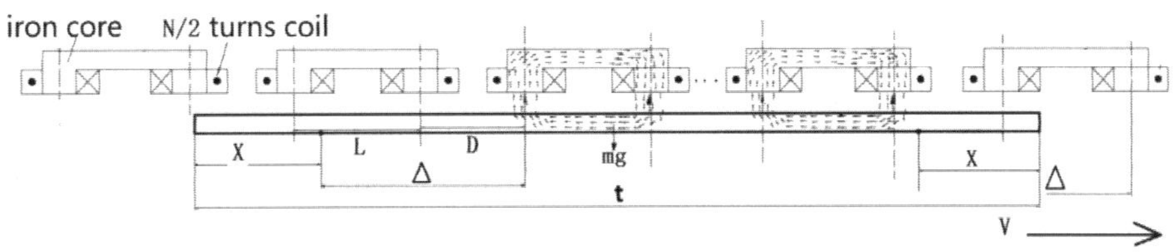

Figure 4. Dynamic filling insertion analysis.

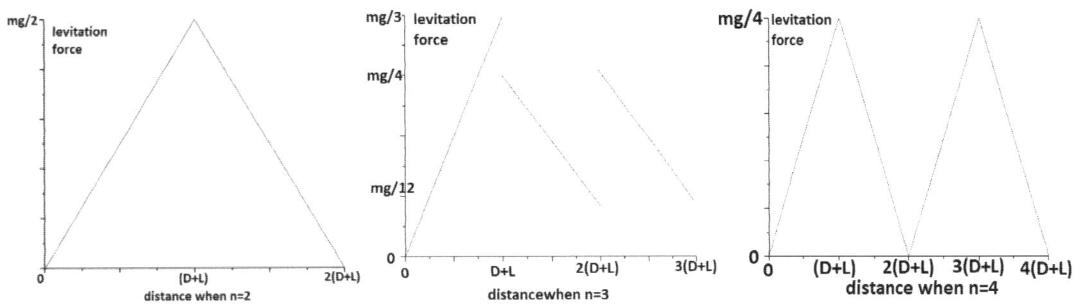

Figure 5. *n* relation with levitation force.

In figure 5, Let the distance between the left node in the left-most excitation source and the left marking point is *K*, and the balance equation is:

$$2 \times (F1 + F2 + \cdots + Fn) = mg \tag{10}$$

IOP Conf. Series: Materials Science and Engineering **629** (2019) 012012 doi:10.1088/1757-899X/629/1/012012

$$\left(\frac{t}{2}-\mathrm{X}-\mathrm{K}\right)\times F1+\left(\frac{t}{2}-\mathrm{X}-\mathrm{K}-\mathrm{L}\right)\times \mathrm{F1}+\cdots+\left(\frac{t}{2}-\mathrm{X}-\mathrm{K}-(\mathrm{n}-1)\mathrm{D}-(\mathrm{n}-1)\mathrm{L}\right)\times \mathrm{Fn} + \left(\frac{t}{2}-\mathrm{X}-\mathrm{K}-(\mathrm{n}-1)\mathrm{D}-\mathrm{nL}\right)\times Fn=0 \tag{11}$$

Due to too many parameters, we cannot be solved it by MATLAB according to equations (9), (10), (11), and we can't get the relationship of them.

Solution: Set n (numbers of drivers working at the same time) as known quantity and consider them one by one. We need consider $n = 2, 3, 4$.

Reason: limited constraint conditions (there are only four equations: force balance, moment balance, initial condition, termination condition).

It can be seen from figure 6 that the required suspension force is continuous when n=2 and n=4, which is convenient to achieve.

We can't solve this problem because we have finite constraints when n>5.

After analyses of the above three cases, the feasible models are n=2 and n=4. Considering that increasing the levitation force of the excitation source requires increasing the current, coil heating is more serious. In this sense, the solution is better when $n = 4$.

3. Electromagnetic simulation

Before the experiment, we need to use software simulation to predict the feasibility of the experiment. Meanwhile, in the simulation process, the best parameters can be simulated to facilitate the experimental debugging. As shown in figure 6, The experimental model consists of coil array and soft iron. and the coil is accelerated continuously by controlling the on-off of the coil sequence.

Figure 7 shows the time-speed relationship simulation diagram.

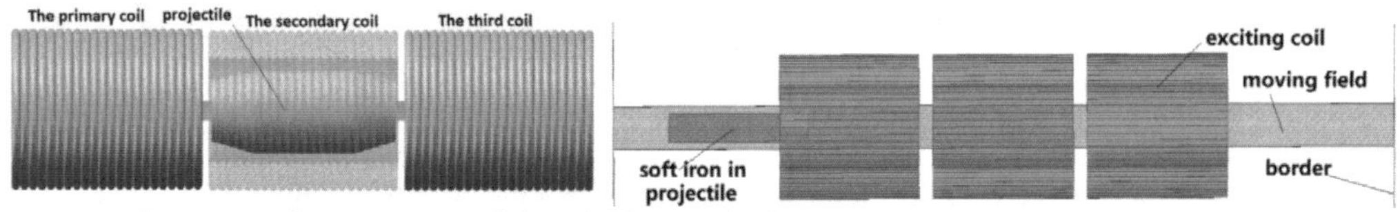

Figure 6. The structure of the picking and Electromagnetic simulation diagram.

Figure 7. Time-Velocity Transient simulation diagram.

Through simulation analysis, the shuttle speed can be up to 6.5 m/s in 0.01 s.

After the projecting is completed, shuttle enters the process of levitation and insertion. At this point, the shuttle should be suspended and flying forward, and a controllable jitter can occur. The control of jitter needs to be adjusted through feedback. In the simulation process, the noise force should be as low as possible, the more stable the suspension force, the better.

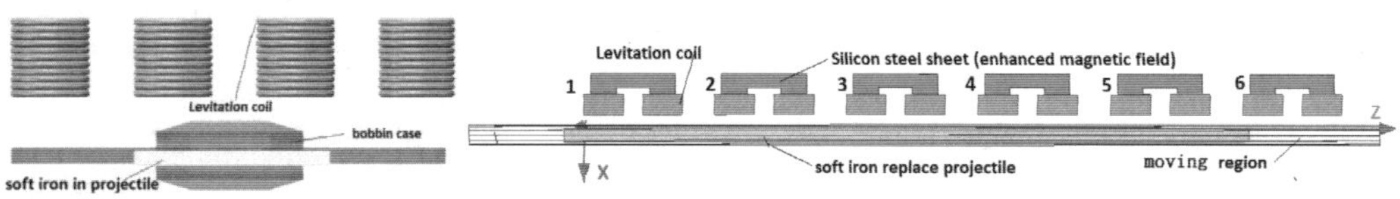

Figure 8. Electromagnetic levitation structure and simulation.

Figure 8 shows the dynamic suspension model and simulation diagram.

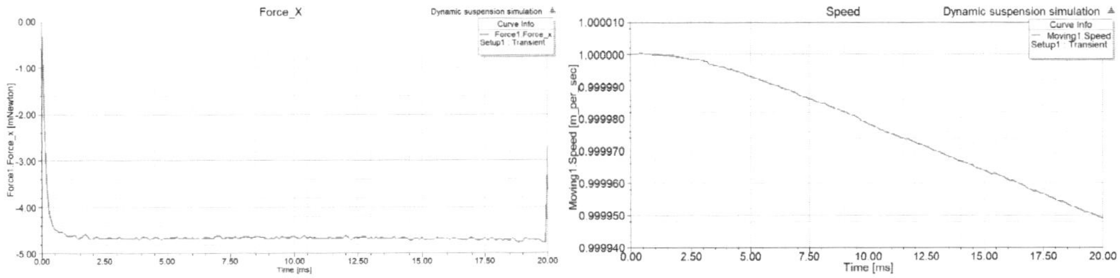

Figure 9. The force diagram of suspension direction and motion direction in software simulation.

As can be seen from the figure 9, the speed of the projectile changes from 1 m/s to 0.99995 m/s, which is in a continuous downward trend. It can be seen that the generation of minimal vortexes has a minimal impact on the speed of the projectile.When the gripper flies in dozens of arrays of levitating coil elements, the speed reduction is still very small. According to the simulation results, the weft insertion scheme is feasible in mechanical mechanism layout and suspension principle, and can be adjusted according to the actual application environment.Meanwhile, the feedback control of the levitation force of the coil can be used for stable and reliable adjustment of a good hardware platform.

4. Simulation of static suspension control system

As shown in figure 10, the excitation source is the excited coil, the control object is the suspended object (soft iron), and the hall sensor is located at the bottom.

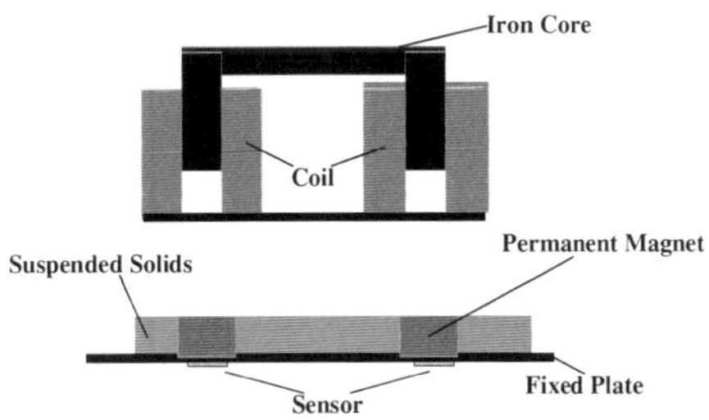

Figure 10. Suspension structure diagram.

Set the lower end of the coil as zero, the initial distance as 50 mm, and the target suspension height as 40 mm (10 mm higher than the initial point).

$$\mathrm{F} = \frac{K1 \times I}{X^3} - mg \tag{12}$$

$$a_x = \frac{F}{m} \tag{13}$$

$$v_x = \int a_x dt \tag{14}$$

$$\mathrm{x} = \int v_x dt \tag{15}$$

In the above formula, F is the total force exerted on the suspension, I is the exciting current, and X is the distance. $K1$ is the parameter that determines the relation between suspension force and I and x, and

Mg is gravity. The driving equation of *I* be:

$$\mathrm{I} = -\mathrm{K2} \times (\mathrm{X} - \mathrm{Xt}) - \mathrm{K3} \times v_x - K4 \times a_x \tag{16}$$

Let the initial conditions be *x0*=50 mm, *M*=0.0125, and omit the *K1* parameter (*K2*, *K3* and *K4* are also parameters). The expected convergence is *x0*=40 mm. MATLAB *ode45* function to solve.

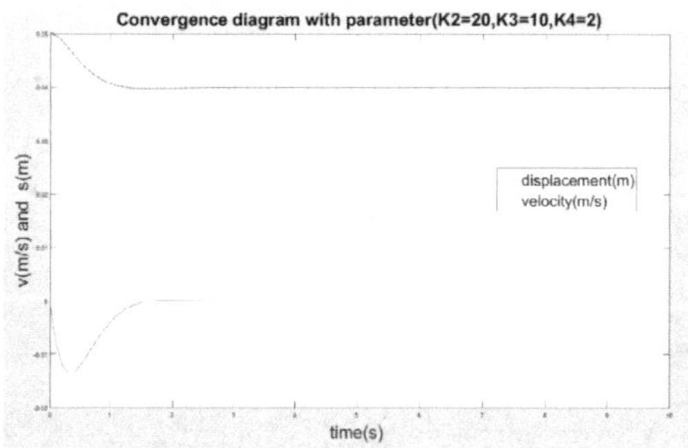

Figure 11. The velocity and displacement of the floating direction projectile (*K2=20, K3=10, K4=2)*.

In figure 11, *K2*, *K3* and *K4* are respectively the parameters of displacement (*X*), velocity (Vx) and acceleration (a_x). Because you can't get the instantaneous Vx and a_x directly. It can only be inferred from position *X* and time. Firstly, the relationship between the magnetic field intensity and the height of the projectile is measured experimentally.

5. Experiment and result analysis

After the software simulation proves that the scheme is feasible, we start the experiment of electromagnetic projection and magnetic levitation. In the process of projection, velocity sensor is used to detect the velocity of the shuttle in real time, and dry battery and large capacitance are used as generating device of the magnetic field in figure 12. Experiments were carried out for different initial positions and different voltages. It can be seen that the experimental results are similar to the simulation results.

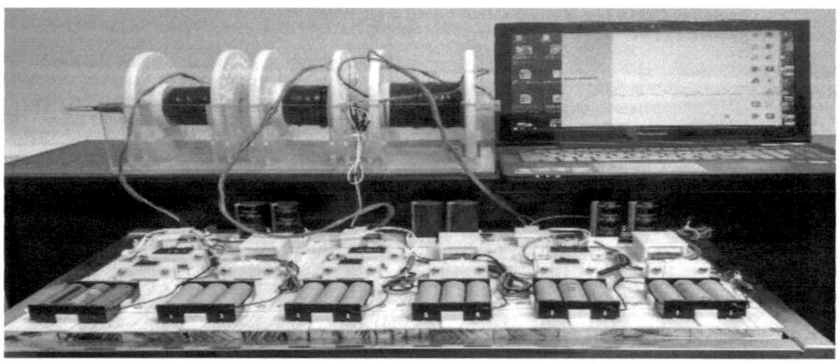

Figure 12. Picking experiment.

Figure 13 shows different exit velocities at different initial positions with the same voltage 250 V and shows exit velocity at the same initial position at different voltage.

In the suspension stage, the real-time magnetic field is monitored through Hall sensor under the shuttle and the coil current is adjusted according to the control requirements. In order to improve response speed, the high speed operational amplifier module is used. Single chip microcomputer is used for voltage regulation responding to a four-way hall sensor, the experimental process is shown in figure

14.

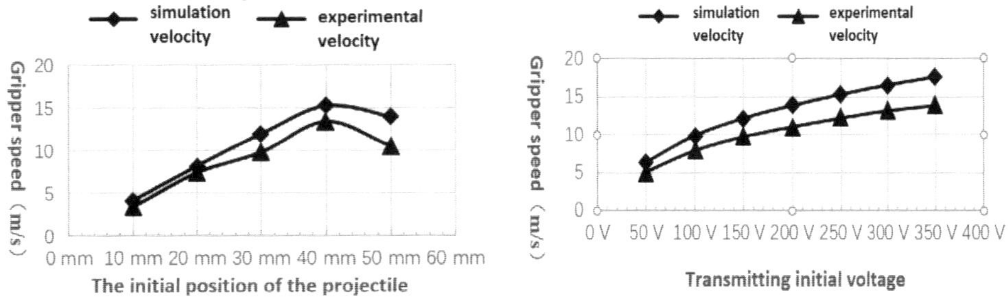

Figure 13. Speed comparison diagram with same voltage and Speed comparison diagram with same voltage position.

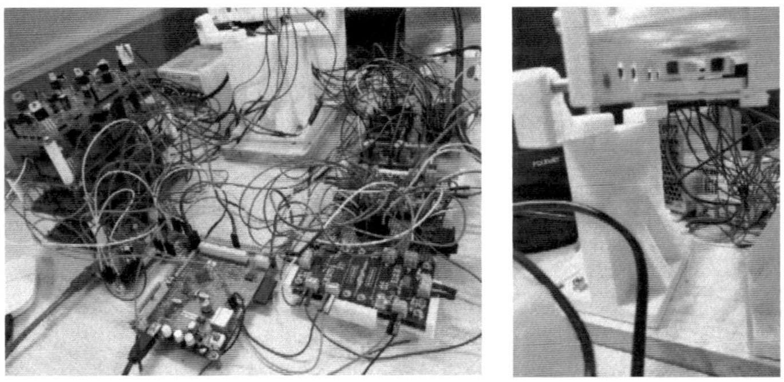

Figure 14. Static suspension experiment diagram.

With the parameters above, as we can see in figure 15, the response time reached 0.1 s, the experimental results show that the shuttle is vibrating up and down (in contact with the sensor and coil).

As we can see in figure 16, by changing the driving mode and subdivision degree, shuttle jitter is reduced by 10 times, which basically meets the experimental requirements. In order to further ensure the stability of the experiment, the high-subdivision, low-deviation DC-DC Analogy adjust circuit needs to be studied. The sensor with high performance and low power consumption is selected at the same time.

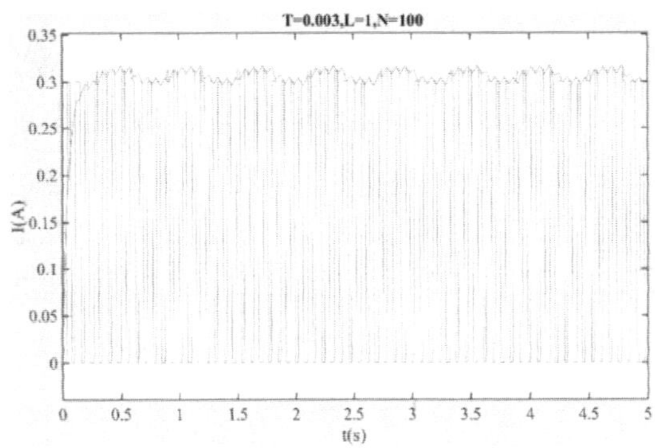

Figure 15. PWM driver response diagram.

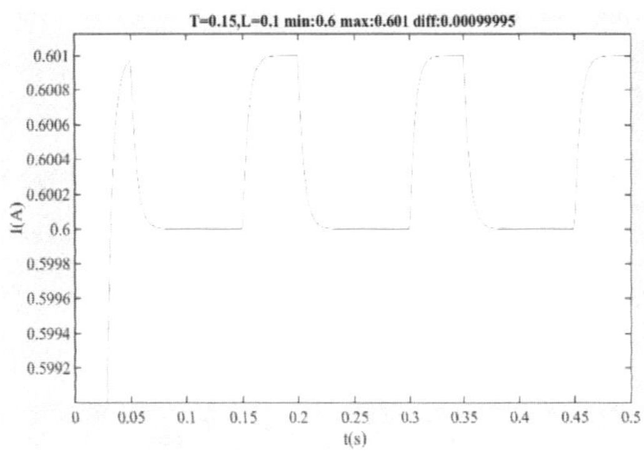

Figure 16. DC driver response diagram.

6. Conclusion

This paper introduces a new electromagnetic-driven dual-directional weft insertion loom, electromagnetic transmitting mechanism and magnetic levitation system, after analyzes their working principle, makes a detailed comparison with the original gripper loom. The analysis results show that the structure of the new type of loom simplifies the structure of the loom, it can significantly improve the rate of weft insertion, and reduces the impact and vibration during the process of weft insertion with high speed. We use the built-in A/D module of stm32F4 to collect the magnetic field data of hall sensor HW101, and use the high-speed D/A module TLV5618 as the control unit of the coil output voltage. With a quick response and feedback, levitation becomes possible. this paper provides a basis for further research.

Acknowledgment

The work was supported by National Natural Science Foundation of China (Grant No. 51305309, 51875414) and Natural Science Foundation of Hubei Province (Grant No. 2017CFB585).

References

[1] Chen M 1981 Torsion bar weft insertion mechanism *Journal of East China Institute of Textile Technology* **4** 48-55
[2] Xu H Y 2003 Technical progress of projectile loom *Beijing Textile Journal* **4** 43-6
[3] Duan J R 2013 Analysis of weft yarn velocity on the projectile weaving machine *Journal of Xi' an Polytechnic University* **6** 711-8
[4] Zhang H, Kou B Q, Jin Y X *et al* 2016 A cylindrical magnetic levitation gravity compensator with Halbach secondary structure *Transactions of Chinese Electrotechnical Society* **6** 30-7
[5] Liu Z W, Chen X M, Dong J *et al* 2015 The optimization of multistage electromagnetic launching system of mini-SHB *Journal of Experimental Mechanics* **1** 9-16
[6] Yan J C *Electromagnetism* (Hefei: Press of University of Science and Technology of China, 1997)
[7] Wang J, Wu H *et al* 2016 Performance analysis and optimization of high speed switch valve based on PWM control mode *Journal of Hefei University of Technology (Natural Science)(JHUT)* **9** 32-5

IOP Conf. Series: Materials Science and Engineering **629** (2019) 012013 doi:10.1088/1757-899X/629/1/012013

Crack propagation in mixed-mode specimens described via multi-parameter fracture mechanics

L Malikova[1,2,3], P Miarka[1,2] and H Simonova[1]

[1]Brno University of Technology, Faculty of Civil Engineering, Institute of Structural Mechanics, Veveri 95, 602 00 Brno, Czech Republic
[2]Institute of Physics of Materials, Czech Academy of Sciences, Zizkova 22, 616 62 Brno, Czech Republic

E-mail: malikova.l@fce.vutbr.cz

Abstract. The main objective of the paper is to employ a multi-parameter fracture mechanics concept to describe crack propagation through a specimen loaded in mixed-mode. This concept in particular was used because it has been shown that application of the generalized fracture mechanics concept can play a key role for materials with specific fracture behaviour, i.e. when fracture processes occur not only in the very vicinity of the crack tip, but also at larger distances from it. Two mixed-mode (I+II) geometries for the investigation of crack behaviour are presented here. The Williams series expansion is used for crack-tip stress field approximation. Then, considering the higher-order terms of the Williams expansion with regard to maximum tangential stress criterion can provide better estimates of the crack deflection angle. The coefficients of the Williams expansion were determined by means of the over-deterministic method for the purposes of this work. This analysis was performed for each cracked configuration, which is very time-consuming and makes the analysis very extensive. The crack propagation angle obtained by means of the generalized fracture criterion is discussed in detail. It was found that single-parameter fracture mechanics is sufficient when applied close to the crack tip and when mode I of loading prevails, while multi-parameter fracture mechanics can be recommended at larger distances from the crack tip and for configurations where mode II becomes dominant.

1. Introduction

It is very important to study crack behaviour under mixed-mode (I+II) loading because it often occurs in practice. It has been observed that cracks try to propagate in such a manner as to minimize loading mode II, and that they attempt to propagate under pure mode I if possible. Nevertheless, before that the crack is mostly subjected to arbitrary mixed-mode loading. Therefore, it is very reasonable to study the crack path of defects that exist in real structures.

Several approaches can be chosen for the investigation of the fracture behaviour of cracked configurations that arise often because of mechanical or thermal loading or just due to the manipulation of these structural elements. The basic approach is single-parameter linear elastic fracture mechanics, which has been studied extensively [1-4]. Nevertheless, this concept is fully valid only for brittle materials, where the singular crack-tip stress field can be described via the stress intensity factor sufficiently, i.e. where plastic deformation does not occur at all, or only to a very small extent. On the other hand, many materials (such as quasi-brittle or elastic-plastic ones) exist in which fracture processes take place at a greater distance from the crack tip and are more complicated. Consequently, the first task before a complex fracture analysis can begin is to describe the crack-tip stress field reliably. It has been shown that multi-parameter fracture mechanics can be a suitable method of doing this. Furthermore, the multi-parameter approximation of the crack-tip stress field can be included in the fracture criteria that decide crack stability and/or crack propagation direction.

In this work, the maximum tangential stress criterion is utilized to estimate the crack deflection angle under mixed-mode (I+II) loading. The analysis is performed on two selected cracked configurations and the results obtained are discussed with regard to the choice of the proper critical distance at which the criterion should be applied, as well as with regard to the suitable choice of the number of the initial terms of the Williams expansion to be considered for the tangential stress approximation.

2. Methodology and basic theoretical terms

Basic terms and methodology used within the fracture analysis conducted for this research are presented and explained in the following subsections.

2.1. The Williams power series

The basic idea behind this work consists in the application of the Williams expansion [5] for the reconstruction of a crack-tip stress field. The Williams power series was derived for a homogeneous elastic isotropic cracked body with arbitrary remote loading and has the following form:

$$\sigma_{ij} = \sum_{n=1}^{\infty} A_n \frac{n}{2} r^{\frac{n}{2}-1} f_{ij}(n,\theta) + \sum_{n=1}^{\infty} B_m \frac{m}{2} r^{\frac{m}{2}-1} g_{ij}(m,\theta)\ , \text{ where } \ i,j \in \{x,y\}\ . \qquad (1)$$

Thus, each stress tensor component can be expressed as a function of the polar coordinates (r, θ) of the corresponding point, the known functions f_{ij}, g_{ij} and coefficients A_n, B_m depending on the cracked geometry that must be calculated numerically. The series represents a summation over indexes n and m and is used in its truncated form, i.e. N and M terms are considered for mode I and II, respectively.

2.2. Determination of the coefficients of the Williams expansion

As is stated above, it is necessary to calculate the coefficients of the Williams power series in order to be able to reconstruct the crack-tip stress field. Although more methods exist [6-8], one of them seems to be easier to use than the others because of its low requirements with regard to software and mathematical definitions and that is the over-deterministic method (ODM). It is based on the direct application of the equation for the description of displacement vector components:

$$u_i = \sum_{n=0}^{\infty} A_n\, r^{\frac{n}{2}} f_i(n,\theta,E,\nu) + \sum_{n=0}^{\infty} B_m\, r^{\frac{m}{2}} g_i(m,\theta,E,\nu)\ , \text{ where } \ i \in \{x,y\}\ . \qquad (2)$$

A set of nodes (usually at a radial distance from the crack tip) are selected in the numerical model of the cracked specimen and their displacements under mixed-mode loading are calculated. The obtained values are then used as inputs for equation (2) together with the polar coordinates of the corresponding nodes. Taking into account the elastic properties of the material (Young's modulus E and Poisson's ratio ν), it is then possible to solve a set of $2k$ equations, where k is the number of nodes selected around the crack tip. It has to hold that $2k \geq N + M + 2$ in order to satisfy the condition about the over-determined system of equations. More details and recommendations on using the ODM can be found for instance in [9,10].

2.3. The multi-parameter maximum tangential stress (MPMTS) criterion

When the coefficients of the Williams expansion are known, an arbitrary stress tensor component can be reconstructed in both Cartesian and polar coordinates. When the Maximum Tangential Stress (MTS) criterion is used, the tangential stress $\sigma_{\theta\theta}$ has to be expressed via the Williams power series. According to the theory behind the MTS criterion, the crack propagates in the direction of the maximum tangential stress [3]. When the MPMTS form of the criterion is considered, it is necessary to derivate the Williams power series expression of the tangential stress and fulfil the following equation to find the proper crack deflection angle:

$$\frac{\partial \sigma_{\theta\theta}}{\partial \theta} = 0 \quad \text{and} \quad \frac{\partial^2 \sigma_{\theta\theta}}{\partial \theta^2} < 0 \,. \tag{3}$$

The procedure of searching for the maximum of the function of the tangential stress is performed numerically using Wolfram Mathematica software [11].

3. Mixed-mode I+II configurations

As mentioned in the previous sections, two mixed-mode configurations have been chosen for the analysis of the crack deflection angle under various levels of mixed-mode conditions: an Eccentric Asymmetric 4-Point-Bending (EA4PB) specimen and a Semi-Circular Bending (SCB) specimen; see figure 1 for more details about the geometry. Both of the cracked configurations make it quite easy to choose a certain level of mixed-mode conditions by changing one selected parameter: in both cases all the following parameters were kept except one: crack eccentricity e in the EA4PB specimen and crack inclination angle β in the SCB specimen.

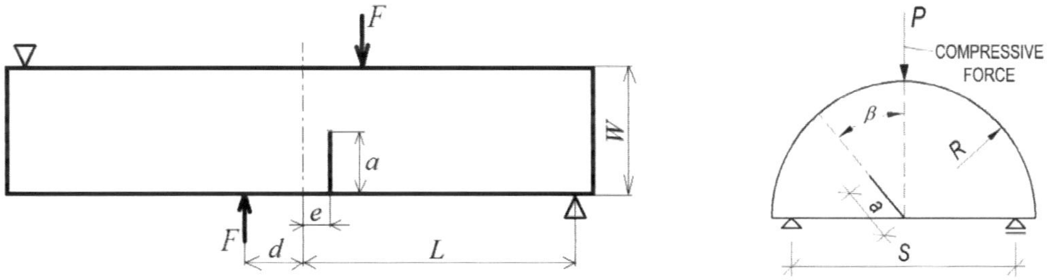

Figure 1. Schema of the mixed-mode I+II cracked configurations under study: (a) Eccentric Asymmetric 4-Point-Bending (EA4PB) specimen; (b) Semi-Circular Bending (SCB) specimen.

The dimensions considered invariable in the numerical models of the specimens presented in figure 1 are as follows (note that the numerical models were modelled as two-dimensional and plane strain conditions were assumed):

- EA4PB:
  - ➢ half distance between supports $L = 100$ mm;
  - ➢ half distance between loading forces $d = 20$ mm;
  - ➢ specimen width $W = 40$ mm;
  - ➢ relative crack length $a/W = 0.3$.
- SCB:
  - ➢ distance between supports $S = 80$ mm;
  - ➢ specimen radius $R = 50$ mm;
  - ➢ relative crack length $a/W = 0.3$.

In contrast, the parameters that were varied in order to study various levels of mixed-mode loading were the crack eccentricity in the EA4PB specimen e ($e/W = 0$, 0.1, 0.2, 0.3 and 0.4) and the crack inclination angle β ($\beta = 10$, 20, 30, 40 and 50°), respectively.

The elastic parameters of the material model in the numerical simulations were set according to the properties of common concrete, i.e. Young's modulus 35 GPa and Poisson's ratio 0.2. ANSYS commercial finite element software [12] was used for the simulations and quadrilateral element types (PLANE183) were applied when meshing the cracked specimen with refinement of the mesh at the crack tip.

Note that the MPMTS criterion was applied at three different critical distances ($r_c = 0.1$, 0.5 and 1.0 mm); the influence of this parameter is discussed in the following section.

4. Results and discussion

IOP Conf. Series: Materials Science and Engineering **629** (2019) 012013 doi:10.1088/1757-899X/629/1/012013

In the following section, the results obtained based on the multi-parameter linear elastic fracture mechanics analysis are presented. In figure 2, dependences of the crack deflection angle γ estimated by means of the MPMTS criterion on the K_I/K_{II} ratio (which represents the level of the mixed-mode loading) are presented. The dependences were obtained at various critical distances (r_c = 0.1, 0.5 and 1.0 mm) at different mixed-mode I+II cracked configurations (EA4PB and SCB) and considering various numbers of the initial Williams expansion terms (N, M = 1, 2, 3, 4 and 5).

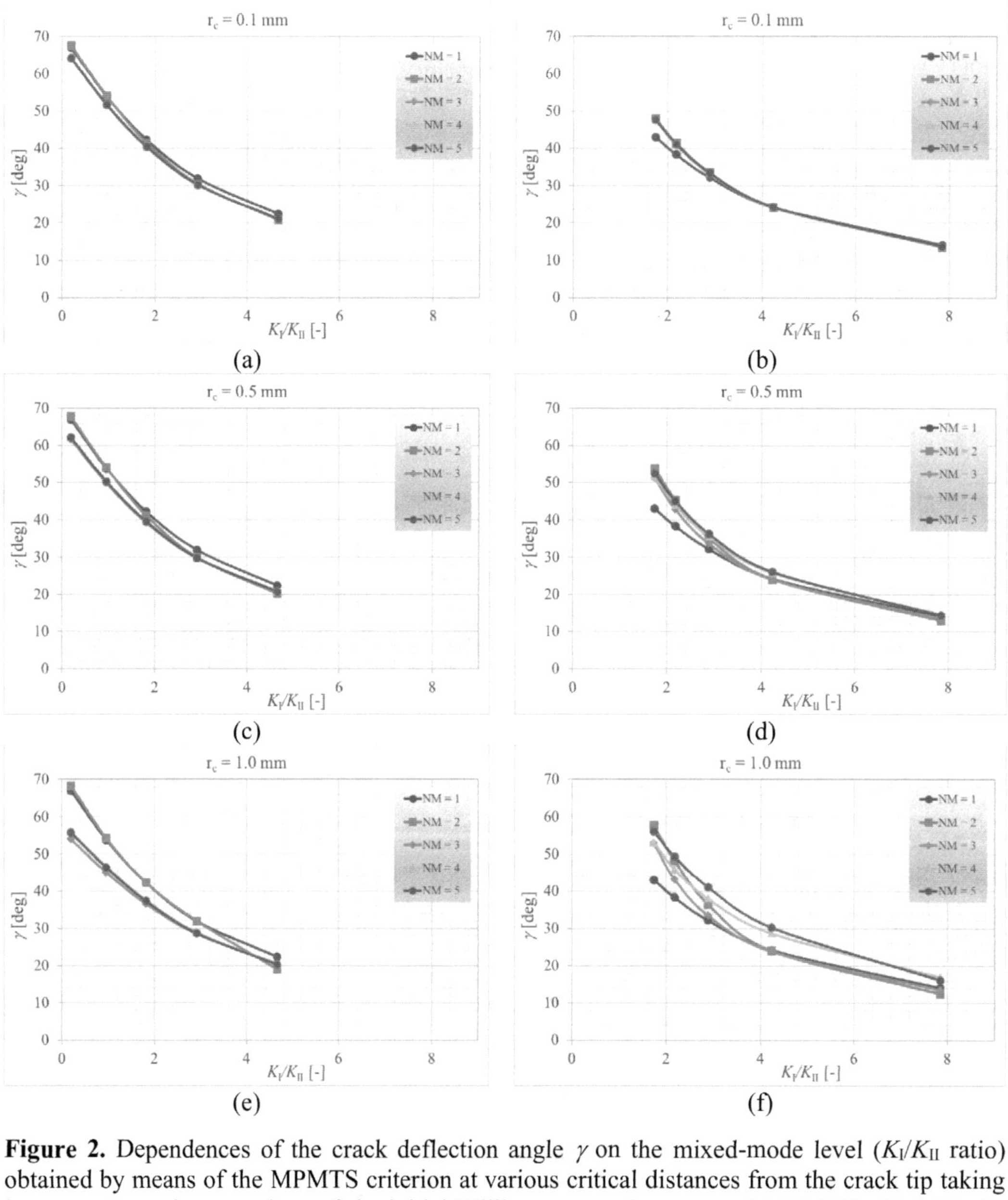

Figure 2. Dependences of the crack deflection angle γ on the mixed-mode level (K_I/K_{II} ratio) obtained by means of the MPMTS criterion at various critical distances from the crack tip taking into account various numbers of the initial Williams expansion terms. (a) EA4PB, (b) SCB, (c) EA4PB, (d) SCB, (e) EA4PB and (f) SCB.

From the dependences presented in figure 2 the following statements can be formulated:

- The mixed-mode I+II conditions modelled via the EA4PB specimen include cases where mode II of loading prevails ($K_{II} > K_I$).
- The resultant crack deflection angles γ obtained for both mixed-mode specimens qualitatively exhibit the same trends.
- At very small distances close to the crack tip ($r_c = 0.1$ mm), the differences between the crack deflection angles γ obtained by considering one or more terms of the Williams expansion are not very significant.
- The differences between the dependences determined by considering various numbers of the Williams expansion terms grow with increasing critical distance r_c at which the MPMTS criterion is applied.
- Similarly, the results of the analysis prove that the higher-order terms of the Williams expansion play a key role when the crack is loaded under mixed-mode I+II conditions with mode II as the dominant mode of loading.

5. Conclusions

Within this work a parametric study was performed on two mixed-mode (I+II) configurations in order to investigate the importance of the higher-order terms of the Williams power expansion used for the approximation of the tangential stress in the multi-parameter maximum tangential stress (MPMTS) criterion. Based on the dependences obtained, it can be concluded that single-parameter fracture mechanics is sufficient when applied close to the crack tip and when mode I of loading prevails, while multi-parameter fracture mechanics can be recommended at larger distances from the crack tip and for configurations where mode II becomes dominant. The choice of the proper distance at which the fracture criterion should be applied can depend on several different factors and has to be made responsibly with regard to material composition and material properties.

Acknowledgments
Financial support from the Czech Science Foundation (project No. 18-12289Y) and from the Faculty of Civil Engineering, Brno University of Technology (project No. FAST-S-19-5896) is gratefully acknowledged.

References

[1] Anderson T L 1995 *Fracture Mechanics: Fundamentals and Applications* (Florida: CRC Press LLC)
[2] Ayatollahi M R and Abbasi H 2001 *Build Res. J.* **49** 167-80
[3] Erdogan F and Sih G C 1963 *J. Basic Engng. Trans. ASME* **85** 519-27
[4] Sih G C 1974 *Int. J. Fract.* **10** 305-21
[5] Williams M L 1957 *J. Appl. Mech. (ASME)* **24** 109-14
[6] Karihaloo B L and Xiao Q Z 2001 *Engng. Fract. Mech.* **68** 1609-30
[7] Tong P, Pian T H H and Lasry S J 1973 *Int. J. Num. Meth. Engng.* **7** 297-308
[8] Xiao Q Z, Karihaloo B L and Liu X Y 2004 *Int. J. Fract.* **125** 207-25
[9] Ayatollahi M R and Nejati M 2010 *Fatigue Fract. Engng. Mater. Struct.* **34** 159-76
[10] Šestáková L 2013 *Appl. Mech. Mater.* **245** 120-5
[11] *Wolfram Mathematica Documentation Center* 2018 (Wolfram Research: Champaign)
[12] *Ansys Program Documentation* 2019 (Swanson Analysis System: Houston)

Thermal response analysis of reinforced concrete box girder during hot-mixed asphalt mixture paving

D Zheng[1], Z D Qian[1,2], D X Liu[1] and Q B Huang[1]

[1]Intelligent Transportation System Research Center, Southeast University, Nanjing 210096, China

E-mail: qianzd@seu.edu.cn

Abstract. Hot-mixed asphalt (HMA) concrete has been extensively used in the concrete box-girder bridge deck pavement due to its excellent performance and driving comfort. Nevertheless, during HMA concrete construction, the high paving temperature may disturb the thermal field of the bridge, thereby causing the significant thermal response throughout the box girder. Therefore, this research aims to investigate the temperature behavior and thermal stress of a reinforced concrete box girder segment during HMA concrete paving. The transient thermal field theory and element deletion method were first adopted to establish a three-dimensional finite element model of a reinforced concrete box girder segment and simulate the dynamic paving process, respectively. Subsequently, the bridge monitoring data was used to validate the thermal field model and then the temperature distribution analysis was conducted. Finally, the thermal stress in the bridge was studied in cross-sectional area. Results demonstrate that the vertical temperature of the concrete box-girder during the paving process is declined with the increasing vertical distance to SFRC leveling layer, and its highest value occurs at 36min after starting to paving. Furthermore, in the vertical direction, the west web-plate is subjected to compressive stress at locations of two ends and tensile stress at locations of the middle. Meanwhile, the thermal stresses undergo a decline trend after short-term growth both in the box and flange slab in the transversal direction.

1. Introduction

In practice, the temperature-induced cracking has become one of common early distresses that occur in the bridge superstructures [1]. Exposed to the thermal actions (air temperature variations, solar radiation and so on), a concrete bridge is ineluctably subjected to the non-uniform thermal field, thereby causing the thermal deformations in the bridge components due to the thermal expansion and contracting of materials [2]. Once these deformations are retrained, the significant thermal stress will occur in the bridge, which might exceed the load-induced stress [3,4]. As a result, excessive thermal stress threatens the serviceability of bridge.

Currently, most researches have focused on the thermal field characteristics of the bridge in the natural environment. For instance, Zhao and Wang [5] studied the thermal field of a concrete bridge against different temperature boundary conditions. Gu *et al* [6] used the numerical simulation method to investigate the effect of wind speed, atmospheric cleanliness, bridge orientation and latitude, on the three-dimensional temperature changes throughout the box girder. Nevertheless, fewer attentions have been devoted to the thermal effects of the concrete box girder when hot-mixed asphalt (HMA) concrete is paving on the bridge deck. With the development of hot-mixed modified asphalt binder, the paving temperature of HMA pavement is higher than that of conventional asphalt pavement, which may cause the significant thermal effects in the box girder. Moreover, steel fiber reinforced concrete (SFRC) has been applied as a leveling layer in many concrete bridges, and its high impact strength, toughness and thermal conductivity may contribute to the temperature changes of the concrete box girder.

In this research, the thermal response of the concrete box-girder segment containing SFRC leveling layer during HMA concrete paving was analyzed using a three-dimensional thermal field model. In detail, the temperature distribution characteristics were studied based on the established finite element (FE) model, followed by the mechanical response (thermal stress) analysis of the bridge during the paving process.

2. Description of the bridge and temperature monitoring

In this research, Dantu main line concrete box-girder bridge (Dantu Bridge) in China was selected for establishing the numerical model. Figure 1 shows the cross section of Dantu Bridge which is elongated in north-south direction. The box girder is 130 cm in height and 135 cm in width. In addition, there are 5 cm-thick SFRC leveling layer, 6 cm-thick sublayer asphalt pavement and 4 cm-thick asphalt concrete surface upon the box-girder section.

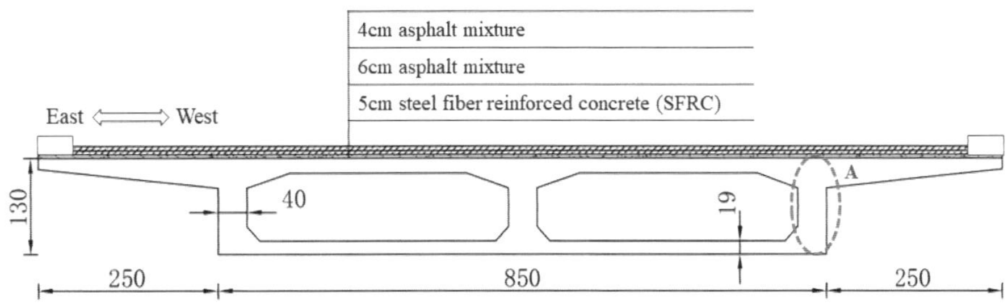

Figure 1. Cross section of Dantu Bridge (Units are centimeter).

In a mid-span cross section of Dantu Bridge, the temperature monitoring was conducted on October 20, 2004 to record the vertical temperature in real time during the asphalt concrete paving, using the resistance thermometers (RT-1) [7,8]. The arrangement of twelve vertical monitoring points in the west web-plate is shown in figure 2.

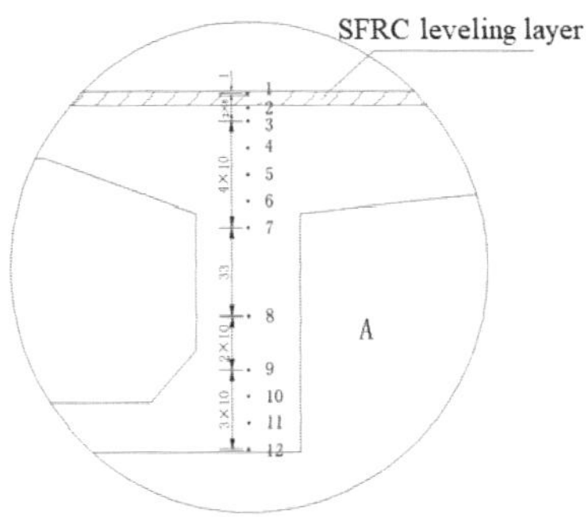

Figure 2. Arrangement of the vertical monitoring points in the west-plate.

3. Numerical simulation

3.1. Finite element model for thermal analysis

According to the geometrical configuration of the bridge in figure 1, a 21 m-long segment model of the concrete box-girder was developed using the commercial FE package ABAQUS 6.11, given that there is no obvious longitudinal temperature change on the non-paved bridge deck [9]. The FE model of concrete box-girder is shown in figure 3.

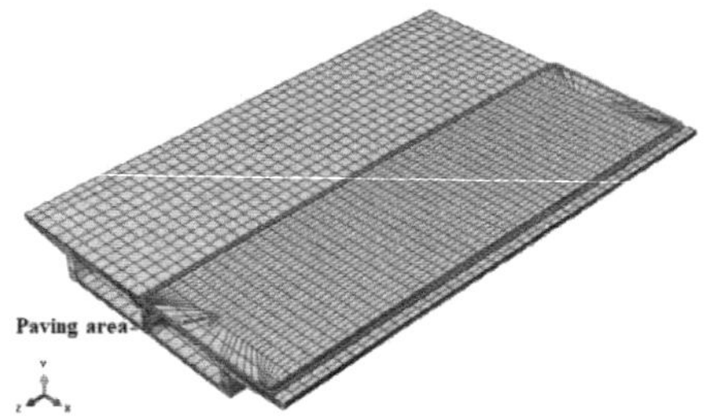

Figure 3. FE model of concrete box-girder.

3.2. Temperature boundary conditions and material parameters

In the paving process of HMA concrete, the paver could be seen as a movable heat producer. Therefore, the temperature variation in the concrete box-girder during HMA concrete construction is treated as the transient temperature field [10]. In addition, the continuous temperature change of the bridge deck and pavement system is subjected to solar radiation, radiation heat transfer and natural convective heat transfer. Furthermore, there is no solar radiation in the sides, bottom and inner of box girder, and negligible natural convection occurs in the boxes [11]. As a result, according to the transient thermal field theory, the temperature boundary conditions of FE model are presented by equation (1).

$$\begin{cases} -k\dfrac{\partial T_s}{\partial y} = h_c f[T_s(t)-T_a] + q(t) + \varepsilon\sigma\left[(T_s(t)-T_z)^4-(T_a-T_z)^4\right] \\ -k\dfrac{\partial T_b}{\partial y} = h_c f[T_b(t)-T_a] + \varepsilon\sigma\left[(T_b(t)-T_z)^4-(T_a-T_z)^4\right] \\ -k\dfrac{\partial T_{in}}{\partial y} = \varepsilon\sigma\left[(T_{in}(t)-T_z)^4-(T_a-T_z)^4\right] \end{cases} \tag{1}$$

where k and ε are thermal conductivity coefficient and emissivity coefficient of the materials, respectively; T_s, T_a and T_z are pavement surface temperature, ambient air temperature and absolute zero, respectively; T_b and T_{in} are the temperature of the sides and bottom of the box and the inner-surface temperature of the box, respectively; $q(t)$ is solar radiation intensity; h_c and σ are natural convective heat transfer coefficient and Stefan-Boltzmann constant, respectively.

According to some investigations [11-15], the required parameters for numerical simulation are listed in table 1, including some parameters in equation (1) and material thermal parameters.

Table 1. Parameters for thermal analysis.

Temperature boundary parameters		Materials			
		Item	Reinforced concrete	SFRC	Asphalt concrete
$q(t)$/W·m^{-2}	$q(t)$=0, t∈[0, 12-c/2]	k (W/m·K)	1.7	2.2	1.0
	$q(t)$=$q_0$cos$m\pi(t$-12)/12, t∈ (12-c/2, 12+c/2)	ε	0.9	0.9	0.81
		Density (kg/m^3)	2500	2600	2100
		Specific heat (J/kg·K)	850	900	810
	$q(t)$=0, t∈[12-c/2, 24]	Modulus (MPa)	4.04E4	3.45E4	1200
h_c/W·(m^{-2}K^{-1})	h_c=4.0v+5.8	Poisson ratio	0.225	0.2	0.35
σ/W/(m^2·K^4)	5.6697E-8	Expansion coefficient (10^{-5}/°C)	1	1	5

Note: q_0 is maximum sunshine intensity at noon, where q_0=1.572Q/c, in which Q and c are daily total solar radiation and effective duration of sunshine, respectively; v is the wind speed, m/s.

3.3. Thermal field model

Before the thermal analysis, the local meteorological data during HMA concrete paving was used to determine the initial thermal field of the model. In addition, the paving temperature and thickness of HMA concrete were chosen as 142°C and 6cm according to the field measuring results in the Master thesis of Zhu [7]. Furthermore, the element deletion method was adopted to simulate the dynamic paving process of HMA concrete and its procedures are summarized below [11].

- Along the longitudinal direction, dividing the 21m-long FE model into 21 parts and deactivating all solid elements of HMA concrete;
- Defining 22 heat transfer steps. In detail, the paving process was simulated by reactivating the solid elements of HMA concrete in the first 21 steps part by part at 30-second interval. Moreover, the temperature variations of the model after paving were simulated by the 22th step. Figure 4 shows the thermal field distributions at different heat transfer steps during the simulation process.

(a) (b) (c)

Figure 4. Simulating the paving process at different heat transfer steps. (a) Step 8, (b) Step 12 and (c) Step 16.

3.4. Model validation

To validate the accuracy of thermal field model, the comparisons between the simulated and measured temperature at the pavement surface, Point ①, Point ③ and Point ④ were conducted, as shown in figure 5. It can be found from figure 5 that the difference between the simulated and measured temperature is less than 5°C. When taking the field measurement errors and simulation simplification into account, the simulated results are considered to agree well with the measured results, indicating the feasibility of the established FE model and simulation method. In addition, as shown in figures 5(b)-5(d), the closer the locations are to the top of web-plate, the greater the influence of HMA concrete paving on the vertical temperature is. In detail, the highest temperature of the box girder occurs at 36 min at Point ①, while there is no obvious influence of HMA concrete construction on the vertical temperature at Point ④. Therefore, the critical time for thermal analysis was chosen as 36 min after the paving begins, due to the highest temperature difference in the vertical direction at this time.

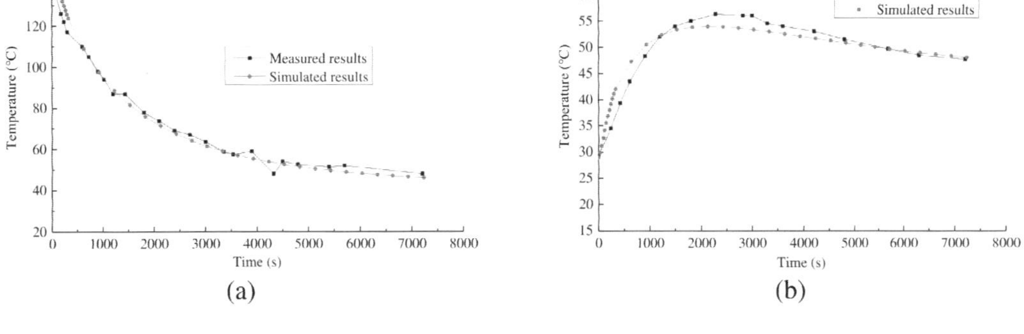

(a) (b)

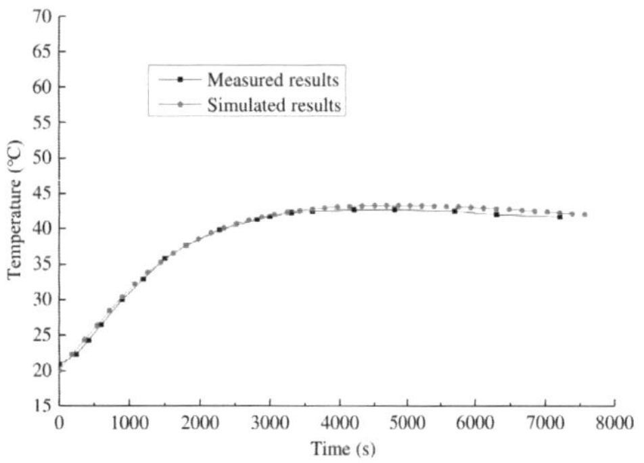

(c)

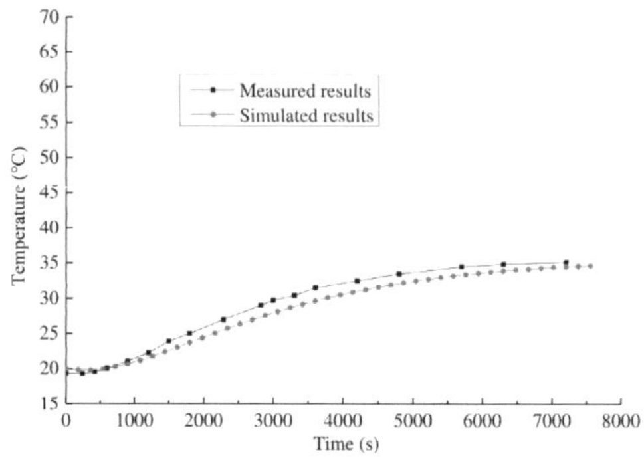

(d)

Figure 5. Comparison between the simulated and measured results. (a) Pavement surface, (b) Point ①, (c) Point ③ and (d) Point ④.

4. Thermal stress analysis

4.1. Mechanical model and data acquisition

The geometry of the mechanical model in this research was same with that of the thermal field model. Additionally, the simply supported constraints were set as the boundary conditions of the mechanical model, which can ensure the longitudinal expansion of the concrete box-girder. Moreover, the mechanical parameters of materials were determined according to the results in table 1. Furthermore, the thermal stress analysis was conducted in the mid-span cross section where the mechanical state is considered to be the most unfavorable.

Subsequently, the data acquisition scheme of the mechanical model was formulated to analyze the thermal effects of the box girder after the paving begins. In detail, the vertical arrangement of the data acquisition scheme was determined according to figure 2. Besides, in the transversal direction, the measuring locations were arranged as shown in figure 6, including Point #1 at the middle of the box and Point #2 at the flange slab.

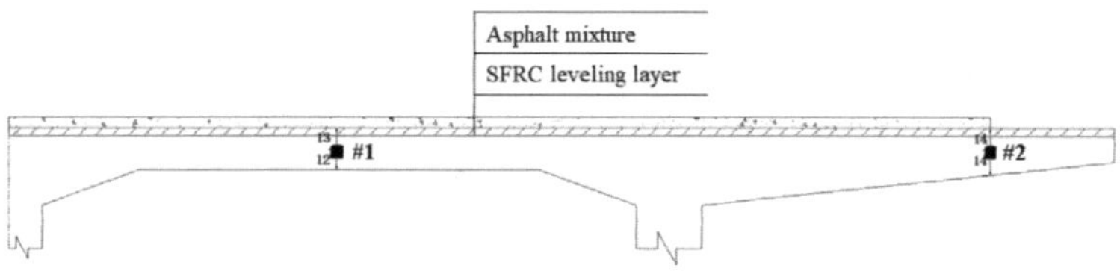

Figure 6. Transversal arrangement of the data acquisition scheme.

4.2. Thermal stress distribution in the vertical direction

The thermal stress distribution in the vertical direction at the critical time is illustrated in figure 7 (X-axis is represented by the vertical distance to the top of SFRC leveling layer). It can be found from figure 7 that both transversal and longitudinal thermal stresses present the compression status (negative number) at two ends and the tension status (positive number) at the middle in the west web-plate. In addition, the most unfavorable tension status occurs between Point ③ and Point ⑤. Due to most of damage in cement concrete occurring in the tension status, Point ④ was chosen to analyze the characteristics of thermal stress distribution during the whole paving process, as figure 8 shows. It indicates that in the tension status, the thermal stress has a trend from rising to decline, and the highest values of transversal and longitudinal stress both are close to 4 MPa, which is considerable and non-negligible in the tress field analysis of the concrete box-girder.

IOP Conf. Series: Materials Science and Engineering **629** (2019) 012014 doi:10.1088/1757-899X/629/1/012014

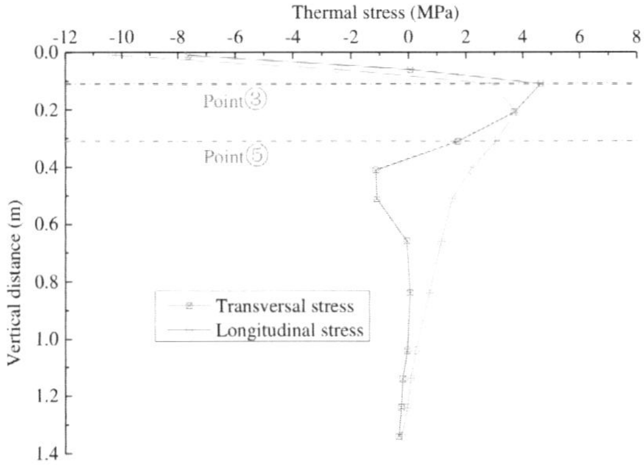

Figure 7. Thermal stress distribution at different vertical locations.

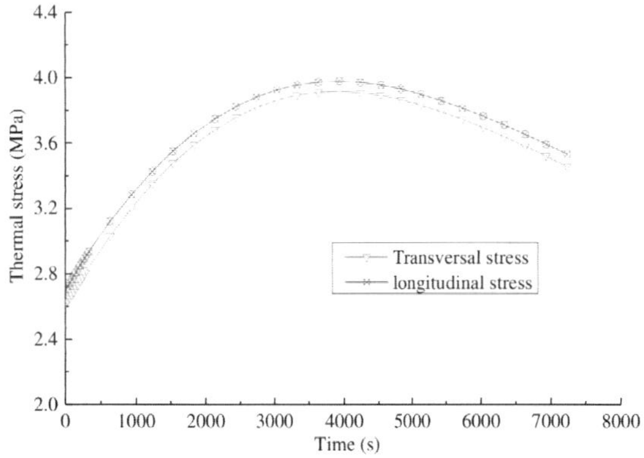

Figure 8. Thermal stress distribution at Point ④.

4.3. Thermal stress distribution in the transversal direction

Figure 9 illustrates the thermal stress distribution of two transversal points during the whole paving process, respectively. As shown in figure 9, two measuring locations, including Point #1 in the box and Point #2 in the flange slab, are in the tension status. Besides, the thermal stress presents a short-term increase, followed by a decline trend on the measuring locations in the box and flange slab. Moreover, transversal stress is larger than longitudinal stress at Point #1 in the box, which is opposed to Point #2 in the flange slab.

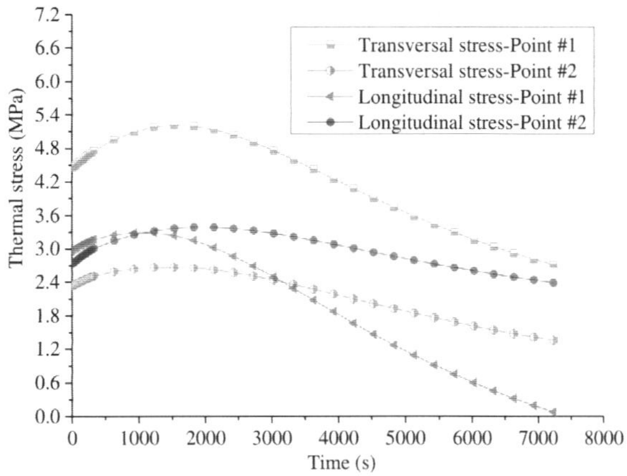

Figure 9. Thermal stress distribution at the transversal Point #1 and Point #2.

5. Conclusions

This paper investigated the thermal effects of the concrete box girder segment containing SFRC leveling layer after the paving begins, and the following conclusions can be drawn.

- The influence of HMA concrete construction on the temperature distribution in the vertical direction is attenuated as the vertical distance to the top of SFRC leveling layer increases. Moreover, the highest vertical temperature difference occurs at 36 min after starting to paving.
- In the vertical direction, the west web-plate is subjected to compressive stress at locations of two ends and tensile stress at locations of the middle. In addition, at Point ④ the unfavorable tensile stresses present a trend from rising to decline during the whole HMA concrete paving, the highest values of which are close to 4 MPa.
- In the transversal direction, the thermal stresses undergo a decline trend after short-term growth

both in the box and flange slab.

Acknowledgments
The authors are grateful to the financial support of National Natural Science Foundation of China (NSFC, No. 51378122).

References

[1] Subramaniam K V, Kunin J, Curtis R *et al* 2010 Influence of early temperature rise on movements and stress development in concrete decks *J Bridge Eng* **15** 108-16
[2] Song Z, Xiao J and Shen L 2012 On temperature gradients in high-performance concrete box girder under solar radiation *Adv Struct Eng* **15** 399-416
[3] Mirambell E and Aguado A 1990 Temperature and stress distributions in concrete box girder bridges *J Struct Eng* **116** 2388-409
[4] Vecchio F 1987 Nonlinear analysis of reinforced concrete frames subjected to thermal and mechanical loads *ACI Struct J* **84** 492-501
[5] Zhao R D and Wang Y B 2016 Studies on temperature field boundary conditions for concrete box-girder bridges under solar radiation *China Journal of Highway & Transport* **29** 52-61 (in Chinese)
[6] Gu Y, Li Y D and Yao C R 2016 Research of temperature field of concrete box girder under solar radiation *Journal of Highway & Transportation Research & Development* **33** 46-53 (in Chinese)
[7] Zhu J 2005 Study on temperature fields under the action of solar radiation and bituminous deck pavement of the reinforced concrete box girder bridge (Jiangsu, China: Southeast University) (in Chinese)
[8] Liu Q W, Zhu J, Tang B H *et al* 2007 Experiment on temperature distribution of reinforced concrete box girder under asphalt high temperature paving *China Journal of Highway and Transport* **20** 96-100 (in Chinese)
[9] Liu Y, Qian Z, Hu J *et al* 2017 Temperature behavior and stability analysis of orthotropic steel bridge deck during gussasphalt pavement paving *J Bridge Eng* **23** 04017117
[10] Liu Y, Qian Z and Hu H 2016 Thermal field characteristic analysis of steel bridge deck during high-temperature asphalt pavement paving KSCE *J Civ Eng* **20** 2811-21
[11] Zheng D, Qian Z, Liu D *et al* 2018 Thermal field characteristics of reinforced concrete box girder during high-temperature asphalt pavement paving *Transport Res Rec* Doi: 10.1177/0361198118768529
[12] Sabbagh Yazdi S R and Saadat Abadi T A 2011 Sequential computations of two-dimensional temperature profiles and thermal stresses on an unstructured triangular mesh by GFVM method *Int J Civ Eng* **9** 171-82
[13] Chen X Q, Liu Q W and Zhu J 2009 Measurement and theoretical analysis of solar temperature field in steel-concrete composite girder *J Southeast Univ (Eng Ed)* **25** 513-7
[14] Peng Y, Zhu X and Qiang S 2009 Three dimensional analyses of thermal stresses in concrete box girders *Journal of the China Railway Society* **31** 116-21
[15] Rodriguez L E, Barr P J and Halling M W 2014 Temperature effects on a box-girder integral-abutment bridge *J Perform Constr Fac* **28** 583-91

IOP Conf. Series: Materials Science and Engineering **629** (2019) 012015 doi:10.1088/1757-899X/629/1/012015

A solution to the double crack's J-integral in the plain plate

H Li[1] and X Y Zhang[1,2]

[1]College of Aerospace and Civil Engineering, Harbin Engineering University, Harbin, China

E-mail: zhangxueyi@hrbeu.edu.cn

Abstract. Based on the relevant criteria and theory of fracture mechanics and the description and definition of corresponding fracture parameters, the J integral factor of double crack propagation in plain plate is analyzed. In order to obtain the J integral factor of double crack propagation, the method of defining seam is used to preset the path of crack's propagation. With the variation of load, the crack would be acquired a propagation forward to the presetting seam, furthermore, and simulated the singularity of crack tip, the stress field, displacement field and strain field at the crack tip are analyzed for the causes and regular changes. The model under the same load is also calculated using the model of the viscoelastic plate, and the corresponding regular changes and discussions are also given. The result is showed that the distribution of stress about crack tip is more larger than other district, concentration of stress is so obvious, the value of stress is more decreasingly less than other distribution; the strain of crack tip would acquire an increasing tendency to the direction of crack propagation, the displacement's magnitude also presents a significantly growth across the central point toward both sides, and the value of displacement is largest of the both sides of plate. Compared to the classical elastic plate, the J integral factor is smaller, and the strain energy according to the crack tip is less.

1. Introduction

The occurrence of a large number of fracture problems in actual engineering, especially for the fracture of the plate and shell structures, is an important measure for the calculation and determination of the corresponding fracture parameters. And the corresponding degree of damage and the form of failure are also reflected. The reference and judgment, especially the stress field, the displacement field and strain field at the crack tip have a certain description and mechanism [1].

2. Basic theory

2.1. The elastic plain plate

2.1.1. The process of the model. According to the size requirements, the geometric model of the corresponding flat plate is established [2-5], as shown in the figure 1. The geometric model is shown in figure 2, the element meshing is shown in figure 3.

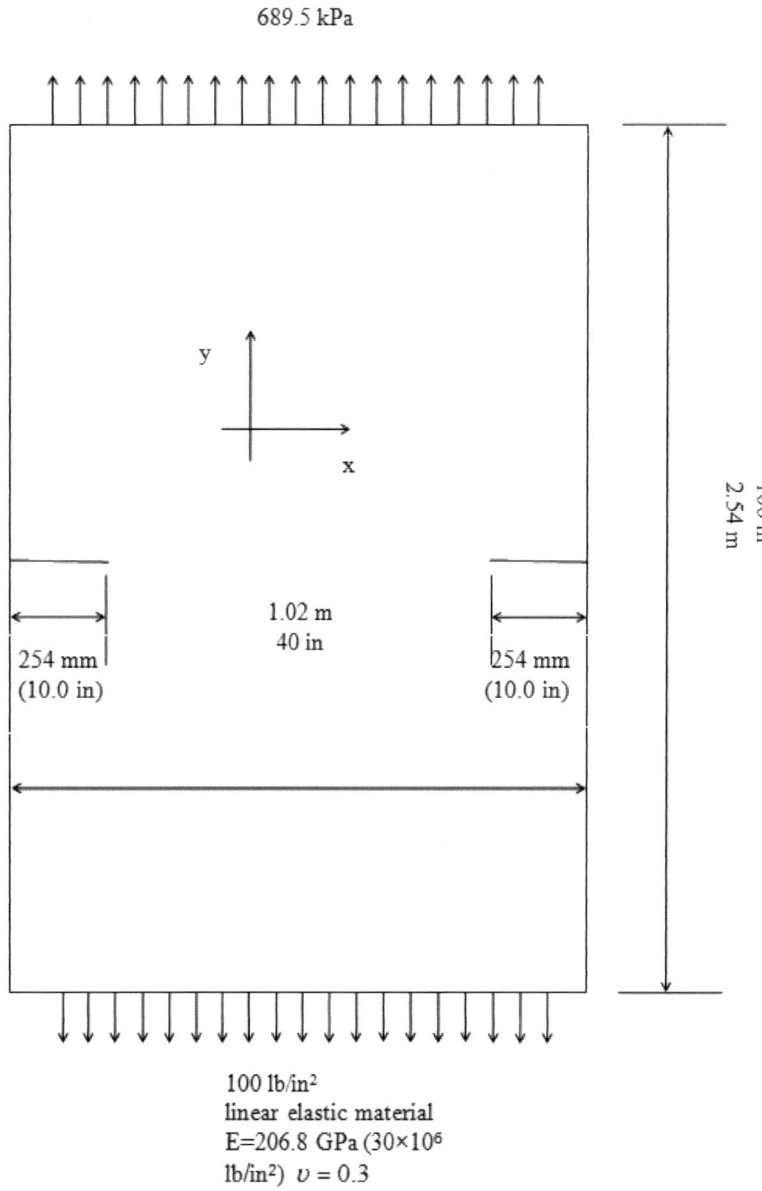

Figure 1. Load diagram of the entire plate.

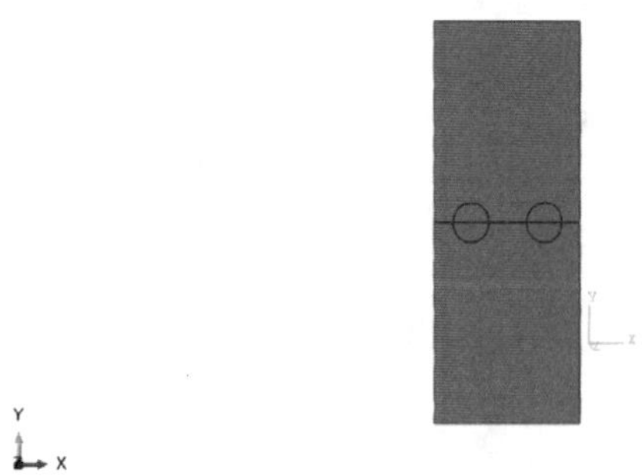

Figure 2. The geometric model of the plain plate.

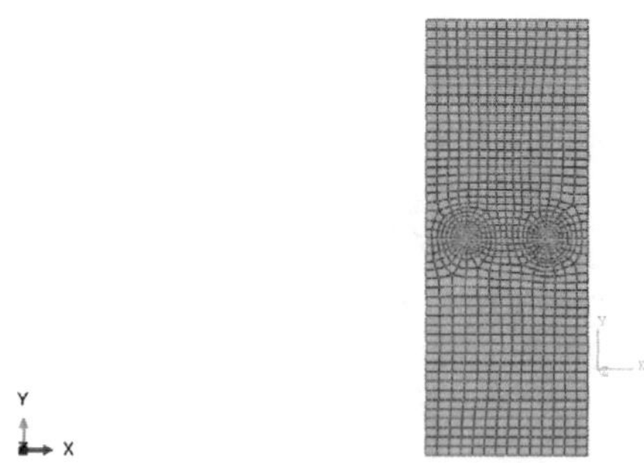

Figure 3. Corresponding finite element meshing.

IOP Conf. Series: Materials Science and Engineering **629** (2019) 012015 doi:10.1088/1757-899X/629/1/012015

2.1.2. Mesh. Divide the finite element mesh and preset the crack model to refine the mesh at the crack.

2.1.3. Result. As a result of the deformation of the plain plate after biaxial stretching, the crack growth has become into a symmetrical shape distribution. As shown in figure 4.

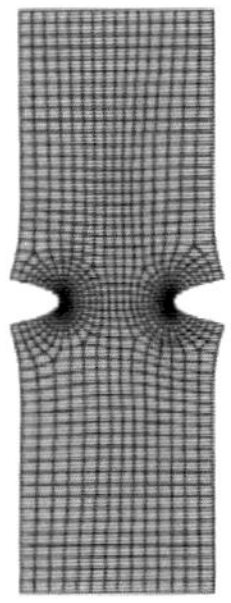

Figure 4. The result of crack propagation.

2.1.4. Crack tip stress field distribution. The simulation results show that stress distribution at the crack tip is extraordinary large, the stress concentration occurs at the same time, the stress value decreases along the crack propagation direction, and at the stress concentration, the value is the largest, it may be a serious concentration at crack tip, in the figure 5, which is about 13.60 MPa.

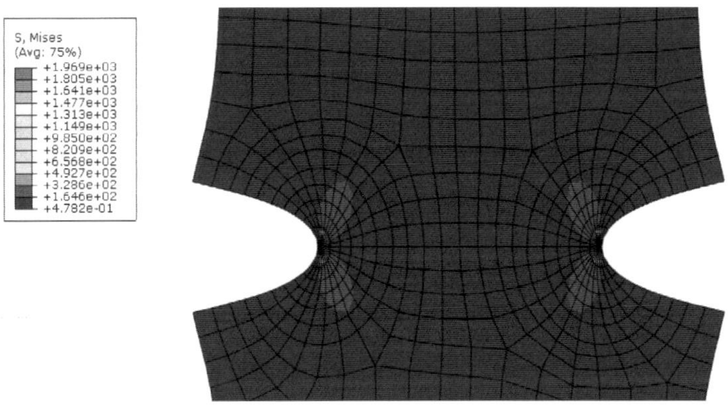

Figure 5. Stress field distribution of corresponding crack.

2.1.5. Crack tip displacement field distribution. The strain value at the tip increases in the crack propagation direction [6-8], and the total displacement gradually increases along the center point toward both sides. From the figure 6, the displacement at the boundary between the two ends of the plate is maximal, and the maximum value is 2.82×10^{-6} m.

On the one hand, the method of defining the seam is used to preset the crack propagation path. As the load is applied, the crack will expand along the seam. It's different from debonding method, the latter is limited in practical application range, the former is convenient to calculate the fracture parameters such as stress intensity factor and J-integral, and in the former method, the approach of defining the crack front is to select the crack tip (2D) or the crack tip line (3D) directly. Using this method to define the crack front does not need to define the next crack tip/line, which is relatively simple.

IOP Conf. Series: Materials Science and Engineering **629** (2019) 012015 doi:10.1088/1757-899X/629/1/012015

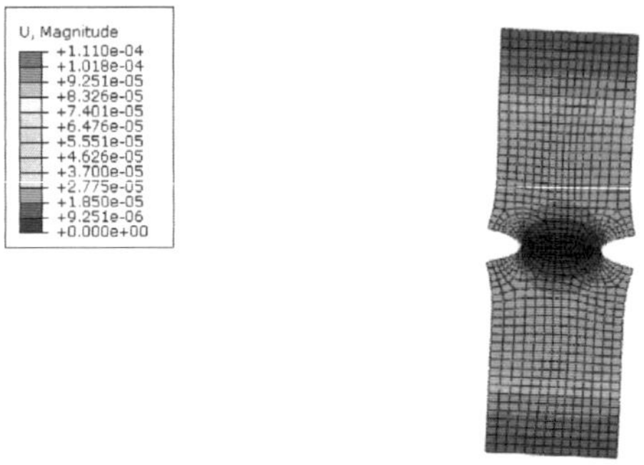

Figure 6. Displacement field distribution of corresponding crack.

2.1.6. J integral factor. Solution of corresponding parameters after crack propagation (J integral factor).
Elastic plain plate:

Table 1. J - Integral estimates.

parameters	1	2	3	4	5	6	7	8
Left crack	1.3042E-02	1.3019E-02	1.3027E-02	1.3028E-02	1.3028E-02	1.3028E-02	1.3028E-02	1.3029E-02
Right crack	1.3043E-02	1.3019E-02	1.3027E-02	1.3028E-02	1.3028E-02	1.3028E-02	1.3028E-02	1.3029E-02

The crack at the left end expands to the right, and the crack at the right end expands to the left, as shown in the table 1, the J integral factor according to the crack propagation at the left and right ends is stable at 1.30e-02.

2.2. The viscoelastic plain plate

2.2.1. Crack tip stress field distribution (static step). The crack tip stress distribution in the visoelastic plain plate law is basically the same as the elastic plain plate, as shown in the figure 7, the largest value is 14.89 Mpa.

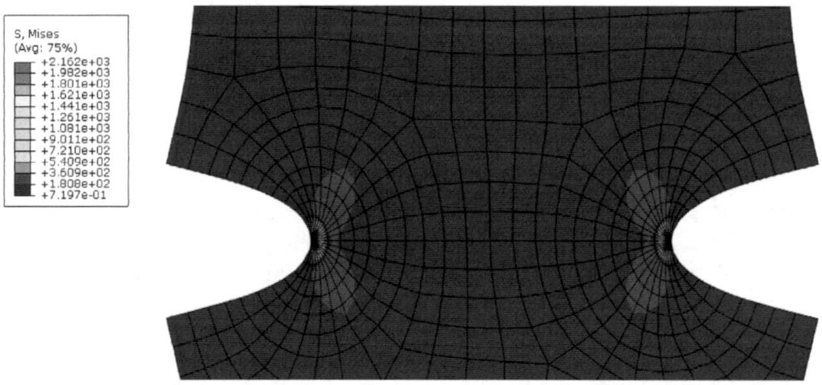

Figure 7. Stress field distribution of corresponding crack.

2.2.2. Crack tip displacement field distribution. The displacement field at the tip is the same as that of the elastic plain plate. The total displacement increases gradually from the center point to the sides, and the displacement at the two ends of the plate is the largest. From the figure 8, the maximum value is

4.67×10^{-6} m.

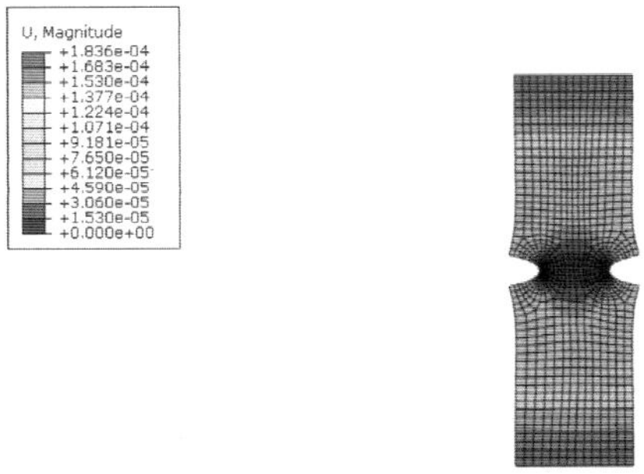

Figure 8. Displacement field distribution of corresponding crack.

2.2.3. J integral factor. The J integral factor according to the crack propagation at the left and right ends are stable at 7.94e-03, as shown in the table 2, the value is less than in the elastic plain plate.

Table 2. J - Integral estimates.

parameters	1	2	3	4	5	6	7	8
Left crack	7.9459E-03	7.9318E-03	7.9365E-03	7.9372E-03	7.9374E-03	7.9373E-03	7.9372E-03	7.9376E-03
Right crack	7.9462E-03	7.9320E-03	7.9366E-03	7.9372E-03	7.9374E-03	7.9373E-03	7.9373E-03	7.9375E-03

2.2.4. Crack tip stress field distribution (visco step). The crack tip stress distribution in the visoelastic plain plate law is basically the same as the plane elastic plate, from the figure 9, the largest value is 14.87 Mpa.

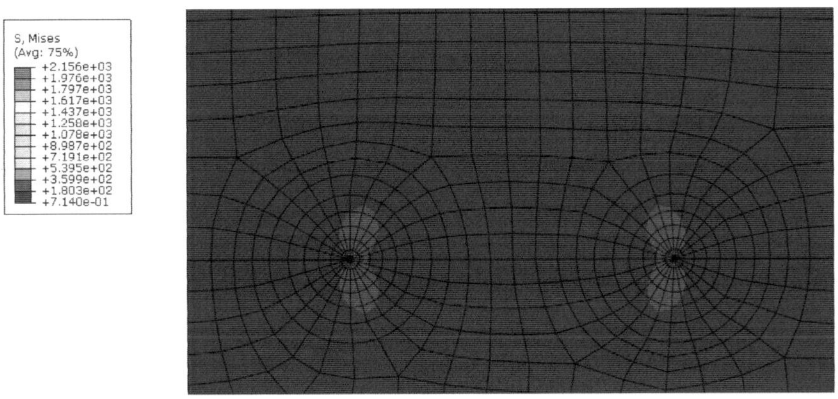

Figure 9. Corresponding crack stress field distribution.

2.2.5. Crack tip displacement field distribution. The displacement field at the tip is the same as that of the elastic plain plate. The total displacement increases gradually from the center point to the sides, and the displacement at the two ends of the plate is the largest. From the figure 10, the maximum value is 2.84×10^{-6} m.

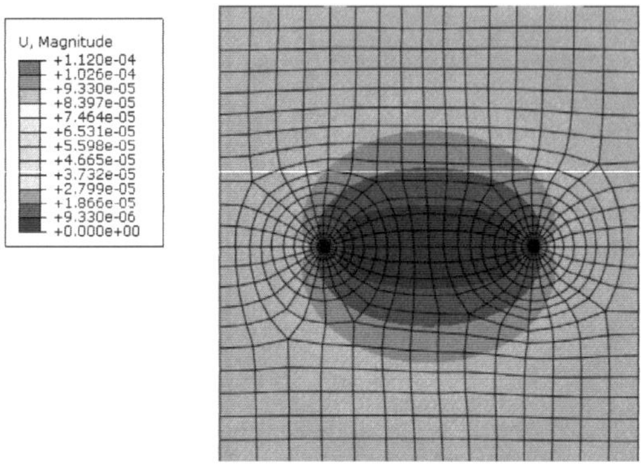

Figure 10. Corresponding crack displacement field distribution.

2.2.6. J integral factor. The J integral factor according to the crack propagation at the left and right ends are stable at 8.00e-03, as shown in the table 3, the value is less than in elastic plain plate and larger than in the static step of viscoelastic plain plate, there is creep during the stretching process of the viscoelastic plate, the plastic deformation, resulting in a small J integral value than before.

Table 3. J - Integral estimates.

parameters	1	2	3	4	5	6	7	8
Left crack	8.0135E-03	7.9992E-03	8.0040E-03	8.0047E-03	8.0048E-03	8.0047E-03	8.0047E-03	8.0051E-03
Right crack	8.0138E-03	7.9994E-03	7.9366E-03	8.0041E-03	7.9374E-03	7.9373E-03	7.9373E-03	7.9375E-03

Table 4. C - Integral estimates.

parameters	1	2	3	4	5	6	7	8
Left crack	6.3790E-03	6.3253E-03	6.2494E-03	6.1369E-03	5.9849E-03	5.7814E-03	5.4987E-03	5.1433E-03
Right crack	6.3800E-03	6.3261E-03	6.2499E-03	6.1371E-03	5.9849E-03	5.7818E-03	5.5813E-03	5.3743E-03

Due to the existence of the creep, as shown in the table 4, C-integral is less than the J integral factor in the viscoelastic plain plate.

3. Conclusion

The simulation of the biaxially stretched plate, whether elastic or viscoelastic, it shows that the stress distribution at the crack tip is relatively large, the stress concentration is obvious, and the stress value decreases along the crack propagation direction [9]; the strain value at the tip decreases first, then increasing along the direction of crack propagation, the total displacement gradually increases along the center point toward both sides, and the displacement at the boundary between the two ends of the plate is the largest [10]. And in the viscoelastic plain plate, the J integral factor is less than in the elastic plain plate, and deformation during load application has appeared and creep affects crack propagation and energy release.

Acknowledgment

This paper is funded by the International Exchange Program of Harbin Engineering University for Innovation-oriented Talents Cultivation and is supported by my supervisor X Y Zhang, and my apprentice Y N Wu. Specific acknowledgement for the help from the Ministry of Industry and Information Technology of the People's Republic of China, Mechanics and Engineering Technology Experimental Teaching Demonstration Center.

References

[1] Geubelle P H and Rice J R 1995 A spectral method for three-dimensional elastodynamic fracture

IOP Conf. Series: Materials Science and Engineering **629** (2019) 012015 doi:10.1088/1757-899X/629/1/012015

problems *J Mech Phys Solids* **43** 1791-824

[2] Su Y 2007 Research on numerical solution method of three-dimensional crack J integral (Shandong, China: China University of Petroleum)

[3] Fan C Y 2010 Study on the theory and numerical method of piezoelectric solid fracture of piezoelectric voltage (Gansu, China: Lanzhou University)

[4] Chen Y Y, Wang L Q, Jiao L and Qu Y P 2003 Solving the J-integration of fracture mechanics by the integral equation of surface force boundary *(CJCM) Chinese Journal of Computational Mechanics* **2003** 583-6

[5] Xiao D Y and Dexter R J 1998 Finite element calculation of applied J-integral for cracked ship structural details *ENG FRACT MECH* **60** 59-82

[6] He W, Cai L X, Chen H and Peng Y Q 2018 Semi-analytical method for solving J-integral of shaped cracked element *(CSTAM) Chinese Journal of Theoretical and Applied Mechanics* **50** 579-88

[7] Yaren M F, Demir O, Ayhan A O and İriç S 2018 Three-dimensional mode-I/III fatigue crack propagation: Computational modeling and experiments *INT J FATIGUE* **24** 124-34

[8] Yang D, Dong W, Liu X F, Yi S H and He X Q 2018 Investigation on mode-I crack propagation in concrete using bond-based peridynamics with a new damage model *ENG FRACT MECH* **199** 567-81

[9] Gao C H and Tang X S 2014 Model of the stress region of the crack tip of type I and its analytical solution *(CJAM) Chinese Journal of Applied Mechanics* **31** 883-8

[10] Zhao G F and Xia K W 2018 A study of mode-I self-similar dynamic crack propagation using a lattice spring model *COMPUT GEOTECH* **96** 215-25

IOP Conf. Series: Materials Science and Engineering **629** (2019) 012016 doi:10.1088/1757-899X/629/1/012016

Effects of process parameters on the microstructure and mechanical properties of medium-carbon steels for improving high-strain rate properties

Y B Song[1,2], J H Ham[1] and H K Kim[1]

[1]Agency for Defense Development, Republic of Korea

E-mail: smain@hanmail.net

Abstract. The effects of accelerated continuous cooling and isothermal heat treatment process on the microstructure and mechanical properties of medium carbon bainitic steels was investigated to find an alternative process of the conventional austempering treatment process. Two kinds of medium carbon steel billets melted by Vacuum Induction Melting (VIM) were hot rolled to the thickness of 6~12mm plates and accelerated-cooled to predetermined temperatures and then isothermal heat treatment was conducted immediately. For quantitative analysis of microstructures, the fraction of constituent phases such as retained austenite, martensite, and bainite was measured by XRD and dilatometry. A series of hardness, tensile, and impact tests was conducted to evaluate the mechanical properties of the plates. As the finish cooling temperature increased, the fraction of retained austenite increased and the fraction of martensite and bainite decreased due to the differences in carbon diffusion and transformation kinetics during isothermal heat treatment process. Microstructural observation results showed that the lath/block thickness and fraction of bainite and martensite were mainly dependent on austempering temperature and time. The thickness of bainite and the fraction of martensite were found to be the major factors influencing mechanical properties of the present medium-carbon nanostructured bainitic steels.

1. Introduction

The use of high hardness steel is increasing for armour protection. Recently there has been a growing interest in a new concept of bainitic steel having high hardness and strength due to high carbon content and fine microstructure. These bainitic steels are designed to have high strength of ~ 2.3 GPa and high hardness in the range of 600–670 Hv [1-3]. The microstructure of the steel consists of nano-sized bainitic ferrite lath and carbon enriched austenite film by isothermal heat treatment just above martensite start temperature – so often called 'nanostructured bainitic steel' [4-10]. In spite of high hardness and strength, the steel may be inapplicable to the commercial products due to their inherently low impact toughness and very long isothermal holding time for bainite transformation.

This study aims at replacing conventional prolonged austempering process by isothermal treatment after continuous accelerated cooling. For this purpose, the static and dynamic mechanical properties and microstructures of two medium carbon bainitic steels were examined under various conditions of the continuous accelerated cooling and subsequent isothermal treatment.

2. Experimental

To evaluate the effects of carbon content on the mechanical properties and microstructures, two kinds of medium-carbon steels were prepared with vacuum-induction-melting and hot rolling in thickness of

6~12 mm. Chemical compositions of the alloys are listed in Table 1.

Figure 1 represents continuous accelerated cooling and isothermal treatment process. The initial slab size for hot rolling was 120x220x290 mm and the finish rolling temperature was about 900°C. After hot rolling, acceleration cooling was performed by laminar cooling or spray cooling with water. The final cooling temperature of plates was controlled between 300°C to 600°C and the cooling rate was about 25°C/s. Then the cooled plates were isothermal-treated at 250°C for 1 hour for bainite transformation. In addition, for the purpose of comparison of high strain rate properties, the alloy B was austempered at various temperatures for 30 min in a salt bath for bainite transformation after austenitizing at 900°C.

The ferrite-austenite transition temperature (Ac_3) and martensite start temperature (Ms) were measured by dilatometric analysis. For quantitative analysis, the fraction of retained austenite and martensite was measured by XRD and dilatometry. A series of hardness, tensile, and impact tests was conducted and the high strain rate properties were also evaluated by performing the Split-Hokinson Pressure Bar (SHPB) test.

Table 1. Chemical composition of the present medium carbon steels. (wt%)

Specimen	C	Si	Mn	Ni	Cr+Mo+V	Fe
alloy A	0.27	0.23	0.97	3.03	0.82	Bal.
alloy B	0.50	1.45	0.68		0.86	Bal.

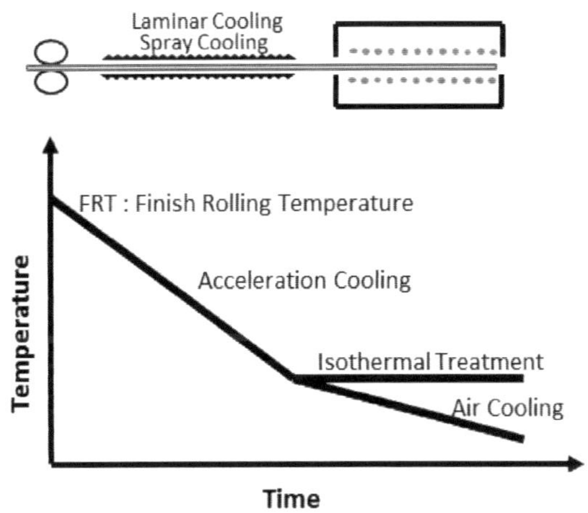

Figure 1. Schematic diagram of accelerated cooling and isothermal treatment process.

3. Results & discussions

3.1. Dilatometry study for analysis temperature of phase transformation

Figure 2 shows the dilatation curves of the alloy A and alloy B. The Ac_3 and Ms temperature were measured as 780°C and 296°C for alloy A and 857°C and 270°C for alloy B, respectively. With these temperatures, the austenitizing and isothermal temperatures were determined.

3.2. XRD analysis

The role of martensite is very important on mechanical properties of steel. Especially, the mechanical properties (hardness, strength, etc.) of the medium-carbon bainitic steel are influenced by the martensite containing highly concentrated carbon. The X-ray analysis was conducted to identify the alloy phases and to evaluate their fraction. As shown in Figure 3, the hot-rolled plated was nearly martensitic and the peaks of retained austenite were not clearly detected, indicating very low retained austenite content if it exists.

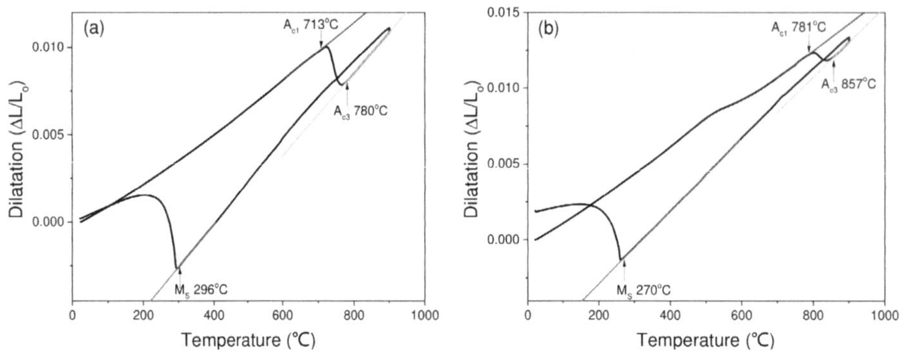

Figure 2. Dilatation curves and the characteristic temperatures of the (a) alloy A and (b) alloy B.

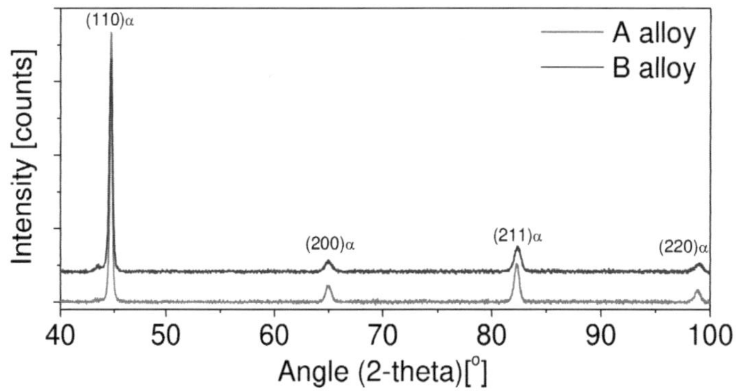

Figure 3. XRD patterns of the hot rolled plate of the alloy A and alloy B.

3.3. Effect of austempering process on the high strain rate properties of the alloy B

In this study, static/quasi-static tensile test and high strain rate compression test with SHPB were conducted to evaluate the mechanical properties of quench-tempered (Q/T) and austempered samples of the alloy B. The strain rate of each test was $1x10^{-3}$ /s, 1 /s, and 10^3 /s, respectively.

In Table 2, the hardness of the alloy B austempered at different temperatures was compared with that of Q/T alloy. The hardness of the Q/T sample is the highest because of the high volume fraction of martensite. As the austempering temperature decreased, the hardness increased because the fraction of tempered martensite increased.

Figure 4 shows the quasi-static and high strain rate properties of the Q/T and austemperd samples. Comparing with the Q/T treated sample, the austempered samples which have lower strength at quasi-static test condition tend to have high strength and also good ductility at high strain rate condition indicating austempering treatment could improve the impact properties of the steels at high strain rate test condition such as ballistic condition.

3.4. Effect of continuous accelerated cooling and isothermal treatment of the alloy B

Two finish cooling temperature of 610°C and 410°C were applied to alloy B for accelerated cooling. Figure 5 and Table 3 represent the cooling curves on the CCT diagram of the alloy B and the phase fraction calculated with JmatPro software. In the alloy B, high temperature phases such as pearlite and ferrite can form in the slow cooling rate because the CCT diagrams of the alloy B is shift to left than that of the alloy A. Microstructure in Figure 6 shows that, while the low cooled plate (air cooling) has pearlite, ferrite and bainite, the major phases of accelerated cooled plates were bainite and martensite. The accelerated-cooled plates cooled to finish cooling temperature were isothermal-treated at 250°C for 1 hr for bainite transformation.

The summary of hardness and tensile properties of the accelerated-cooled and isothermal heat-treated plates is represented in Figure 7. The air-cooled plate had low hardness and strength and high ductility because the main phase was pearlite as shown in Figures 6(a) and (b). The accelerated cooled plates had higher hardness and strength because of high volume fraction of martensite (Figures 6(c) and (d)) but the ductility sharply decreased as the finish cooling temperature decreased because of formation of fresh martensite. The accelerated-cooled and isothermal treated plates had high strength and reasonable ductility because of the presence of bainite formed during isothermal treatment.

Table 2. Hardness of three-austempered conditions of B alloy plates investigated in this study.

Conditions	Austempered			Q/T
	245°C 30min	255°C 30min	265°C 30min	200°C 4hr
Hardness (H_RC)	56.0±0.09	55.7±0.08	53.5±0.09	59.0±0.15

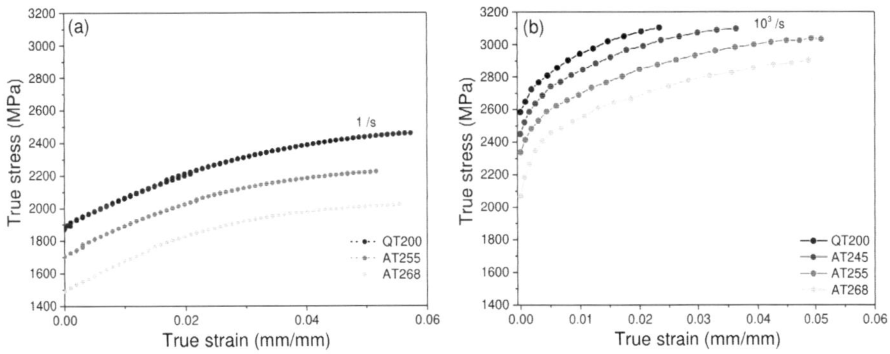

Figure 4. (a) Quasi-static and (b) high strain rate properties of austempered and Q/T samples of the alloy B.

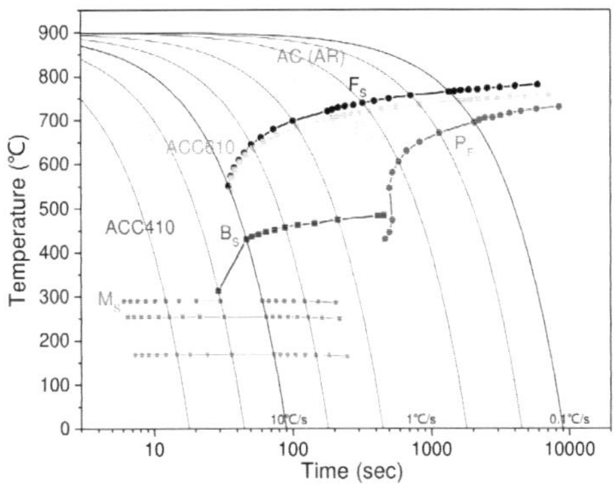

Figure 5. Accelerated cooling curves plotted in the continuous cooling transformation diagram of the alloy B.

Table 3. Calculated phase fractions of the alloy B samples with controlled cooling conditions.

Conditions	Fraction of phases (%)				
	f_M[a]	f_B[b]	f_F[c]	f_P[d]	f_A[e]
Air cooling		14.15	4.04	81.8	
ACC610	13.81	74.55	4.38	7.15	0.15
ACC410	79.3	20.05	0.06	0.02	0.58

[a] Martensite [b] Bainite [c] Ferrite [d] Pearite [e] Austenite

IOP Conf. Series: Materials Science and Engineering **629** (2019) 012016 doi:10.1088/1757-899X/629/1/012016

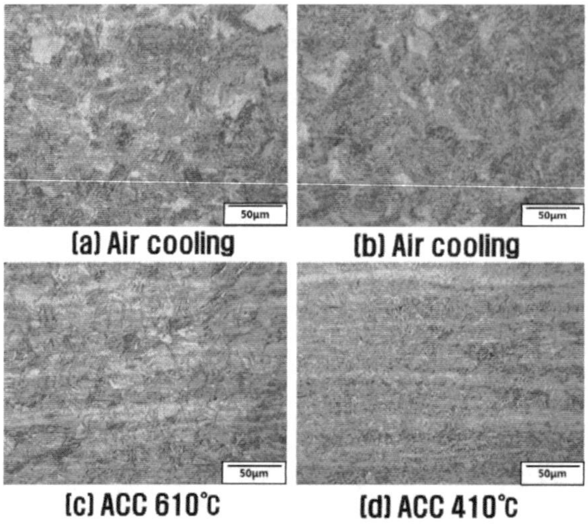

Figure 6. Microstructure of the air cooled and accelerated cooled samples of the alloy B.

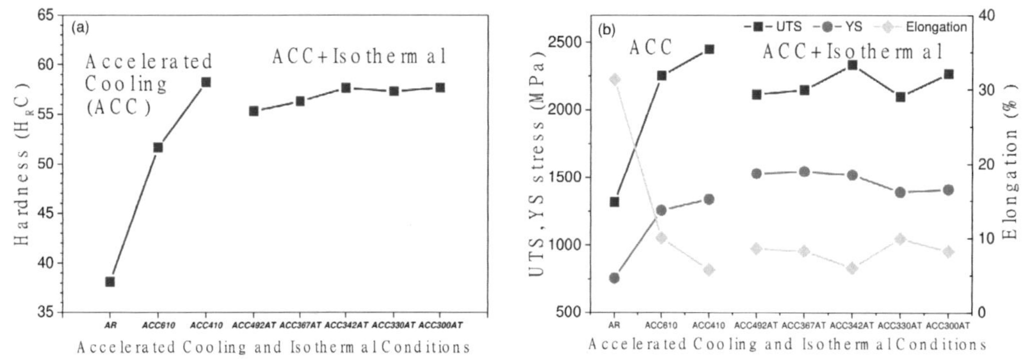

Figure 7. (a) hardness and (b) tensile properties of the alloy B with accelerated cooling only and accelerated cooling + isothermal treatment.

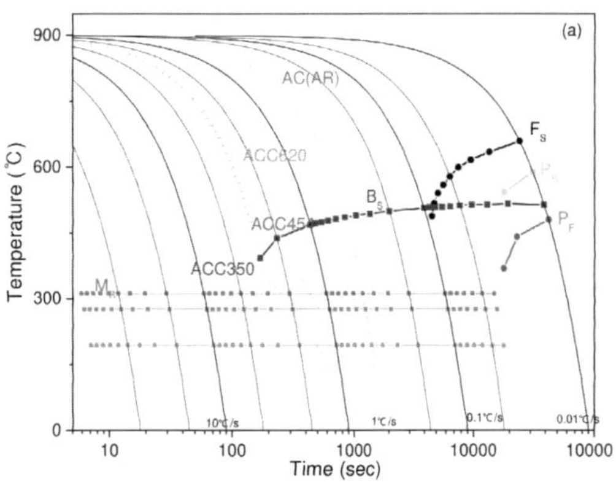

Figure 8. Accelerated cooling curves plotted in continuous cooling transformation(CCT) diagram of the alloy A.

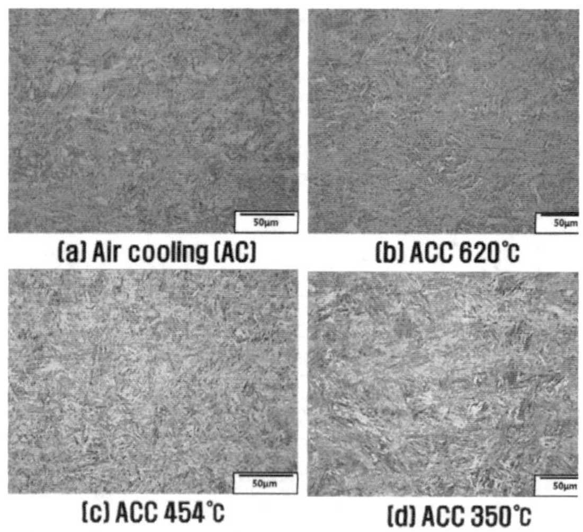

Figure 9. Microstructure of air-cooled and accelerated-cooled alloy A.

3.5. Effects of continuous accelerated cooling for alloy A

Accelerated cooling of the hot-rolled plates of the alloy A was conducted from the finish rolling temperature of 900°C to the finish cooling temperatures of 620°C, 454°C, and 350°C followed by air cooling to room temperature. Figure 8 and Table 4 show the cooling curves on the CCT diagram of the alloy A and the phase fraction after cooling to RT calculated with JmatPro software. The accelerated plates of the alloy A consisted of martensite, bainite, and retained austenite. As the finish accelerated cooling temperature decreased from 620°C to 350°C, the martensite fraction increased and the bainite fraction decreased while the retained austenite fraction only slightly increased. The optical microstructure of the accelerated-cooled alloy A is shown in Figure 9.

Figure 10 shows the hardness and tensile properties of accelerated cooling samples of the alloy A. The direct quenched plate had the highest hardness and tensile strength due to the highest fraction of martensite. The air-cooled plate had lower hardness and strength because of higher fraction of bainite. For the accelerated plates, the yield strength decreased and elongation increased with decreasing the finish cooling temperature because the retained autenite fraction increased correspondingly.

Table 4. Calculated phase fractions of the alloy A samples with controlled cooling conditions.

Conditions	Volume fraction of phases (%)			
	f_M[a]	f_B[b]	f_F[c]	f_A[d]
Air cooling	75.13	24.56	0.01	0.29
ACC620	86.42	13.25		0.32
ACC454	95.07	4.57		0.36
ACC350	99.38	0.25		0.37
Direct quench	99.46	0.15		0.39

[a] Martensite [b] Bainite [c] Ferrite [d] Austenite

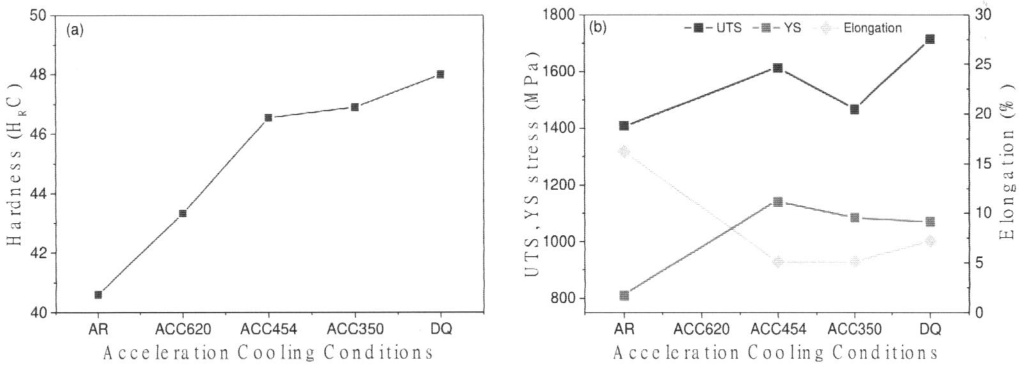

Figure 10. (a) hardness and (b) tensile properties of the accelerated-cooled plates of the alloy A.

4. Conclusions

- The austempered alloy B had better high strain rate properties than the Q/T sample due to high strength and good ductility.
- By adapting the accelerated cooling and isothermal process to substitute the austempering process, 0.5 wt% medium carbon bainitic steel (alloy B) showing the high hardness and over 2000 MPa tensile strength was achieved. These excellent mechanical properties are originated from a mixed microstructure of martensite and bainite.

IOP Conf. Series: Materials Science and Engineering **629** (2019) 012016 doi:10.1088/1757-899X/629/1/012016

- In the case of the accelerated cooling process for 0.23 wt% low carbon bainitic steel (alloy A), the hardness and tensile strength increased as the finish cooling temperature decreased because the volume fraction of martensite increased but that of bainite decreased.
- The present accelerated cooling and isothermal process can substitute the austempering process with salt bath in bainitic steel.

References

[1] Bhadeshia H and Edmonds D 1983 *Met. Sci.* **17** 420–5
[2] Caballero F, Bhadeshia H, Mawella K, Jones D and Brown 2001 P *Mater. Sci. Technol.* **17** 512–6
[3] Gong W, Tomota Y, Harjo S, Sua Y H, Aizawa K 2015 *Acta Mater.* **85** 243-9
[4] Caballero F, Bhadeshia H, Mawella K, Jones D and Brown P 2002 *Materials science and technology* **18** 279-84
[5] Sung H K, Shin S Y, Hwang B, Lee C G, Kim N J and S Lee 2010 *Korean Journal of Metals and Materials* **48** 798-806
[6] Bhadeshia H and Edmonds D 1983 *Metal Science* **17** 411-9
[7] Lee D J, Kim M S, Ku G E, Heo S H, Kim N Y and Lee J M 2018 *Korean Journal of Metals and Materials* **56** 221-6
[8] Caballero F, Roelofs H, Hasler S, Capdevila C, Chao J, Cornide J and Garcia-Mateo C 2012 *Materials Science and Technology* **28** 95-102
[9] Garcia-Mateo C, Jiménez J A, Yen H W, Miller M, Morales-Rivas L, Kuntz M, Ringer S, Yang J R and Caballero F 2015 *Acta Materialia* **91** 162-73
[10] Kim G H, Jang J H, Kim S H, Kim B J, Sohn K Y and Nam D G 2017 *Korean Journal of Metals and Materials* **55** 559-65

IOP Conf. Series: Materials Science and Engineering **629** (2019) 012017 doi:10.1088/1757-899X/629/1/012017

Modification of the surface layer of magnesium with 1050A aluminum alloy using electrospark deposition

S Spadło[1,2], W Depczyński[1] and P Młynarczyk[1]

[1]Faculty of Mechatronics and Machine Building, Kielce University of Technology, Poland

E-mail: sspadlo@tu.kielce.pl

Abstract. This paper discusses the results of the preliminary investigations into the use of electrospark deposition (ESD) for the surface enrichment of magnesium with 1050A aluminum alloy. The experiments focused on the effects of the capacitor's value on the surface layer microstructure. Four capacitances were considered. The modified material was analyzed at the macro and micro scales, using optical and electron scanning microscopy. The microhardness of the modified layer was also determined.

1. Introduction

Over the recent years, much attention has been paid to surface modification of soft and reactive materials in order to improve their surface properties, for example, their resistance to corrosion and abrasive wear. Surface modification is particularly important when the production of machine components needs to be cost-effective.

Although electrodischarge machining (EDM) [1-4] is generally classified as a removal process, it can be used to surface treat materials whose properties are unsatisfactory. In such cases, the technique is called electrospark deposition (ESD) [5-8] or electrical discharge alloying (EDA), and various materials can be used as electrodes. The method is also employed to precision repair expensive components [9-13]. The ESD process consists in depositing the material from the electrode in the molten state onto the surface of the solid substrate. The layer produced by electric discharges is generally metallurgically bonded after solidification. The main parameters of the ESD process are voltage, capacitance and discharge frequency. To select optimal parameters requires analyzing the properties of the electrode material, including its melting temperature, thermal conductivity and density. This study focused on the effects of the capacitor's value on the surface layer formation.

The use of thermally sensitive materials such as aluminum as an electrode in ESD to be deposited onto magnesium for protective purposes have not been investigated so far. Researchers dealing with the subject have applied other techniques to improve the surface properties of magnesium, for example, metal inert gas (MIG) welding [14], laser welding [15, 16], electron beam welding (EBW) [17], friction stir welding [18], diffusion bonding [19, 20], hot rolling [21, 22], explosive cladding [23], twin-roll casting [24], and other casting techniques [25, 26].

ESD involves rapid solidification of the deposited material. Research in this area has been done for superalloys, used as the electrode [27, 28], and Fe-based alloys and magnesium alloys [29, 30], employed as the substrate. However, the ESD process has rarely been applied to produce Al or Al alloy coatings that improve the microstructure and the cavitation erosion behavior of the magnesium substrate.

Magnesium is a highly reactive material, which makes it susceptible to corrosion attack. Numerous solutions are being looked into to protect magnesium in mechanical components from the interaction with other chemicals found in the environment, e.g. chloride in sea water. Some applications of Mg require plating it with Al, which is justified when large surface areas are to be protected. When parts complex in shape are involved, it is impossible or too costly to use this method of protection. ESD could be an alternative solution as the method is relatively inexpensive and does not require complicated equipment.

This paper presents the results of the preliminary research into the use of ESD to produce thick surface layers of 1050A aluminum alloy on the magnesium substrate. Different capacitances were analyzed. The macro- and microstructure of the surface layers was characterized and their hardness was measured.

2. Experiment

2.1. Surface layer fabrication

The experiments involved enriching the Mg surface with 1050A aluminum alloy using the electrospark deposition technique. The aluminum alloy in the form of wire (2 x 3 mm) was used as the anode, while the Mg substrate (40 mm x 20 mm x 5 mm plates) acted as the cathode. The chemical composition of the aluminum alloy is shown in table 1.

Table 1. Chemical composition of 1050A aluminum alloy (according to EN 573-1).

Mg [%]	Mn [%]	Fe [%]	Si [%]	Cu [%]	Zn [%]	Ti [%]	other, each [%]	Al [%]
≤0.05	≤0.05	≤0.40	≤0.25	≤0.05	≤0.07	≤0.05	≤0.03	≥99.50

A specially designed ESD system was used for the experiments. The parameters assumed to be optimal for this surface modification process were selected on the basis of the authors'earlier studies, and they were as follows: voltage $U = 600$ V, capacitance $C = 50$ µF–250 µF, and discharge frequency f=50 Hz. The four values of the capacitance tested were: 50 µF, 150 µF, 200 µF and 250 µF. The electrospark deposition was carried out at room temperature using a hand-held gun.

2.2. Microstructure and hardness analysis

The surface-modified specimens were sectioned and placed in acrylic cold setting resin (VariDur 10). Metallographic polishing was then performed using a STRUERS machine, with the final polishing suspension being the Microdiamant 0.05 µm Al_2O_3 suspension. The specimen preparation for the microstructural analysis did not include etching. The examinations involved optical microscopy (OM) using a Nikon Eclipse 200 with NIS Elements AR imaging software and scanning electron microscopy (SEM) applying a JEOL JSM-7100F coupled with an Oxford Instruments X-MAX energy dispersive spectrometry (EDS) detector. The Vickers microhardness of the material tested was measured at aload of 0.1961 N applied for 15 s. an INNOVATEST Nexus 4504 was used for this purpose. The indentations were made at regular intervals across the specimen cross-section to compare the hardness of the surface layer with that of the bulk.

3. Results and discussion

The SEM images in figure1 show the microstructures of the Mg specimens modified at the surface with 1050A aluminum alloy through ESD at four different capacitances. As can be seen, the microstructure and thickness of the surface layer were affected by the capacitor's value. Melting the Mg surface with lower value capacitors (50 µF and 150 µF) led to the formation of thicker layers characterized by relatively more uniform morphology and smaller porosity (figures 1(a) and 1(b)), respectively). When higher value capacitors (200 µF and 250 µF) were used, the enriched surface layer of the Mg substrate was thinner, exhibiting more heterogeneous microstructure with larger porosity (figures 1(c) and 1(d)), respectively). At the highest capacitance, extensive cracking and delamination were observed. This was due to the fact that the pulse energy was too high, part of the deposited material evaporated, and the stress relaxation occurred too rapidly. At lower capacitances, more thermal energy could be absorbed by the material so the cooling (stress relaxation) was slower.

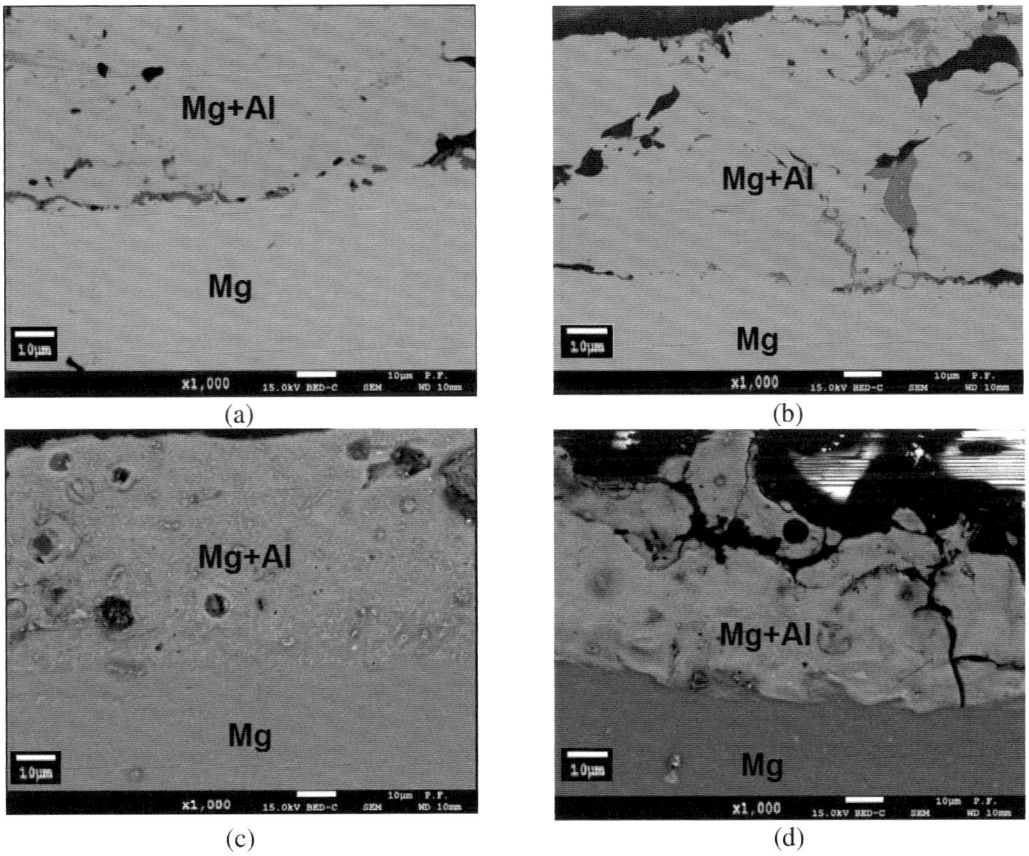

Figure 1. SEM images of the Mg specimens with the surface modified through ESD at (a) 50 μF, (b) 150 μF, (c) 200 μF, (d) 250 μF (magnification x1000).

Figure 2 shows a panoramic OM image of the Mg specimen surface-modified with 1050A through ESD capacitance of 200 μF. The surface layer is not uniform in thickness; locally, there are pores differing in size and numerous precipitates.

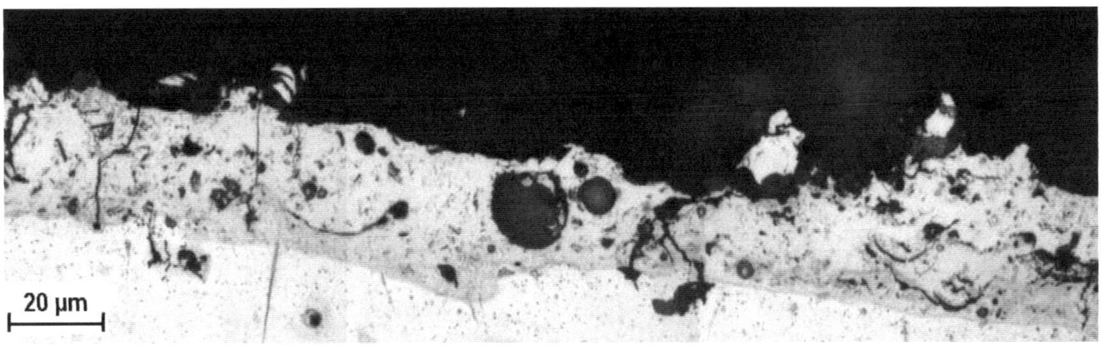

Figure 2. Microstructure of the surface layer fabricated using a 200 μF capacitor (magnification x500).

The specimen microstructure was further analyzed using higher magnification (figure 3).

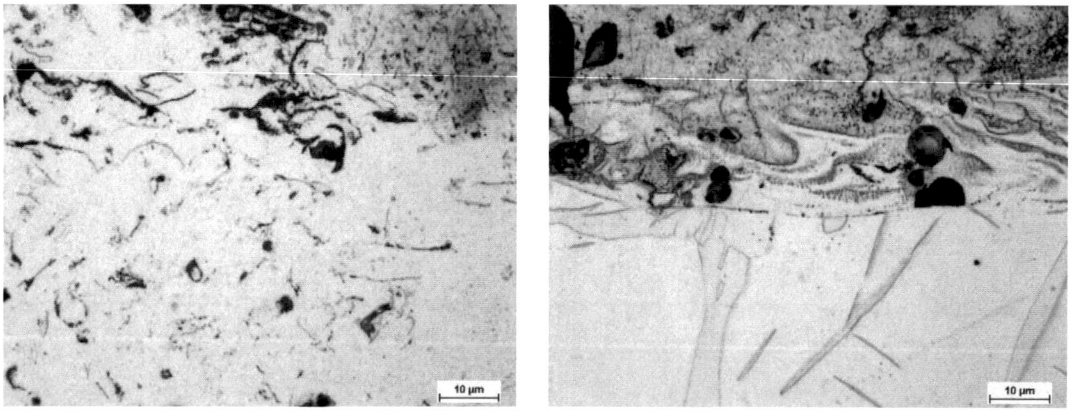

(a) (b)

Figure 3. OM images of the surface layer microstructure fabricated using a 200 μF capacitor: (a) upper zone and (b) lower zone (magnification x1000).

The OM images in figure 3 reveal a relatively uniform microstructure with numerous precipitates at the top and dendrites at the interface between the modified surface layer and the substrate.

Figures 4(a) and 4(b) present the SEM and EDS results obtained for the surface layers produced at 200 μF and 50 μF, respectively.

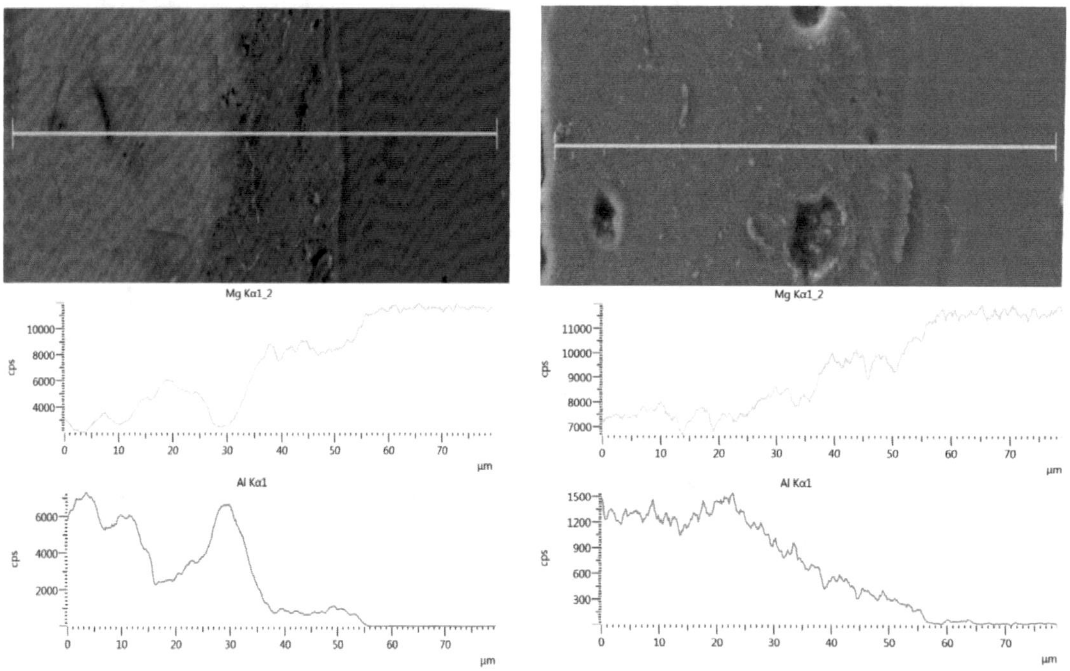

(a) (b)

Figure 4. SEM cross-sections and the corresponding EDS line scans obtained for the Mg specimens enriched at the surface with 1050A at (a) 200 μF and (b) 50 μF.

The EDS analysis indicated diffusion bonding of the two materials. As can be seen from figure 4, the aluminum alloy used to enrich the Mg surface diffused deep into the substrate, with the interface being about 30 μm in thickness. From figure 4 (b), it is clear that the composition of the material modified at 50 μF changed less rapidly.

Figure 5 provides the SEM-EDS data recorded for the surface layer produced with a 200 μF capacitor. The SEM image in figure 5 (b) shows the specimen microstructure with two characteristic zones: the thicker, lighter zone at the top and the thinner, darker zone adjacent to the Mg substrate.

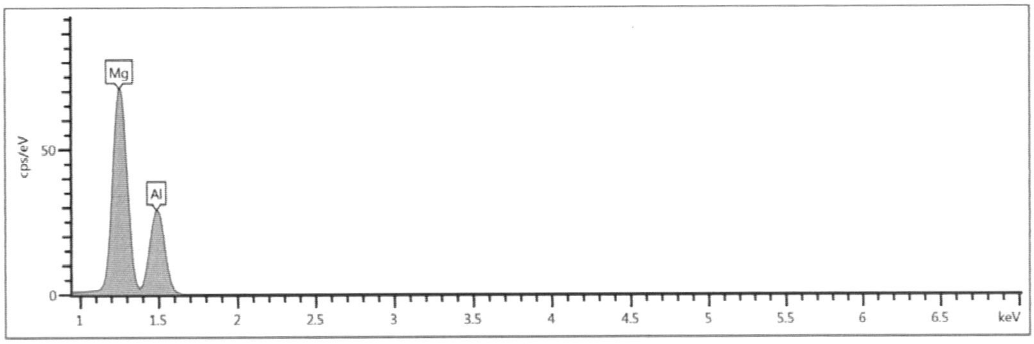

(a)

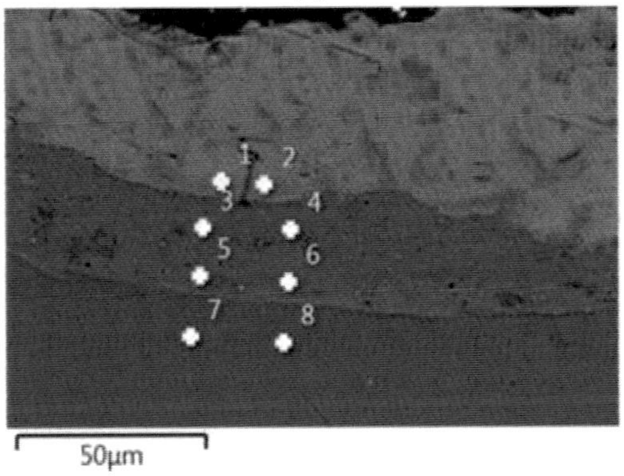

(b)

Figure 5. (a) EDS spectrum showing the elements present at point 1 and (b) SEM image showing details of the microstructure of the surface layer fabricated using a 200 μF capacitor with points of the EDS quantitative analysis.

Table 2. EDS results at the points shown in figure 5(b).

Point	Mg at. %	Al at. %
1	55.95	44.05
2	55.55	44.45
3	83.76	16.24
4	85.59	14.41
5	84.14	15.86
6	84.01	15.99
7	100.00	0.00
8	100.00	0.00

These observations are similar to those described by other researchers for Al-enriched surface layers on Mg fabricated through welding [25, 26], with the difference that the structure of layers obtained via ESD is much finer and its characterization requires the use of relatively large magnifications. The EDS quantitative analysis was carried out at six points, as marked in figure 5(b). The content data are given in table 2. The microstructures of the surface layers were characterized using the Mg-Al binary phase diagram. The results obtained for the material at points 1 and 2 indicate the occurrence of the intermetallic phase ($Mg_{17}Al_{12}$). The surface layer contained both Mg and Al in similar amounts. The contents of Mg at points 3,4,5, and 6 ranged from 83.76% to 85.59%, which suggests a solid solution of Al in Mg. Full identification of the phases present in the material after the ESD would be required, using the XRD technique.

In all the cases considered, the microhardness of the surface layer was higher than that of the Mg substrate. The substrate hardness was about 40 $HV_{0.01}$, whereas that of the ESD-modified surface varied between 200 and 220 $HV_{0.01}$.

4. Conclusions

The experimental data suggest that ESD can be used to modify the surface layer of the Mg substrate using 1050A aluminum alloy. The microstructure and thickness of the enriched layer depends on the process parameters. In this study, four different capacitances were considered. The layers produced with lower value capacitors were thicker and more uniform in structure. The surface enrichment achieved with higher value capacitors resulted in thinner layers. There were two characteristic zones: the upper zone containing similar amounts of Al and Mg ($Mg_{17}Al_{12}$ intermetallic phase), and the lower zone, i.e., the interface, with a much higher content of Mg, which suggests a solid solution of Al in Mg. The highest capacitance used (250 μF) led to the formation of a layer characterized by extensive cracking and delamination. In all the cases studied, the surface layer had much higher microhardness than the Mg substrate.

References

[1] Davydov A D and Kozak J 1991 Physico-chemical principles of electrochemical discharge machining *Surf. Eng. Appl. Electrochem.* **3** 3-13

[2] Schumacher B 2004 After 60 years of EDM the discharge process remains still disputed *Proc. XIV. ISEM* 149 (Edinburgh, UK) 376-81

[3] Świercz R and Oniszczuk-Świercz D 2017 Experimental investigation of surface layer properties of high thermal conductivity tool steel after electrical discharge machining *Metals* 7 550

[4] Yan B H, Lin Y C, Huang F Y and Wang C H 2001 Surface modification of SKD 61 during EDM with metal powder in the dielectric *Mater. Trans. Jim* **42** 2597-604

[5] Młynarczyk P and Spadło S 2016 The analysis of the effects formation iron - tungsten carbide layer on aluminum alloy by electrical discharge alloying process *Proceedings of 25th International Conference on Metallurgy and Materials* (Brno, Czech Republic) 1109-14

[6] Rastkerdar E, Aghajani H, Kianvash A and Sorrell C C 2018 Microstructural characterization of AA5183 aluminum clad AISI 1018 steel prepared by electro spark deposition *Mater. Res. Express* **5** Doi: 10.1088/2053-1591/aab81f

[7] Spadło S, Młynarczyk P and Depczyński W 2015 Investigation of the selected properties of the superficial layer alloying with the tungsten electrodes *Proceedings of 24th International Conference on Metallurgy and Materials* (Brno, Czech Republic) 863-7

[8] Tsuekawa Y, Okumiya N, Mohri N and Takahashi I 1994 Surface modification of aluminum by electrical discharge alloying *Mater. Sci. Eng. A***174** 193-8

[9] Kovacik J, Baksa P and Emmer S 2016 Electro spark deposition of TiB_2 layers on Ti6Al4V alloy *Acta Metallurgica Slovaca* **22** 52-9

[10] Lee H G, Simao J, Aspinwall D K, Dewes R C and Voice W 2004 Electrical discharge surface alloying *J. Mater. Process Technol.* **149** 334-40

[11] Młynarczyk P, Krajcarz D and Bańkowski D 2017 The selected properties of the micro electrical discharge alloying process using tungsten electrode on aluminum *12th International Scientific Conference of Young Scientists on Sustainable, Procedia Engineering* **192** 603-8

[12] Spadło S 2002 Comparative studies of brush electrodischarge machining with electrodes of alloy

steel and tungsten *Advanced Manufacturing Systems and Technology, International Centre for Mechanical Sciences (CISM)* Vol **437** (New York: Springer – Veralag Wien) 515-24

[13] Spadło S and Młynarczyk P 2016 Analysis of the mechanical interactions of the filament brush electrode on the formation of the surface roughness *Proceedings of 25th International Conference on Metallurgy and Materials* (Brno, Czech Republic) 1169-74

[14] Zhang H T and Song J Q 2011 Microstructure evolution of aluminum/magnesium lap joints welded using MIG process with zinc foil as interlayer *Mater. Lett.* **65** 3292-4

[15] Borrisutthekul R, Miyashita Y and Mutoh Y 2005 Dissimilar material laser welding between magnesium alloy AZ31B and aluminum alloy A5052-0 *Sci. Tech. Adv. Mater.* **6** 199-204

[16] Mola R, Dziadoń A and Jagielska-Wiaderek K 2016 Properties of Mg laser alloyed with Al or AlSi20 *Surf. Eng.* **32** 908-15

[17] Spadło S, Młynarczyk P, Depczyński W, Skowron E and Mijas R 2017 Research on the influence of electron beam welding of titanium alloy microstructure connection *Proceedings of 26th International Conference on Metallurgy and Materials* (Brno, Czech Republic) 1939-44

[18] Morishige T, Kawaguchi A, Tsujikawa M, Hino M, Hirata T and Higashi K 2008 Dissimilar welding of Al and Mg alloys by FSW *Mater. Trans.* **49** 1129-31

[19] Li Y, Liu P, Wang J and Ma H 2008 XRD and SEM analysis near the diffusion bonding interface of Mg/Al dissimilar materials *Vacuum.* **82** 15-9

[20] Dziadoń A, Mola R and Błaż L 2011 Formation of layered Mg-eutectic composite using diffusional process at the Mg-Al interface *Arch. of Metall. and Mater.* **56** 677-84

[21] Zhang XP, Yang TH, Castagne S and Wang JT 2011 Microstructure; bonding strength and thickness ratio of Al/Mg/Al alloy laminated composites prepared by hot rolling *Mater. Sci. Eng. A* **528** 1954-60

[22] Wierzba A, Mróz S, Szota P, Stefanik A and Mola R 2015 The influence of the asymetric ARB process on the properties of Al-Mg-Al multi-layer sheets *Arch. of Metall. and Mater.* **60** 2821-5

[23] Mróz S, Stradomski G, Dyja H and Galka A 2015 Using the explosive cladding method for production of Mg-Al. bimetallic bars *Arch. Civ. Mech. Eng.* **15** 317-23

[24] Bae JH, Prasada Rao AK, Kim KH and Kim NJ 2011 Cladding of Mg alloy with Al by twin-roll casting *Scripta Mater.* **64** 836-9

[25] Mola R, Bucki T and Dziadoń A 2016 Formation of Al-alloyed layer on magnesium with use of casting techniques *Arch. of Foundry Eng.* **16** 112-6

[26] Mola R, Bucki T and Dziadoń A 2017 Microstructure of the bonding zone between AZ91 and AlSi17 formed by compound casting *Arch. of Foundry Eng.* **17** 202-6

[27] Lachowicz M B and Dudziński W 2012 Characteristics of GTA fusion zones and heat affected zones in superalloy 713C *Mater. Sci.* **30** 159-69

[28] Młynarczyk P, Spadło S and Depczyński W 2017 Investigations into the effects of spot welding on thin sheet of superalloys Hastelloy X and Haynes 230® *Proceedings of 26th International Conference on Metallurgy and Materials* (Brno, Czech Republic) 1881-6

[29] Dziadoń A, Mola R and Błaż L 2016 The microstructure of the surface layer of magnesium laser alloyed with aluminum and silicon *Mater. Charact.* **118** 505-13

[30] Mola R 2015 The properties of Mg protected by Al- and Al/Zn-enriched layers containing intermetallic phases *J. Mater. Res.* **30** 3682-91

IOP Conf. Series: Materials Science and Engineering **629** (2019) 012018 doi:10.1088/1757-899X/629/1/012018

Fatigue crack growth analysis of surge line under thermal stratification conditions

P Tang[1,2], J C Luo[1], L Sun[1] and P Z Li[1]

[1]Nuclear Power Institute of China, Chengdu, China

E-mail: tangpengnpic@163.com

Abstract. As the connection between the pressurizer and the main pipeline of nuclear reactor, thermal stratification occurs in the pressurizer surge line during the operation. The thermal stratification through the pipe wall increases the possibility of plastic deformation and fatigue failure in the nozzle, tee, elbow and weld. So it is of great significance to carry out research on the fracture mechanics of surge line under thermal stratification conditions. Therefore, in this paper, based on the results of finite element analysis and ASME specifications, and with the results of pipeline stress analysis under temperature and pressure transient loads, the fatigue crack growth analysis of the surge line under thermal stratification conditions is carried out, and the technical means of suppressing crack propagation is explored, which provides a reference for the safe operation of the surge line of the nuclear power plant.

1. Introduction

In the primary loop system of nuclear power plant, the pipeline system, as the transportation artery of the coolant, plays an important role in the safe operation of the reactor equipment. The pressurizer surge line is an important link for connecting the pressurizer and the primary pipe. Inside the surge line, the fluid from the hot section of the primary pipeline has a low temperature and a high density, occupying the lower part of the pipeline section; while the fluid from the pressurizer has a high temperature and low density, and occupies the upper part of the pipe section, thus forming a thermal stratification phenomenon. The corresponding temperature gradient results in local stresses, and the pipe support constraints lead to global stress, making global bending inconsistent with the design and reducing pipe fatigue life [1,2]. During the long-term service of the surge line, cracks or defects may occur at the inner surface of the pipe. Affected by the thermal stratification conditions for a long time, the crack may continue to expand and eventually penetrate through the wall thickness, causing the pipeline to rupture and the leakage of coolant, which seriously affect the safe operation of the surge line and other primary loop equipment. Therefore, in order to ensure the safe operation of the nuclear power plant, it is of great engineering value to carry out the analysis of the fatigue crack propagation of the surge line, understand the crack propagation mechanism and explore the crack arrest technology.

Since the large value of stress intensity and fatigue usage factor of the surge line always occur at the joint of the straight pipe and the elbow, it is meaningful to carry out the research on the fatigue crack propagation of the weld structure. Zheng [3] discussed the crack initiation mechanism and early crack propagation behavior of TC21 titanium alloy electron beam welded joints. Ihara [4] estimated the residual stress distribution generated by butt welding and surface machining, and performed thermal elastoplastic analysis on the welding model. The surface machining simulation was performed on the thermo-mechanical coupling analysis based on the Johnson-Cook material model. The calculated crack propagation analysis based on the stress intensity factor (SIF) was performed using the calculated residual stress distribution generated by welding and surface machining. Gotoh [5] investigated the fatigue crack propagation behavior of the out-of-plane welded joint under biaxial loading and two different phase conditions. The phase difference effect of the fatigue crack shape evolution under biaxial loading was confirmed by measurement. Arora [6] studied the fatigue crack propagation (FCG) behavior of austenitic stainless steel pipe welds. Compact tension (CT) specimens have been tested from actual narrow gap pipe welds. The slit is located in the weld centerline (WCL), heat affected zone (HAZ), HAZ-Fusion line (FL) interface and base metal. The FCG rates at different locations of the crack were compared. Testing was performed using the standard procedure of ASTM E647. The FCG efficiency of narrow gap welds (HAZ, WCL, HAZ-FL interface and parent metal) was compared to conventional VW weld seam welds (SMAW).

In order to reduce the crack propagation and slowdown the crack further expansion, the surfacing technology is often used in the engineering. By welding the dissimilar metal on the surface of the pipeline, the crack surface is in the state of compressive stress, which resists the crack propagation. Huang [7] conducted a structural integrity analysis of the pilot weld overlay on different metal welds (DMW) of pressurized nozzles in pressurized water reactors (PWR). Based on MRP-169 and ASME code case N-504-2, the analysis of fatigue crack growth and fatigue usage was performed. Figure 1 shows the axial stress after application of the overlay welding procedure. Surfacing provides further compressive residual stress in the DMW, with more than 75% of the original weld thickness being compressed. According to the analysis, the weld overlay layer has the effect of significantly improving the compressive residual stress field of the DMW region.

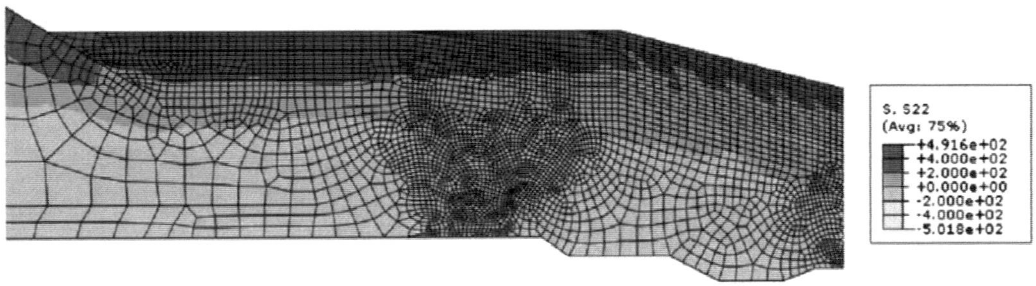

Figure 1. Axial stress after overlay procedure.

Based on the above method, Wei [8,9] constructed the failure assessment curve of the structure after surfacing repair by ABAQUS finite element analysis and fracture mechanics analysis method. The effects of weld overlay thickness, crack depth and crack location on the failure assessment curve and LBB safety margin were analyzed. Jiang [10] proposed a new method of laminating welding on the inner surface of the nozzle to reduce the residual stress of the through joint. The finite element method was used to study the residual stress of the weld before and after surfacing. Based on the size of the spray tube nozzle of the Qinshan I nuclear power plant, Zhang [11] established the axisymmetric finite element model, and made used of the life-and-death element to simulate the transient heat input of the welding process, and investigated the temperature field of the structure during the surfacing repair process. According to the variation distribution characteristics, the calculated structural heat distribution history is taken as the thermal load, and then the variation characteristics of the structural residual stress during the surfacing repair process are obtained. The results show that after the surfacing repair is completed, the residual stress on the inner wall of the DMW weld zone can play the role of controlling the stress corrosion cracking (PWSCC). Liu [12] analyzed the welding residual stress of dissimilar metal welds in the water supply nozzle of boiling water reactor in Taiwan. In order to reduce the degradation of the nozzle weld, weld coverage repair was implemented to ensure its structural integrity. An axisymmetric model of the feedwater nozzle was established, taking into account the on-site welding procedure and detailed weld simulation to predict the weld residual stress during weld coverage repair.

Yusa [13] evaluated the applicability of eddy current testing for fatigue crack detection and size in embedded Inconel weld overlays. Jun [14] performed a series of failure analysis and crack propagation analysis on fractured weld repair rails to determine the cause of rail failure and the effect of residual stress on crack growth rate.

Antunes [15] developed a numerical model to predict plasticity induced crack closure (PICC), and isolated basic micromechanisms of crack closure and developed qualitative and quantitative relations between individual plastic wedges and PICC. García [16] conducted a numerical analysis of PICC based on a cohesive zone model (CZM) with irreversible damage that employs a crack growth method coupled with cyclic plasticity at crack tip and phenomenological fracture. The results show the ability of the proposed CZM to capture the influence of crack tip plasticity in the evaluation of crack closure phenomenon. Funari [17,18] introduced a numerical model to predict crack propagation phenomena in sandwich structures, and proposed numerical approaches which are able to compute the fracture variables (i.e. J integral approach). And the research results mentioned above are essentially meaningful and inspirational for the research of this paper.

According to the overlay analysis method, this paper carried out the analysis of the fatigue crack propagation of the surge line and explored the pipe arrest technology, which provides the method and technical basis for the operation and maintenance of the surge line.

2. Method description

2.1. Overlay stress analysis method for surge line

By depositing a layer of metal on the outer wall of the pipeline, the distribution of the stress field of the pipeline with and without overlay welding was compared. By selecting the relevant path on the pipe section, the stress variation from the inner wall to the outer wall was obtained under different paths.

Table 1. Material property for surge line.

Temperature T(°C)	20	50	100	150	200	250	300	350
Elastic modulus E(GPa)	197.0	195.0	191.5	187.5	184.0	180.0	176.5	172.0
Density ρ(kg/m^3)	7910							
Poisson ratio υ	0.3							
Heat expansion coefficient α(10^{-6}/°C)	15.54	16.00	16.49	16.98	17.47	17.97	18.46	18.95
Specific heat C(J/(kg.°C)	455.0	468.0	494.0	507.0	526.0	538.7	542.7	552.0
Heat conductivity λ(W/(m.°C)	14.0	14.4	15.2	15.8	16.6	17.3	17.9	18.6
Allowable stress intensity S_m(MPa)	358	358	358	358	354	346	339	334
Yield stress S_y(MPa)	199	199	176	160	148	138	131	127
Tensile strength S_u(MPa)	517	517	506	495	495	495	495	495

Table 2. Material property for weld.

Temperature T(°C)	20	50	100	150	200	250	300	350
Elastic modulus E(GPa)	209	208	207	204	202	199	196	191
Density ρ(kg/m^3)	7810							
Poisson ratio υ	0.3							
Heat expansion coefficient α(10^{-6}/°C)	11.1	11.4	11.7	12.0	12.4	12.8	13.1	13.4
Specific heat C(J/(kg.°C)	459.0	465.0	469.0	475.0	481.0	490.0	502.0	512.0
Heat conductivity λ(W/(m.°C)	18.2	32.1	48.1	47.2	46.5	45.1	44.3	43.5
Allowable stress intensity S_m(MPa)	284	270	255	240	226	201	186	172
Yield stress S_y(MPa)	178	169	160	150	141	129	117	107
Tensile strength S_u(MPa)	569	569	569	569	569	569	569	569

The material properties of the surge line and the weld are shown in tables 1 and 2. The load cases, transient load of temperature and pressure for fatigue crack propagation analysis are shown in table 3

IOP Conf. Series: Materials Science and Engineering **629** (2019) 012018 doi:10.1088/1757-899X/629/1/012018

and figure 2.

Table 3. Load cases for fatigue crack propagation analysis.

Load Case	Number	Pressure(MPa)	Low T(°C)	High T(°C)
1	700000	0.1	30	35
2	10700000	15.6	285	345
3	12200000	15.6	290	345
4	4200000	3.4	130	240
5	3000000	15.6	295	345
6	4000000	16.6	240	350
7	4000000	12.9	220	330
8	3000000	16.6	290	350
9	12000000	15.6	320	345
10	10000000	15.6	315	315

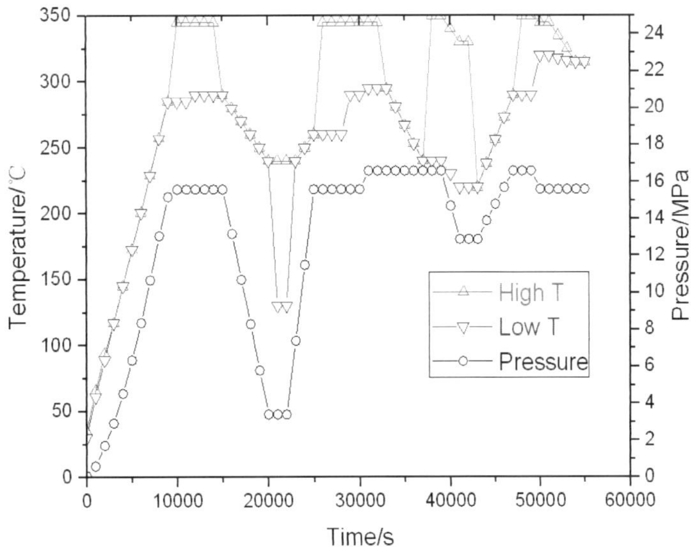

Figure 2. Transient load of temperature and pressure.

The two-dimensional section models with and without a weld overlay layer for surge line were established, as shown in figure 3. According to the 16-division temperature assumption, the inner wall surface of the pipe is divided into 16 equal parts, and different temperature loads are applied to different parts, as shown in figure 4. Transient thermal load was applied to the inner wall surface of the pipeline, and thermal transient analysis was carried out to obtain the temperature distribution of the pipeline section.

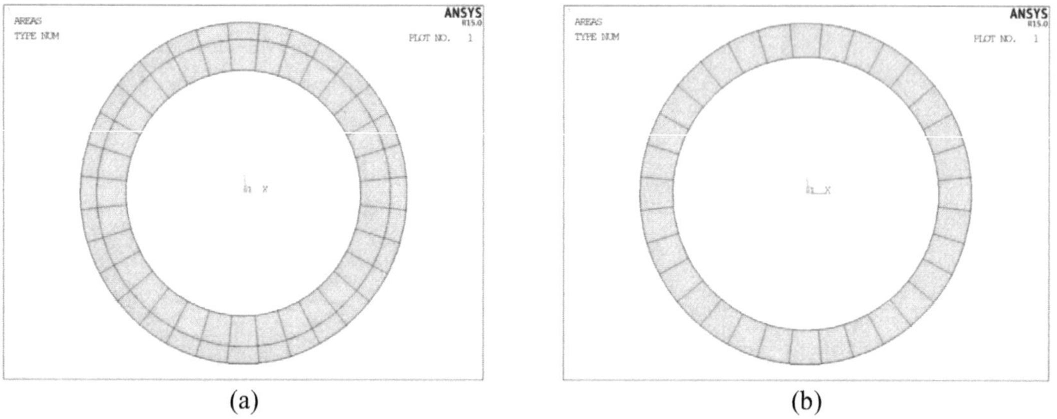

Figure 3. Finite element model of pipe section for the surge line with (a) and without weld overlay (b).

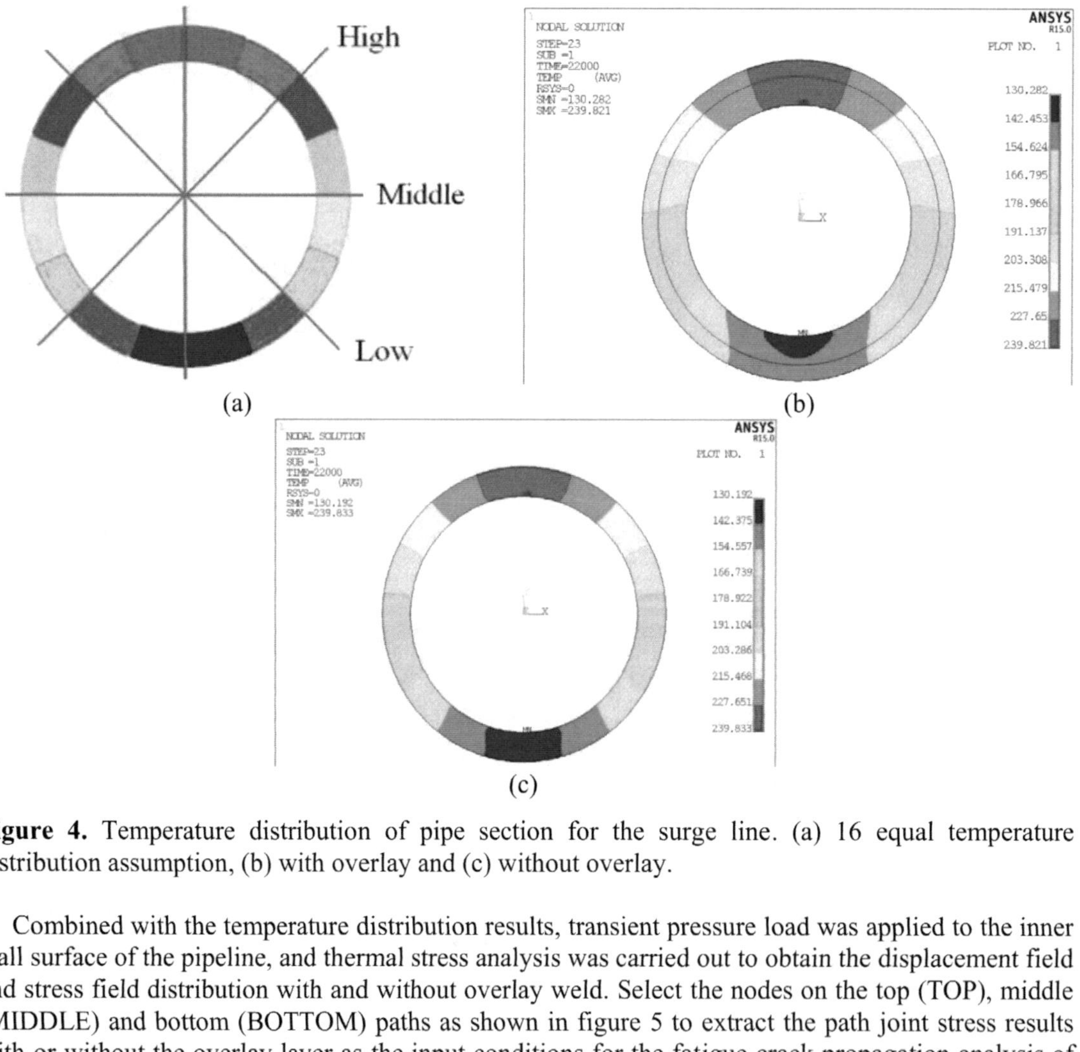

Figure 4. Temperature distribution of pipe section for the surge line. (a) 16 equal temperature distribution assumption, (b) with overlay and (c) without overlay.

Combined with the temperature distribution results, transient pressure load was applied to the inner wall surface of the pipeline, and thermal stress analysis was carried out to obtain the displacement field and stress field distribution with and without overlay weld. Select the nodes on the top (TOP), middle (MIDDLE) and bottom (BOTTOM) paths as shown in figure 5 to extract the path joint stress results with or without the overlay layer as the input conditions for the fatigue crack propagation analysis of the surge line.

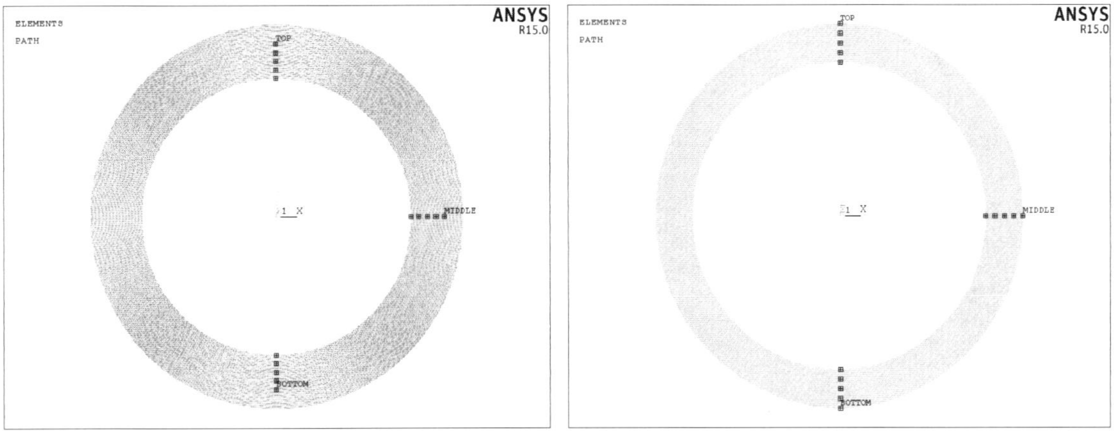

(a) (b)

Figure 5. Path selection of surge line (a) with overlay and (b) without overlay.

2.2. Fatigue crack growth analysis method for surge line

Combined with the stress analysis results of the surge line with or without the overlay layer obtained above, the fatigue crack propagation analysis of the wave tube can be carried out. In this paper, the fatigue crack growth calculation under the thermal stratification condition of the surge line is mainly based on the relevant specifications. The relevant methods are as follows:

Refer to the ASME XI volume IWB-3640 requirements for crack propagation analysis of the overlay design. Due to the subcritical expansion analysis of cracks, a lot of numerical calculations are needed. In order to complete the crack propagation analysis effectively, accurately and quickly, the crack subcritical expansion calculation program was compiled by Matlab software according to the crack propagation analysis method in ASME Code XI volume. With the program, the fatigue crack propagation of the crack for the pressurizer surge line can be calculated. Combined with the transient load of the surge tube operation, the crack propagation amount at the end of the analysis period was obtain.

The surface crack stress intensity factor is calculated according to ASME XI Volume Appendix A-3320 formula:

$$K_I = \left[\left(A_0 + A_p \right) G_0 + A_1 G_1 + A_2 G_2 + A_3 G_3 \right] \sqrt{\pi a / Q} \quad (1)$$

Where, a is crack depth; A_0, A_1, A_2, A_3 are coefficients of equation A-3200(b). G_0, G_1, G_2, G_3, as the free surface correction factors of Point 1 and Point 2, were determined according to Tables A-3320-1 and A-3320-2(Detailed table can be seen in ASME code). The locations of Point 1 and Point 2 are shown in figure 6.

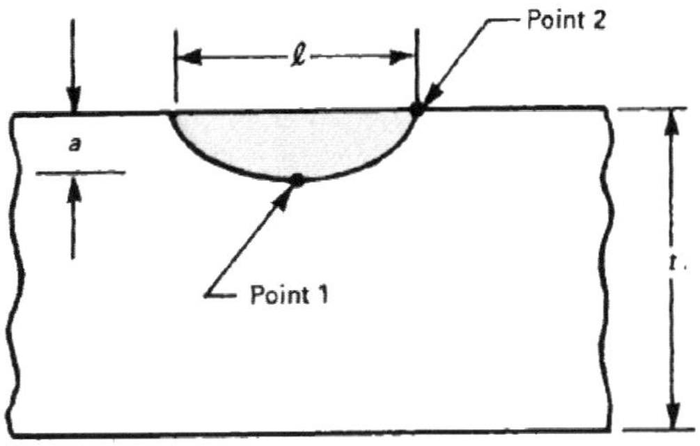

Figure 6. Semi-elliptical crack model.

Q is the crack shape parameters calculated in equation A-3310, equation A-3310(3) is as follows:

$$Q = 1 + 4.593\left(a/l\right)^{1.65} - q_y \tag{2}$$

Where, l is crack length; q_y is the plastic zone correction factor, which is calculated by Equation A-3320(6):

$$q_y = \left[\left(A_0 G_0 + A_p G_0 + A_1 G_1 + A_2 G_2 + A_3 G_3\right)/\sigma_{ys}\right]^2 /6 \tag{3}$$

Where, σ_{ys} is yield strength.

For surface cracks, the relationship between the normal stress and the crack depth of the crack surface at the crack location is determined by the following ASME XI Appendix A Equation A-3200(b)(1)

$$\sigma = A_0 + A_1\left(\frac{x}{a}\right) + A_2\left(\frac{x}{a}\right)^2 + A_3\left(\frac{x}{a}\right)^3 \tag{4}$$

Where, σ is circumferential or longitude stress; x is the distance between the inner surface and crack front; A_i is coefficient, and i=0, 1, 2, 4.

The fatigue crack growth rate CGR(da/dN) is determined by the applied stress intensity factor range ΔK_I and related materials and environmental constants. In ASME XI Vol. Appendix C-3210(1), the formula for calculating the fatigue crack growth rate is as follows:

$$\frac{da}{dN} = C_0\left(\Delta K_I\right)^n \tag{5}$$

Where, n is the slope of log(da/dN) and log(ΔK_I); C_0 is the proportional constant of the environment and material; ΔK_I is the range of applied stress intensity factors; n=3.3; and

$$C_0 = CS \tag{6}$$

Where A is the proportional parameter considering temperature:

$$C = 10^{\left[-8.714+1.34\times10^{-3}T-3.34\times10^{-6}T^2+5.95\times10^{-9}T^3\right]} \tag{7}$$

Where T is the metal temperature (°C), S is a proportional parameter considering the R ratio(K_{Imin}/K_{Imax}) and

$$S = \begin{cases} 1.0 & R \le 0 \\ 1.0 + 1.8R & 0 < R \le 0.79 \\ -43.35 + 57.97R & 0.79 < R < 1.0 \end{cases} \tag{8}$$

3. Results and discussion

3.1. Stress distribution analysis results of overlay weld

The transient temperature load was applied to the inner surface of the surge line section with and without the overlay layer, and the temperature field distribution of the surge line section under different transient loads was obtained. Then, considering the temperature distribution result as an input condition, a transient pressure load was applied to the inner surface of the pipe to carry out thermal stress analysis of the surge line. The stress and displacement field distribution results for the cross section of the surge line with and without the overlay layer were obtained.

Figure 7 shows the maximum displacement values for surge line with and without weld overlay under different operating conditions. It can be seen that under all operating conditions, the maximum displacement value of the surge line containing the weld overlay is less than that associated with the surge line without overlay, which indicates that the displacement level of the surge line is reduced under the condition of the overlay layer.

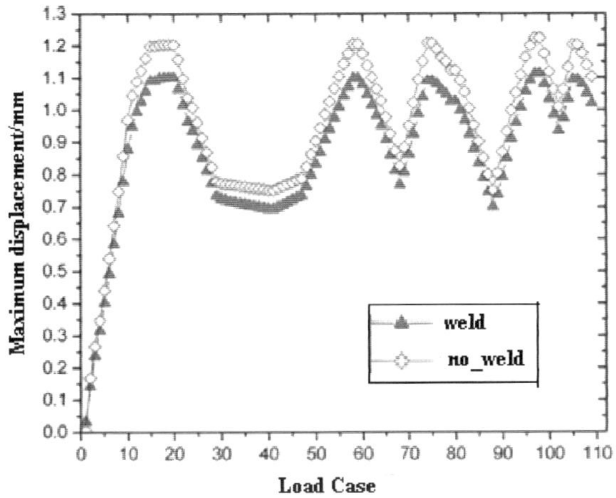

Figure 7. Maximum displacement of surge line with and without weld overlay.

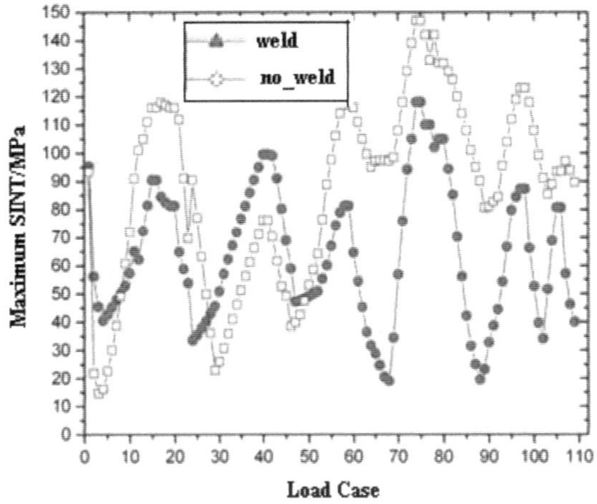

Figure 8. Maximum stress intensity of surge line with and without weld overlay.

Figure 8 shows the maximum stress intensity values for surge line with and without weld overlay under different operating conditions. It can be seen that under most working conditions, the maximum stress intensity value of the surge line containing the weld overlay is smaller than that of the surge line without weld overlay, thus indicating that in the presence of the weld overlay, the stress level of the surge line is reduced.

For the stress distribution of the pipe section, taking the results of 70000s in figure 8 as an example, the stress distribution of the surge line with or without the overlay layer are shown in figure 9. It can be seen that under the condition of no-overlay layer, the stress distribution of the section of the pipeline is: the top is the largest, the middle is the smallest, and the stress is gradually increased from the middle to the both ends. Under the condition of surfacing layer, the maximum stress intensity appears on the top and bottom inner wall surface, the middle stress intensity is relatively small, and the stress gradually decreases from the top to the bottom.

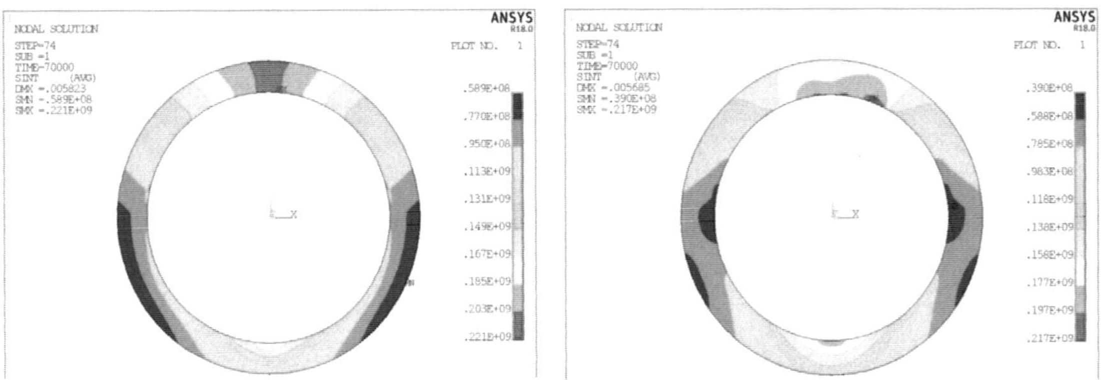

(a) (b)

Figure 9. Stress distribution of the surge line section with (a) or without overlay layer (b) at 70000s.

3.2. Analysis result of fatigue crack expansion

According to the analysis results of the overlay stress, the stress value in the thickness direction of the relevant path is extracted, and the fatigue crack propagation analysis under the transient cyclic load of the surge line is carried out by using the fatigue crack propagation calculation method introduced in Section 2.2.

Assume an initial longitude semi-elliptical surface with a depth of 17.85 mm and a length of 53.55 mm. According to the stress results at the BOTTOM path, MIDDLE path and TOP path of the section, fatigue crack propagation under transient loading were calculated.

The calculated crack propagation profile is shown in figure 10. For the case without overlay layer, the crack at the location of BOTTOM-PATH, MIDDLE-PATH and TOP-PATH were expanded by 3.82 mm, 4.07 mm and 7.15 mm, respectively; for the case with overlay layer, the corresponding results of BOTTOM-PATH, MIDDLE-PATH and TOP-PATH are 3.52 mm, 1.15 mm and 0.87 mm, respectively.

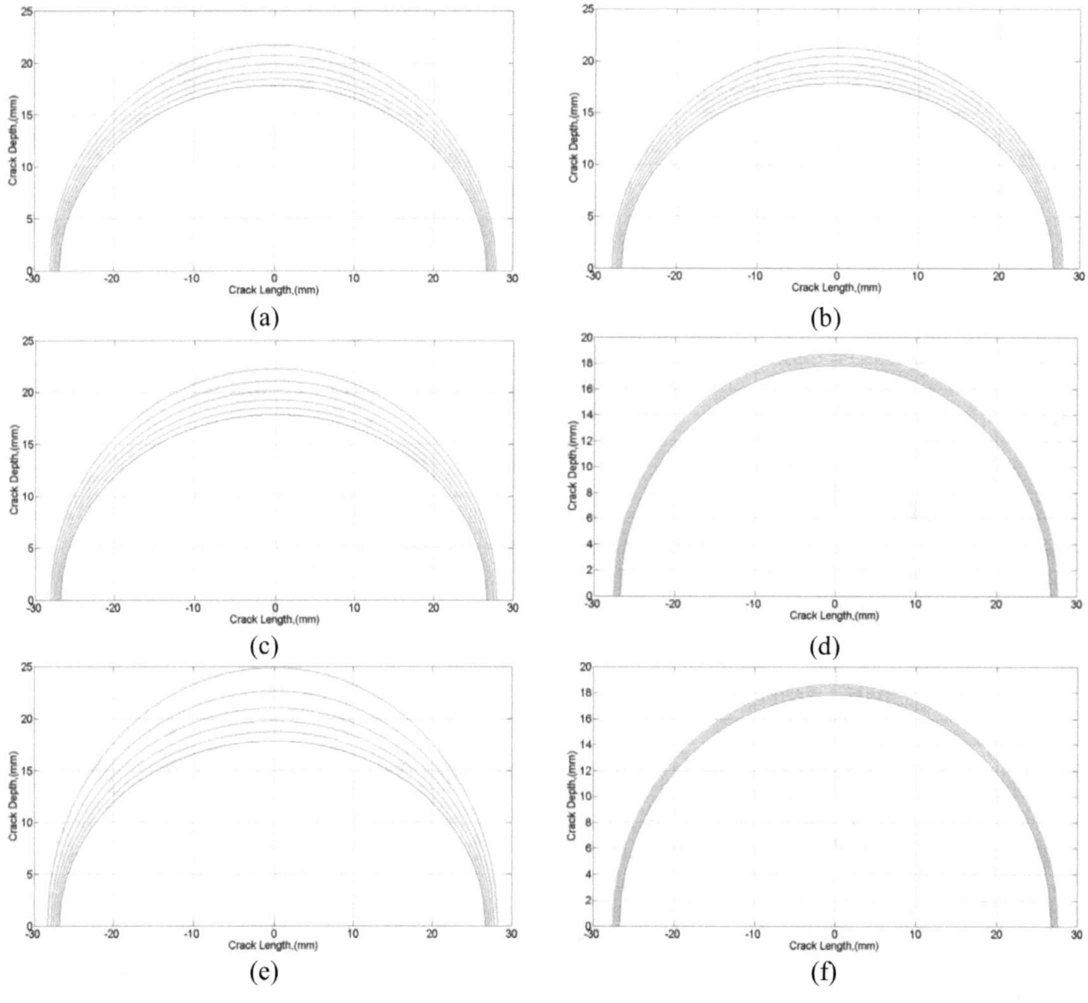

Figure 10. Crack profile of crack growth. (a) Without overlay-BOTTOM-PATH, (b) with overlay-BOTTOM-PATH, (c) without overlay-MIDDLE-PATH, (d) with overlay-MIDDLE-PATH, (e) without overlay-TOP-PATH, (f) with overlay-TOP-PATH.

3.3. Crack arrest evaluation result

The amount of crack propagation obtained with and without the weld overlay were compared, as shown in figure 11. For the cracks at BOTTOM-PATH, MIDDLE-PATH and TOP-PATH, the crack depth extension results under the conditions of the surfacing layer are 0.3 mm, 2.92 mm and 6.28 mm smaller than those without overlay layer, respectively. For the result of the crack length increasement, the former is 0.1 mm, 1.6 mm and 2.2 mm smaller than the latter. It can be seen that the results for the crack located in BOTTOM-PATH is similar. Difference get bigger for the crack located in MIDDLE-PATH, and the crack located at TOP-PATH has the largest difference. Therefore, the crack arresting ability is gradually increased as crack moves from the bottom to the top.

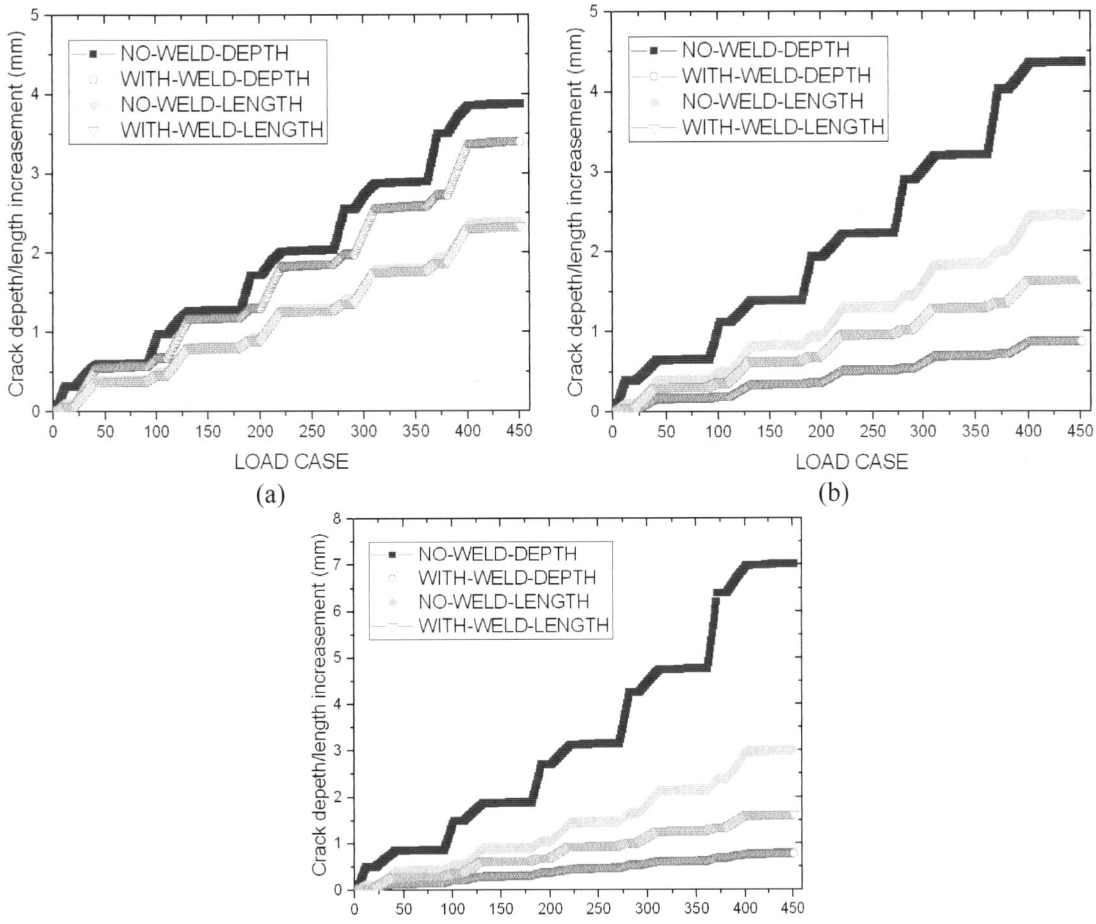

(a) (b) (c)

Figure 11. Comparison of crack depth and length growth with or without overlay layer. (a) BOTTOM-PATH, (b) MIDDLE-PATH and (c) TOP-PATH.

The above results can be explained by figure 12: For the surge line containing the weld overlay, the X-Y direction stress distribution of the pipe section when weld overlay was completed and cooled down to nearly room temperature was examined. It can be found that there is a X-direction compressive stress 1.75 MPa at the left and right ends of the pipe section, and a 1.15 MPa Y-direction compressive stress at the upper and lower ends of the pipe section, thus making the two regions under compressive stress. Therefore, the crack growth ratio at the middle and the top part of pipe section are significantly suppressed. While for the crack at the bottom, despite the compressive stress, the crack arrest effect is weaker due to the relatively lower temperature (See figure 4(a)).

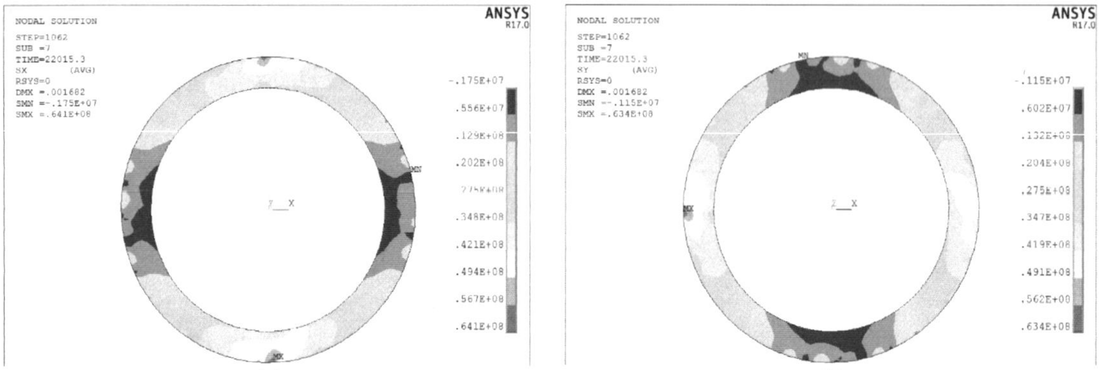

(a) (b)

Figure 12. Stress filed of X-Y-direction. (a) X-direction stress and (b) Y-direction stress.

4. Conclusion

In this paper, the analysis of fatigue crack propagation and crack arrest of the pressurizer surge line were carried out. The main conclusions are as follows:

- Apply transient temperature load to the inner wall surface of the pipe section for the surge line with and without the overlay layer, the temperature field distribution of pipe section under different transient loads were obtained, and then considering the temperature distribution results as the input condition. The transient pressure load was applied to the wall surface, and the thermal stress analysis of the surge line was carried out to obtain the stress and displacement field distribution results of the cross-section of the surge line under two conditions.
  - ➢ Under the condition of overlay layer, the displacement level of the surge line is reduced;
  - ➢ Under the condition of surfacing layer, the stress level of the wave tube is reduced;
  - ➢ Under the condition without overlay layer, the stress distribution of the pipeline section is: the largest value occurs in the top, the smallest appears in the middle, and the stress is gradually increased from the middle to the both ends. Under the condition of overlay layer, the maximum stress intensity appears on the top and bottom inner wall surfaces, the stress intensity in the middle is relatively small, and the stress gradually decreases from top to bottom.
- With the results of surfacing stress analysis, the stress values in the wall thickness direction of the bottom, middle and top paths were extracted, and the fatigue crack propagation analysis under the transient cyclic load of the surge line was carried out.
  - ➢ Obtaining crack propagation profile results on three paths under two conditions;
  - ➢ Comparing the crack propagation of cracks at different locations with or without overlay layer, it can be found that the crack propagation amount for the crack at the top has the largest difference, and the bottom crack propagation amount is similar;
  - ➢ Because of the compressive stress in the top and middle of the pipeline, the crack arresting effect for the cracks at these location are better, and cracking arrest effect for the bottom is the weakest due to the lower temperature. In general, the addition of overlay weld layer to the surge line can effectively decrease the rate of fatigue crack growth, thereby achieving the purpose of crack arrest.

References

[1] Chattopadhyay S 2009 Structural evaluation of a piping system subjected to thermal stratification *Nuclear Engineering and Design* **239** 2236-41
[2] Liang B B, Li G and Wang G Y 2008 Fatigue evaluation of pressurizer surge line concerning thermal stratification effect *Atomic Energy Science and Technology* **42** 448-53 (in Chinese)
[3] Zheng Y C, Zhao Z H, Zhang Z *et al* 2017 Internal crack initiation characteristics and early growth behaviors for very high-cycle fatigue of a titanium alloy electron beam welded joints *Materials*

Science & Engineering A **2017** 311-8

[4] Ihara R, Katsuyama J, Onizawa K *et al* 2011 Prediction of residual stress distributions due to surface machining and welding and crack growth simulation under residual stress distribution *Nuclear Engineering and Design* **2011** 1335-44

[5] Gotoh K, Niwa T and Anai Y 2014 Fatigue crack growth behavior of an out-of –plane gusset welded joints under biaxial tensile loadings with different phases *Procedia Materials Science* **2014** 1536-41

[6] Arora P, Kumar S, Singh P K *et al* 2014 Fatigue crack growth studies on narrow gap pipe welds of austenitic stainless steel material *Procedia Engineering* **2014** 203-8

[7] Huang C C and Liu R F 2012 Structural integrity analyses for preemptive weld overlay on the dissimilar metal weld of a pressurizer nozzle *International Journal of Pressure Vessels and Piping* **2012** 77-83

[8] Wei M, Wang G Z, Xuan F Z *et al* 2013 Effect of dissimilar metal weld overlay between pipe-nozzle of nuclear pressure vessel and safe end on failure assessment curves *Reactor Pressure Vessel* **30** 58-63 (in Chinese)

[9] Wei M, Wang G Z, Xuan F Z *et al* 2014 Leak-before-break analysis of a dissimilar metal welded overlay structure for connecting pipe-nozzle of nuclear reactor pressure vessel to safe end *Nuclear Technology* **37** 1-5 (in Chinese)

[10] Jiang W C, Luo Y, Wang B Y *et al* 2014 Residual stress reduction in the penetration nozzle weld joint by overlay welding *Materials and Design* **60** 443-50

[11] Zhang S W, Chen X D, Zhang Y *et al* 2016 Analysis of weld overlay residual stress for dissimilar metal welds in Qinshan nuclear power plant *Nuclear Power Engineering* **37(S2)** 4-6 (in Chinese)

[12] Liu R F and Huang C C 2013 Welding residual stress analysis for weld overlay on a BWR feedwater nozzle *Nuclear Engineering and Design* **256** 291-303

[13] Yusa N, Janousek L, Rebican M *et al* 2006 Detection of embedded fatigue cracks in inconel weld overlay and the evaluation of the minimum thickness of the weld overlay using eddy current testing *Nuclear Engineering and Design* **236** 1852-9

[14] Jun H K, Seo J W, I Jeon *et al* 2016 Fracture and fatigue crack growth analyses on a weld-repaired railway rail *Engineering Failure Analysis* **59** 478-92

[15] Antunes F V, Rodrigues D M and Branco R 2010 An analytical model of plasticity induced crack closure *Procedia Engineering* **2** 1005-14

[16] García-Collado A, Vasco-Olmo J M and Díaz F A 2017 Numerical analysis of plasticity induced crack closure based on an irreversible cohesive zone model *Theoretical and Applied Fracture Mechanics* **89** 52-62

[17] Funari M F, Greco F, Lonetti P and Spadea S 2019 A numerical model based on ALE formulation to predict crack propagation in sandwich structures *Frattura Ed Integrita Strutturale* **13** 277-93

[18] Funari M F, Lonetti P and Spadea S 2019 A crack growth strategy based on moving mesh method and fracture mechanics *Theoretical and Applied Fracture Mechanics* DOI: 10.1016/j.tafmec.2019.03.007 (accepted manuscript)

Research and application of micro-bead ultra-high performance concrete

N Q Feng[1] and W Yang[2,3]

[1]Department of Civil Engineering, Tsinghua University, Beijing 100084, PR China
[2]China Construction Ready Mixed Concrete Co., Ltd., Wuhan 430205, PR China

E-mail: youngsun8228@qq.com

Abstract. Micro-bead is an ultra-fine fly ash of spherical vitreous with an average particle size of no more than 1.2 μm. In this paper, the physicochemical properties of micro-bead, the preparation method of micro-bead ultra-high performance concrete (MUHPC) were described. Furthermore, the characteristics and slump-loss retarding of fresh MUHPC, the strength and durability of MUHPC and the change of mechanical properties and fracture characteristics of MUHPC after incorporating fiber were studied. Results indicated that UHPC with the cement–micro-bead–silica powder as the main binder and incorporated with the appropriate water reducer had a strength reaching up to 150 MPa and a plasticity reaching up to 3 h and no bleeding segregation phenomenon. The C120 UHPC has a low Cl^- diffusion coefficient and is resistant to freeze-thaw damage and has a high durability. And for the C120 UHPC incorporating polypropylene fiber, as the fiber content increased, the fracture energy, the brittle characteristic length and the toughness of the concrete all increased. C120 concrete has been applied in the Jingji Building in Shenzhen and the East Tower project in Guangzhou and the pumping height is 500 m or more.

1. Introduction

UHPC is defined as a kind of concrete incorporating fiber or fiber web, has a higher strength and durability than HPC and a compressive strength of more than 120 MPa at 28 days [1]. UHPC combined into self-compacting concrete (SCC) is also a new type of concrete which has supper high strength, high durability and good construction properties [2].

Judicious choice of chemical and mineral admixtures can reduce the cement content and result in economical HPC [3].When micro-bead was incorporated into concrete as a replacement of cement, the water consumption of concrete would be reduced, and the fluidity, strength and durability of concrete would be significantly improved [4].

UHPC pushes concrete technology to a new height in terms of strength and durability [5]. In this paper, the workability, deformation performance, strength and durability UHPC with micro-bead, polypropylene fiber or other special materials will be studied.

2. Characteristics of micro-bead

2.1. Scanning electron microscopy (SEM)

As shown in figure 1, micro-bead is a spherical glass bead with a maximum particle size of 1.45 um and a minimum particle size of 2.38 nm.

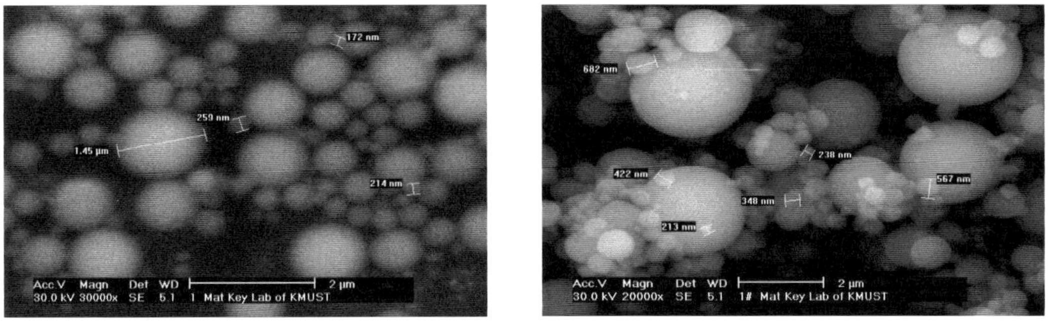

Figure 1. Scanning electron microscopy images of micro-bead.

2.2. Average particle size distribution

The micro-bead with an average particle size of no more than 0.2 µm accounted for 27.23%, 0.2-1.0 µm accounted for 42.43% and no more than 1.0 µm accounted for 69.66%.

The cement with a specific surface area of about 4000 cm^2/g had an average particle diameter of about 10-20 µm. The SF with a specific surface area of about 2000000 cm^2/g has an average particle diameter of about 0.1 µm. As the average particle diameter of the micro-bead is no more than 1.0 µm, in the composite powder of the three components of cement, micro-bead and SF, the micro-beads fill the pores between the cement particles, and the SF particles fill the pores between the micro-beads to obtain a densely packed powder.

2.3. Chemical compositions of micro-bead

The chemical compositions of micro-bead are shown in table 1. It can be seen from table 1 that the main chemical composition of micro-bead is SiO_2 and Al_2O_3 with a total content of 83%, the soluble silicon is about 8.67% of the total silicon, the soluble aluminum is about 15.42% of the total aluminum, and the soluble iron is about 3.49% of the total iron. Therefore, the chemical reactivity of the micro-bead is high.

Table 1. Chemical compositions of micro-bead (%).

SiO_2	CaO	MgO	Al_2O_3	Fe_2O_3	Na_2O	K_2O	SO_3	Carbon content
56.5	4.8	1.3	26.5	5.3	1.4	3.3	0.7	<1

2.4. Loss on ignition

The loss on ignition of micro-bead was about 1.5%, which meet the product quality requirements.

2.5. Standard water demand ratio

The test method and mix proportions are shown in table 2.

Table 2. Mix proportions of standard water demand ratio test (g).

	Cement	Micro-bead	Standard sand	Water requirement
Reference sample	250	\	750	Fluidity 130-140 mm, water requirement 110-113
Contrast sample	175	75	750	Fluidity 130-140 mm, water requirement 115

When the content of micro-bead was 30% and the fluidity of the paste was 130-140 mm, the requirement was about 98 g. According to the calculation formula: the standard water demand ratio of the micro-bead was 98/115*100%=85%.

2.6. Strength of the paste

The pastes in which the cement was partially replaced by 15%, 20%, 25% and 30% micro-bead were cast according to the cement mortar test method [6] and standardly cured to 28d. The flexural strength and compressive strength of the samples were tested and the results are shown in table 3.

Table 3. Flexural strength and compressive strength of the samples (MPa).

Group	Micro-bead content	Flexural strength 1	2	3	Average	Compressive strength 1	2	3	4	Average
1	0%	8.2	8.1	7.8	8.0	54.8	52.7	56.8	55.7	55.0
2	15%	10.8	10.2	10.8	10.6	68.7	69.9	71.8	68.4	69.7
3	20%	10.7	10.8	10.2	10.6	73.9	70.9	70.2	73.1	72.0
4	25%	10.0	9.4	9.7	9.7	68.1	70.4	70.6	71.0	70.0
5	30%	9.9	9.6	10.1	9.9	68.2	70.5	69.5	70.8	69.8

It can be seen from table 3 that compared with the flexural strength of the first set of samples, the flexural strength of the samples containing micro-bead under various dosages is improved, and the increase is between 15% and 20%. The variety of the strength of samples with different dosages of micro-bead is little. The compressive strength of the samples containing micro-bead is significantly improved with a peak appearing in the sample with a content of micro-bead between 15% and 25%.

2.7. Comparison of characteristics of different mineral powders

The comparison of the characteristics of grounded blast furnace slag (GBFS), fly ash (FA), SF and micro-bead are shown in table 4. The micro-bead has the advantages of micro-aggregate filling of SF, and also has strong physical water-reducing and high activity. In the preparation of HPC, not only the water demand is reduced, but also the hydration heat is reduced. Because of its high activity coefficient, the strength of concrete was not reduced after the micro-bead was incorporated. On the contrary, the compactness of concrete was further improved because of its unique micro-aggregate characteristics. Micro-bead provides a new and more effective choice for formulating high performance concrete.

Table 4. Comparison of characteristics of several mineral powders.

Characteristics	FA	GBFS	Micro-bead	SF
Main particle size distribution (μm)	5～30	10～40	0.1~5	0.1～0.5
Grain shape	Mostly spherical	Mostly angular	Completely spherical	Completely spherical
Fillability	General	Poor	Good	Good
Water reducing	Good	Poor	Excellent	Poor
Fluidity	Good	Good	Excellent	Poor
Activity	Poor	Good	Good	Good
Cracking resistance	Good	General	Excellent	Poor

3. Micro-bead ultra-high performance concrete

In China, HPC with a strength more than 100 MPa is called UHPC. It has been tested and applied in some super high-rise buildings. UHPC with cement, micro-bead and SF compounded as cementitious material is better than that with other ultrafine powders.

3.1. Raw materials

The raw materials used in this study included PII52.5R cement, GBFS with a specific surface area of 8500 cm^2/g, micro-bead which was ultrafine FA with a specific surface area of 12540 cm^2/g, a loss on ignition of 1.7%, and a water requirement ratio of 92%, SF with a specific surface area of 2×10^5 cm^2/g and a SiO_2 content of 90%, polycarboxylic acid high-efficiency water reducing agent (PA), gravel (G) with the two-stages of 5~9 mm, 9~16 mm being 3:7 and crushing value less than 6%, washed sea sand (SS) with a fineness modulus of 2.8 and mud content less than 0.6%, natural anhydrous anhydrite (NAA) with a specific surface area of 5200cm^2/g and composition of $CaSO_4 \cdot 0.022H_2O$, and sulpho-aluminate expansion agent (HCSA).

3.2. Mix proportions of concrete

The composition of MUHPC is shown in table 5. In No. 601 concrete, the BASF naphthalene water reducing agent with a solid content of 40% was used. In No. 602, 603, 604, 606 concrete, the naphthalene-ammonia compound water reducing agent (NA) was used. In No. 605 concrete, Sika polycarboxylic acid water reducing agent (SPA) with a solid content of 40% was used.

Table 5. Mix proportions of C120 UHPC (kg/m^3).

Variable	No.	W/B	Cement	Micro-bead	SF	SS	G	Water	Water reducing agent
3% BASF naphthalene water reducing agent	**601**	0.20	500	200	50	750	1000	130	22.5
3.2% naphthalene-ammonia compound water reducing agent	**602**	0.187	500	170	80	700	1000	140	24.0
4% naphthalene-ammonia compound water reducing agent	**603**	0.187	500	170	80	700	1000	140	30.0
3.5% naphthalene-ammonia compound water reducing agent	**604**	0.187	500	200	50	700	1000	140	26.3
2% BASF naphthalene water reducing agent	**605**	0.187	500	200	50	700	1000	140	15.0
4% Sika polycarboxylic acid water reducing agent	**606**	0.187	500	200	50	700	1000	140	30.0

3.3. Performance of fresh concrete

The performance of fresh concrete is shown in table 6.

Table 6. Performances of fresh concrete.

No.	Slump (mm)				Expansion (mm)				Collapse time (s)			
	Initial	1 h	2 h	3 h	Initial	1 h	2 h	3 h	Initial	1 h	2 h	3 h
601	130											
602	270	260	230	225	680×700	640×700	590×590	540×560	5	7	9	12
603	260	250	245	255	660×690	660×650	650×650	670×620	4	5	7	5
604	265			265	710×740			670×710	4			5

Note: For No. 604 concrete, the 4 h slump is 260 mm, the expansion is 670 mm*730 mm, and the collapse time is 7 s. For No. 602 concrete, the admixture dosage is low (3.2%), and after 3 h, the collapse time is too long which is 12 s. For No. 603 and 604 concrete, the slump, expansion and collapse time are in line with the requirements of this study.

The self-developed sulfamic acid-based water reducing agent is blended with BASF's naphthalene water-reducing agent (containing 40% solids), and a small amount of zeolite powder mixed with a water-reducing agent can be used to arrange C120 concrete with excellent plasticity. And the C120 concrete can meet long-distance transportation and ultra-high pumping requirements.

3.4. Self-shrinkage of concrete

In the mix proportion of C120 concrete, NAA ($CaSO_4 \cdot 0.022H_2O$) with a specific surface area of 5200 cm^2/g is added. The early age shrinkage of concretes was compared. The mix proportion of concrete is shown in table 7. The performance index of each group of concrete is controlled at: slump of more than 250 mm, slumping extension of 680 mm or more, and collapse time of less than 15 s. Self-shrinkage measurement results are shown in the table 8 and figure 2.

Table 7. Mix proportions of C120 concrete for self-shrinkage test (kg/m^3).

Type of concrete	W/B	Cement	GBFS	Micro-bead	SF	NAA	HCSA	PA	SS	G
Reference concrete	0.185	550	42	84	24	0	0	3.8%	750	950
Containing 10% anhydrous anhydrite	0.185	550	42	84	24	70	0	4.2%	750	950
Containing 6% HCSA	0.185	550	42	84	24	0	42	4.2%	750	950
Containing 10% HCSA	0.185	550	42	84	24	0	70	4.4%	750	950

Table 8. Self-shrinkage of C120 concrete (10^{-6}).

Time	Reference concrete	Containing 10% anhydrous anhydrite	Containing 6% HCSA	Containing 10% HCSA
0 h	0	0	0	0
1 h	13	17	5	30
2 h	14	30	56	124
3 h	17	30	92	162
4 h	21	69	82	142
5 h	31	136	71	95
6 h	40	181	69	54
7 h	48	196	68	23
8 h	57	204	66	2
9 h	66	210	69	-14
10 h	74	216	71	-26
11 h	81	222	73	-40
12 h	90	226	72	-47
13 h	96	230	74	-55
14 h	103	234	74	-63
15 h	111	237	75	-70
16 h	117	241	77	-76
17 h	124	245	79	-82
18 h	129	244	80	-86
19 h	132	248	81	-89
20 h	139	250	83	-93
21 h	141	253	85	-95
22 h	149	256	86	-98
23 h	150	259	88	-101
24 h	157	261	90	-103
2d	231	312	115	-116

3d	269	339	138	-122
4d		348	157	-103
5d		355	166	-76
7d		369	172	-34
9d		380	176	-22
11d		400	179	-55
14d		425	196	-90
17d		442	202	
19d		To be measured	207	

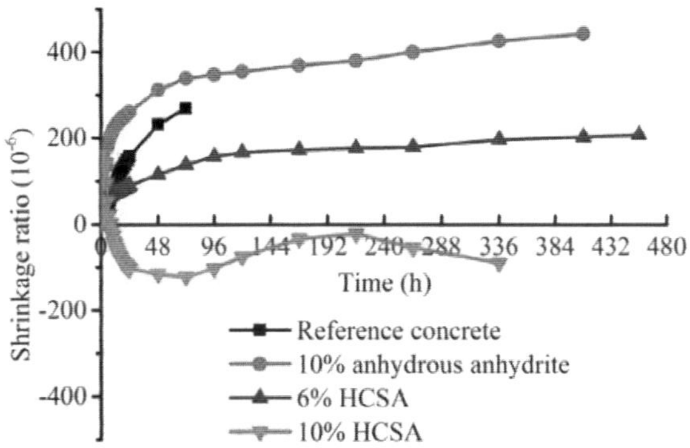

Figure 2. Shrinkage curves of C120 concretes.

It can be seen from table 8 and figure 2 that HCSA can efficiently inhibit the self-shrinkage of concrete. After the incorporation of 5% HCSA into concrete, the 3d self-shrinkage decreased from 269×10^{-6} to 138×10^{-6}, which was reduced by about 50%. After the dosage was increased to 10%, the concrete in 14d was still in a micro-expansion state which was about 90×10^{-6}. When the HCSA dosage was no more than 5%, the self-shrinkage of UHPC was inhibited. After the incorporation of 10% natural anhydrous anhydrite, the self-shrinkage of concrete in 3d increased from 269×10^{-6} to 339×10^{-6}, which was increased by 26%.

3.5. Compressive strength of hardened concrete

The compressive strength of concrete is shown in table 9.

Table 9. Compressive strength of concrete.

No.	3d	7d	28d	56d
601	85.8	97.5	130	140
602	87.1	102	135	145
603	84.3	102	140	150
604	75.1	96	120	130

4. Influence of fiber on the mechanical properties of MUHPC

The fiber was blended into the original C120 concrete to improve the mechanical properties. The mix proportions of polypropylene fiber were 1.0 kg/m^3 and 2.0 kg/m^3, respectively. And the mix proportion of the organic fiber of Shenzhen was 2.0 kg/m^3.

4.1. Pole body compressive strength and static elastic modulus

The static compression modulus test is shown in figure 3.

Figure 3. Static compression modulus test.

The static compression modulus test results are shown in table 10. It can be seen from table 10 that after the fiber is incorporated, the compressive strength of the prism is low, but the static elastic modulus is high.

Table 10. Prismatic compressive strength and static compressive modulus of concrete.

No.	Age (d)	Size of sample (mm)	Prismatic compressive strength		static compressive modulus	
			Individual	Average	Individual	Average
Reference	28	100×100×300	115.2	138.0	5.20×10^4	5.20×10^4
			138.0		5.18×10^4	
			138.8		5.22×10^4	
Fiber 1 kg	28	100×100×300	127.2	128.7	5.10×10^4	5.23×10^4
			126.8		5.21×10^4	
			132.0		5.37×10^4	
Fiber 2 kg	28	100×100×300	137.2	136.6	5.22×10^4	5.30×10^4
			139.0		5.22×10^4	
			133.6		5.45×10^4	
Fiber 2 kg Shenzhen	28	100×100×300	104.8	116.7	5.18×10^4	5.25×10^4
			116.4		5.23×10^4	
			128.8		5.33×10^4	

4.2. Fracture toughness test of concrete

The fracture parameter test of C120 concrete is shown in figure 4.

The load-crack opening displacement (P-CMOD) curves of four groups of concrete with reference, fiber 1 kg, fiber 2 kg and fiber 2 kg (Shenzhen) were determined. And the corresponding fracture parameter values, including cracking strength, tensile strength, bending strength, fracture energy and brittleness parameters were calculated. There were three test pieces for each concrete. Figures 5-8 show the P-CMOD curves of each group of concrete and the calculated concrete softening relationship (σ-w) curve.

IOP Conf. Series: Materials Science and Engineering **629** (2019) 012019 doi:10.1088/1757-899X/629/1/012019

(a) (b)

Figure 4. Fracture parameter test of C120 ultra-high performance concrete. (a) Closed-loop hydraulic servo test machine and (b) Placement of the deformation sensor.

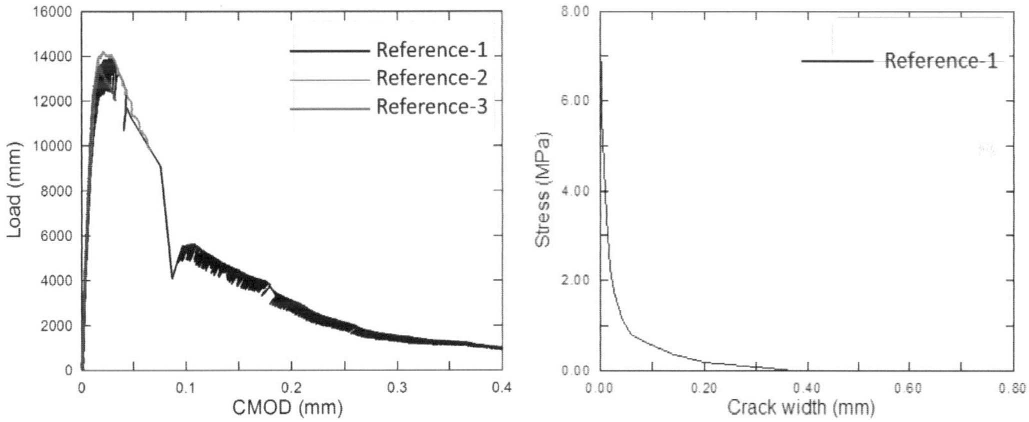

(a) (b)

Figure 5. P-CMOD curve and σ-w curve of reference concrete. (a) P-CMOD and (b) σ-w.

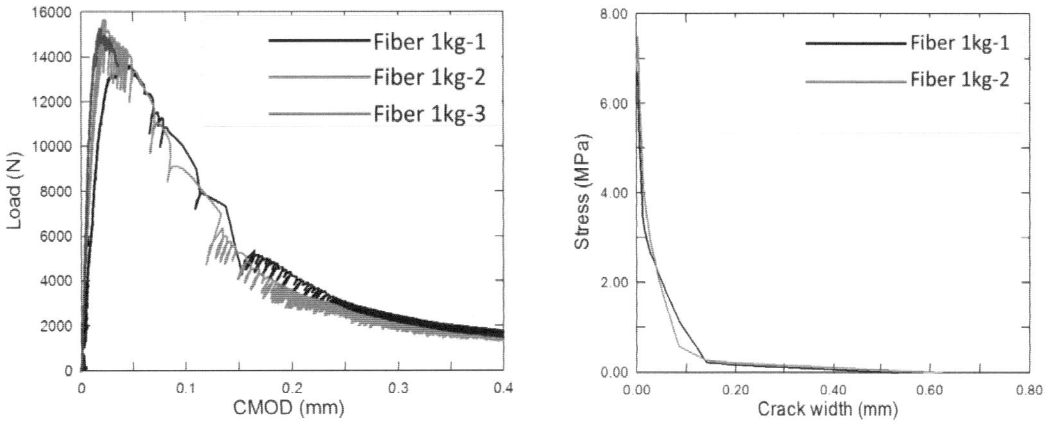

(a) (b)

Figure 6. P-CMOD curve and σ-w curve of concrete with 1 kg fiber. (a) P-CMOD and (b) σ-w.

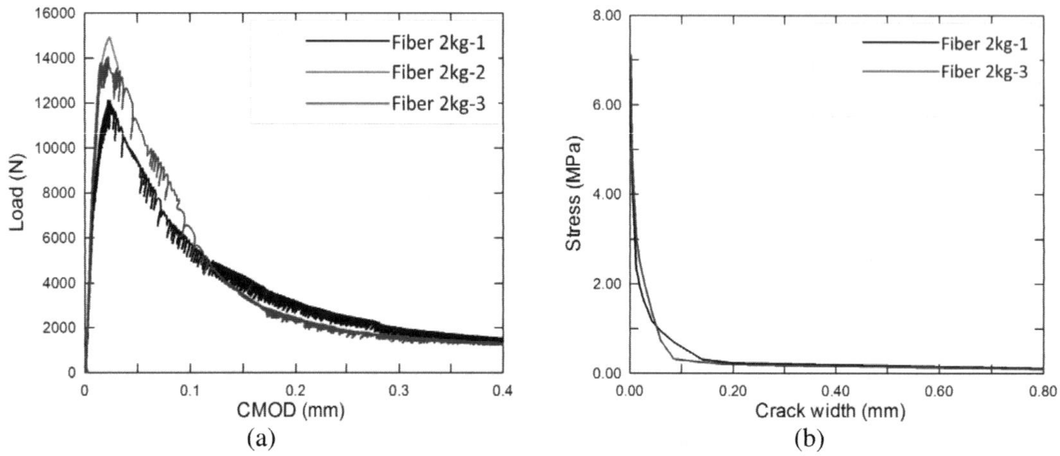

Figure 7. P-CMOD curve and σ-w curve of concrete with 2 kg fiber. (a) P-CMOD and (b) σ-w.

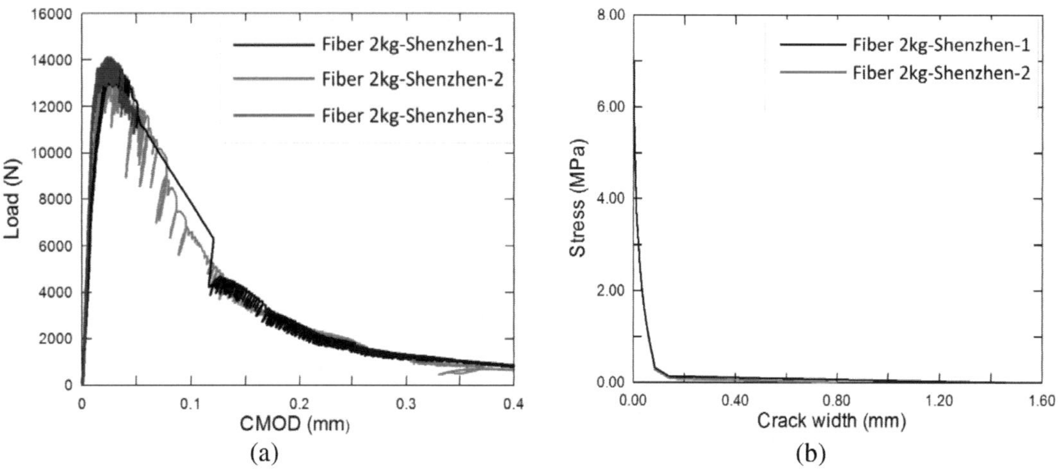

Figure 8. P-CMOD curve and σ-w curve of concrete with 2 kg fiber (Shenzhen). (a) P-CMOD and (b) σ-w.

It can be seen from figures 5-8 that the P-CMOD curve and the σ-w curve are similar in shape between different test pieces of the same kind of concrete, and the difference is not large, indicating that the dispersion between the test pieces is not big.

According the complete P-CMOD curve, its corresponding fracture parameters can be calculated. Table 11 shows the fracture parameter values of different types of concrete. It can be seen from table 11 that the cracking loads of different types of concrete are basically the same. This is because the matrix material is the same, and the fiber is mainly used after the concrete is cracked. As the fiber content increases, the tensile strength of the concrete is substantially unchanged, and the fracture energy of the concrete increases. When the same amount is used and the fiber types are different, the concrete fracture energy is also different. The magnitude of the fracture energy is based on the reference <fiber 1 kg < fiber 2 kg (Shenzhen) < fiber 2 kg. The brittle characteristic length is a physical quantity characterizing the degree of brittleness of concrete. The brittleness of the concrete increases when the characteristic length of concrete decreases. According to the test results, it can be seen that as the fiber content increases, the toughness of the concrete increases. The length of the brittle characteristic length is based on the benchmark <fiber 1 kg < fiber 2 kg (Shenzhen) < fiber 2 kg, and the corresponding concrete toughness is based on the benchmark < fiber 1 kg < fiber 2 kg (Shenzhen) < fiber 2 kg.

Table 11. Summary of concrete fracture parameters.

	No.	Elasticity modulus (GPa)	Cracking load(N)	Bending load (N)	Cracking strength (MPa)	Bending strength (MPa)	Tensile strength (MPa)	Fracture energy (J/m^2)	Characteristic length (cm)
Reference	1#	51.9	5000	14000	6.82	9.07	6.86	213.8	23.5
	2#		4600	14500	6.27	9.39	-	-	-
	3#		4700	13800	6.41	9.59	-	-	-
	Averag		4767	14100	6.5	9.35	6.86	213.8	23.5
Fiber 1 kg	1#	52.2	4800	13800	6.55	8.94	6.67	291.8	33.8
	2#		5000	15500	6.82	10.04	7.08	302.9	27.9
	3#		5000	15300	6.82	9.91	-	-	-
	Averag		4933	14867	6.73	9.63	6.88	297.35	30.9
Fiber 2 kg	1#	52.2	4800	12000	6.55	7.78	6.56	322.8	39.2
	2#		5000	15000	6.82	9.73	-	-	-
	3#		5200	14000	7.09	9.08	7.22	322.1	33.4
	Averag		5000	13667	6.82	8.86	6.89	322.5	36.3
Fiber 2 kg Shenzhen	1#	52.0	5100	14000	6.95	9.07	6.97	300.3	30.0
	2#		5000	14000	6.82	9.07	6.84	321.8	30.7
	3#		5200	14000	7.09	9.07	-	-	-
	Averag		5100	14000	6.95	9.07	6.91	311.1	30.4

5. Durability of concrete

The Cl^- diffusion coefficient and freeze-thaw resistance of C120 concrete were tested.

5.1. Cl^- diffusion coefficient

For the reference concrete, the concrete prepared by river sand (RS), and the concrete prepared by washed SS, the chloride ion diffusion was tested by the NEL method established by Tsinghua University. The test results are shown in table 12. According to the durability design and construction standard evaluation of concrete structure [7], the permeability grade is IV and the permeability is very low.

Table 12. Results of Cl^- diffusion coefficient of C120 UHPC.

No.	Sample type	Age (d)	Cl^- diffusion coefficient (m^2/s)	Test basis
1	Reference	28	0.62×10^{-12}	NEL method for rapid detection of chlorie ion diffusion coefficient in concrete
2	RS	28	0.57×10^{-12}	
3	SS	28	0.64×10^{-12}	

5.2. Frost resistance test

According to the standard requirements [8], C120 concrete was subjected to 300 freeze-thaw cycles. The results are shown in table 13.

Table 13. Results of frost resistance of C120 UHPC.

Content	Result	Standard requirement
Relative dynamic elastic modulus after	99.9	When the concrete is freezed and thawed to a

110 freeze-thaw cycles (%)		predetermined number of cycles, and the relative dynamic elastic modulus is not less than 60%, or the mass loss rate is not more than 5%, it can be considered that the frost resistance of tested concrete has met the design requirements.
Mass loss rate after 110 freeze-thaw cycles (%)	0	

6. Conclusion

Through the detection of micro-bead performance, the preparation and detection of MUHPC, the durability test and the detection of the mechanical properties of MUHPC with the incorporation of fiber, the conclusions of the research work are as follows:

- Micro-bead is an ultrafine powder of spherical glass body with a maximum particle size of 1.45 um and a minimum particle size of 2.38 nm. The cement–micro-bead–silica powders can be combined in appropriate proportions to make the porosity of the powder to reach the minimum. And the concrete has the highest fluidity under the same water consumption, or the water demand is the lowest under the same fluidity.
- UHPC with the cement–micro-bead–silica powder as the main binder and incorporated with the appropriate water reducer has a strength reaching up to 150 MPa and a plasticity reaching up to 3 h (the initial slump being 265 mm–265 mm, expansion being 710 mm×740 mm–670 mm×710 mm, the collapse time being 4 s–5 s) and no bleeding segregation phenomenon.
- On the basis of the reference C120 UHPC, with the incorporation of 5% sulpho-aluminate bulking agent, the self-shrinkage is reduced from 269×10^{-6} to 138×10^{-6}, which is reduced by about 50%. And when a small amount of sulpho-aluminate salt expansion agent is added, the self-shrinkage cracking of UHPC can be effectively inhibited.
- The C120 UHPC has a low Cl^- diffusion coefficient and is resistant to freeze-thaw damage and has a high durability.
- For the C120 UHPC incorporating polypropylene fiber with contents of 1 kg/m^3 and 2 kg/m^3, the shape of the load-crack opening displacement curve (P-CMOD) or the softening relationship curve (σ-w) are nearly the same, and the cracking load is basically the same. This is because the base material of the concrete is the same, and the fiber can play a role mainly after the concrete is cracked. As the fiber content increases, the fracture energy, the brittle characteristic length and the toughness of the concrete all increase.

Reference

[1] Perry V H 2018 What really is ultra-high performance concrete-towards a global definition C J Shi and B C Chen eds. *Proc. the 2nd Int. Conf. on UHPC Materials and Structures* (Fuzhou: RILEM Publications)
[2] Feng N Q, Ye H W *et al* Study on UHP-SCC and its ultra high-rise pumping technology *Our World in Concrete & Structures* (25-27 August 2010, Singapore)
[3] Sarkar S L and Ghosh S N 1993 *Mineral Admixtures in Cement and Concrete* (New Delhi, India: ABI Books)
[4] Li H, Feng N Q, Xu K F, Zhu Z and Li Y H 2008 Properties of micro-bead high performance concrete *Cross-Strait Academic Exchanges*
[5] Feng N Q 2016 *Technology of HPC & UHPC* (Beijing, China: China Architecture & Building Press) (in Chinese)
[6] GB/T 17671-1999 1999 *Cement Mortar Strength Test Method (ISO method)* (Beijing, China: China Standard Press) (in Chinese)
[7] GB/T 50476-2008 2008 *Code for Durability Design of Concrete Structures* (Beijing, China: China Building Industry Press) (in Chinese)

[8] ASTM C 666-2003 2003 *Standard Test Method for Resistance of Concrete to Rapid Freezing and Thawing* (West Conshohocken: ASTM International)

IOP Conf. Series: Materials Science and Engineering **629** (2019) 012020 doi:10.1088/1757-899X/629/1/012020

Effect of the length-diameter ratio on the initial fragment velocity of cylindrical casing

Y G Gao[1], S S Feng[1], B Zhang[1] and T Zhou[1,2]

[1]State Key laboratory of Explosion Science and Technology, Beijing Institute of Technology, Beijing 100081, PR China

E-mail: zhoutong@bit.edu.cn

Abstract. The initial velocity of fragment from cylindrical casing, which detonates at one end along the central of the casing, is the key issue in the field of explosion technology and its protecting. Most of the formula available can predict the initial velocity or the velocity of fragments at middle part of the cylindrical casing with greater length-dimeter ratio (L/d>2). However, when the length-diameter ratio is less than two, the initial velocity of the cylindrical casing filled with explosives will have a big difference. In the present work, a numerical simulation model acknowledged by X-ray radiography experiments was used to determine the influence of the length-diameter ratio. The formula was built on top of the Gurney formula and made use of a correctional function to account for the effect of the length-dimeter ratio. The formula was further acknowledged with the established numerical simulation model. The results indicate that the calculation formula can accurately predict the initial fragment velocity with different length-diameter ratio (L/d≤ 2).

1. Introduction

The cylindrical casing filled up with explosive is a usual used structure in the design of warhead. There are a large amount of numerical simulation and experiments about cylindrical casing. It is critical to calculate the initial velocities for design in explosion techniques and protective devices.

The fragmentation is a very interest dynamic issue to the researchers who focus on explosion drive and protective applications. During the past decades, there are many formulas calculating the initial fragment velocity which have been covered. Assuming the velocity of detonation products is linearly distributed at any time and the shell is infinite in length, Gurney [1] proposed a typical formula to calculate the initial velocity, and it is shown as:

$$v_0 = \sqrt{2E} \cdot (1/\beta + 0.5)^{-1/2} \tag{1}$$

where $\sqrt{2E}$ is Gurney constant of explosive, v_0 is Gurney velocity, and β is the mass ratio of explosive to casing(β=C/M), C and M is the mass of explosive and cylindrical casing. Gurney found that the initial velocity of fragment was only related to the mass ratio and explosive energy. However, the influence of length-dimeter ratio was not taken into account. It performs well when the length-diameter ratio is greater than 2, but when the ratio is less than 2, there will be a large deviation. Charron [2] found that Gurney's method is not suitable well for allocations where length-diameter ratio is no more than 2, even if local C/M's were took into account. According to a lot of experiments, There is a conclusion that the range of the initial fragment velocity would change about $1.7AL/r_0\sqrt{C/M}$ because of changes in the casing length, which A is cross-sectional area, L is the length of the cylindrical casing, and r_0 is the initial radius of the casing. According to conservation of energy, A formula was proposed to count the initial velocity of fragment, and it is shown as:

$$v_0 = \frac{D_e}{2}(Q_v / \frac{1}{\alpha_0} - \frac{1}{2})^{1/2} \tag{2}$$

where D_e is the explosive detonation velocity, Q_v is the heat of detonation of the charge, α_0 is the loading coefficients. Considering the motion equation of the cylindrical casing, an equation to calculate the initial velocity was proposed [3], which can be expressed as:

$$v_0 = \frac{D_e}{2} \cdot \left\{ \frac{\beta}{2+\beta} \left(1 - \left(\frac{r_0}{r} \right)^4 \right) \right\}^{1/2} \tag{3}$$

where r is the expanding radius in one moment. The equations mentioned above are put forward considering that the cylindrical casing is infinite in length, However, according to experimental data of some articles [4], their calculated results are inconsistent with them. The initial velocity is less than that calculated by the equations mentioned above for small length-diameter ratio. The kinetic energy (E_k) is one of the important scales to measure the fragment killing power. $E_k = 1/2mv^2$, so it is very important to predict the velocity as accurately as possible for the warhead structure design. Because the length of the warhead is always limited, especially when the length-diameter ratio is less than 2, such as grenades, tandem warhead, etc.Thus, a formula predicting the initial velocity of cylindrical casing with different length-diameter ratio (L/d$\leq$ 2) is in great request.

2. Validation of simulation model

The SPH method was used to investigate the fragmentation process in the present work [5], which can be used to simulated material when there was large deformation , and it is widely used in the explosive field, To acquire a precise and reliable simulation model, Feng's work initiated at one end [6,7] was used to calibrate the model.

2.1. Characterization of the experiment: cylindrical metal casing initiated at one end

The experiments were carried out to figure out the axial distribution of initial velocity of the casing, which was filled up with composition B (COMP-B). AISI 1045 steel was used as the casing, which is widely used in engineering. The length of casing and charge(L) is 77.3mm, the exterior and interior dimeter of casing and charge are 29.68mm and 23.60mm, and the density of casing and COMP-B are 7.85 and 1.717 $g \cdot cm^{-3}$.

2.2. Simulation model of Feng's trial

Due to the symmetry of experiment devices, it only need model a quarter of them (**Figure 1**). To obtain the most accurate output possible, a great number of evenly distributed particles are needed. Therefore, particles need to be as small as possible. But smaller size requires more computer resources. Li [8] discovered that 0.4 mm is precise enough in the SPH model. The 0.4 mm particle radius was used in our present work, which showed that the results of simulation results are consistent with the experiment.

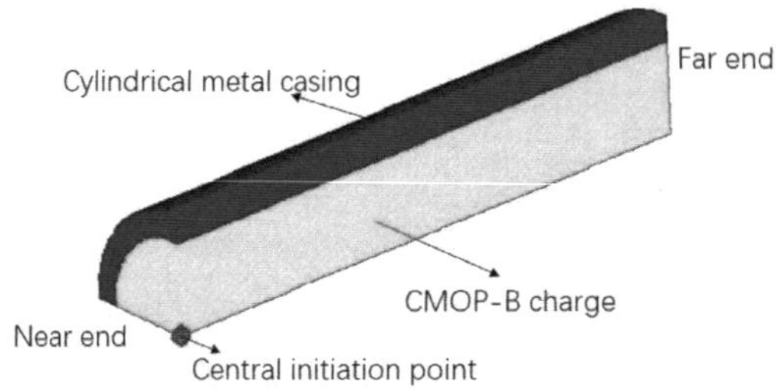

Figure 1. The simulation model of a quarter of cylindrical casing detonated at the near end along the central of the casing.

2.3. Material model

In our present work, the Johnson and Cook (JC) model (equation (4)) is employed to simulated the material properties of the AISI 1045 steel [9]. The model can be expressed as:

$$\sigma = \left(A + B\varepsilon_{ep}^{n}\right)\left(1 + C\ln \dot{\varepsilon}\right)\left(1 - T^{*m}\right) \tag{4}$$

where A, B, C, n and m are the constants of the material, which are 507, 320, 0.064, 0.28, and 1.06 for the casing respectively[10]. Failure model is applied to model the fragmentation [11-13]. As a result, the material involves weak points where the failure occurred [14]. The failure model [13] would be described as:

$$dp = (1 - p)Ce^{\gamma\varepsilon}d\varepsilon \tag{5}$$

where γ and C are constants, (1-p) is the probability that there is not rupture when the strain values is less than ε. γ can be expressed as:

$$\gamma \approx 160\frac{\sigma_2}{\sigma_f\left(1 + \varepsilon_f\right)} \tag{6}$$

where σ_2 is the proportionality coefficient of the law, σ_f is the true stress at the fracture and ε_f are the plastic strain at fracture. According to Li's stand tensile tests [8], the value of γ is 53.8.

The material properties of COMP-B are simulated using the JWL equation of state (EOS), which can be shown as:

$$P_b = C_1\left(1 - \frac{\omega}{r_1 v}\right)e^{-r_1 v} + C_2\left(1 - \frac{\omega}{r_2 v}\right)e^{-r_2 v} + \frac{\omega E}{v} \tag{7}$$

where P_b is the blast pressure, v is the initial relative volume, E is the internal energy every initial volume, ω r_1, r_2, C_1, and C_2 are the constants of the material. The parameters of the material are shown in Table [15].

Table 1. Parameters of explosive model.

Material	$\rho_0/g\cdot cm^{-3}$	$D/m\cdot s^{-1}$	P_{CJ}/GPa	E_0/KJ$\cdot m^3$	C_1/GPa	C_2/GPa	R_1	R_2	ω
COMP-B	1.717	7980	29	8.5×10^6	542	7.68	4.2	1.1	0.24

2.4. Simulation results

The experimental photograph and the numerical image of the fragment outline at the almost same moment (52.6 μs and 50 μs) are shown in Figure 2. From Figure 3, it shows up clearly that the fragment distributions in simulation and trial are in good agreement, which shows that the simulation model is capable to provide responsible predictions and the size of the particles are adequate to provide relatively accurate prediction.

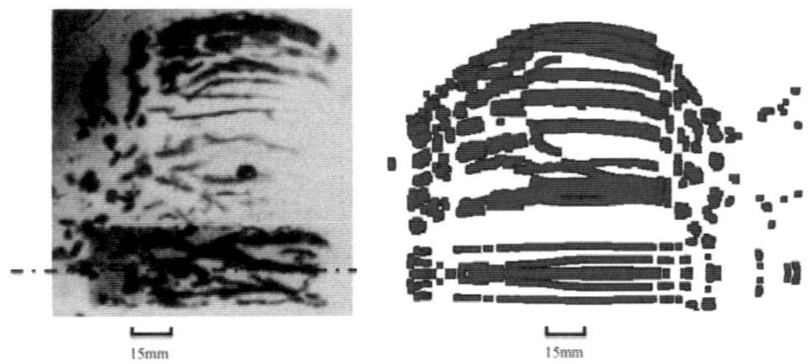

Figure 2. The X-ray image (52.6 μs) (left) and numerical simulation image (50 μs) (right) of a cylindrical casing.

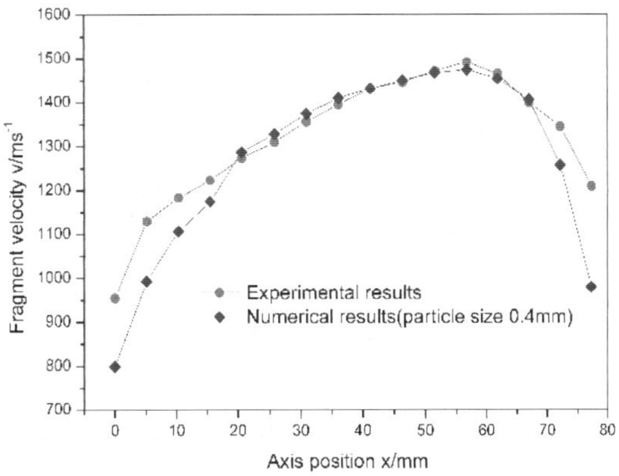

Figure 3. The congruency between the experiment and numerical simulation of fragment velocity.

3. The influence formula of the length-diameter ratio

3.1. Theoretical analysis

In cases where length-diameter ratio is small, the rarefaction waves get to farther inside the casing and the pressure would decrease over a wider axial distance. Hence, the influence formula of the length-diameter ratio can be modelled by multiplying a correctional function with β, which is shown as:

$$v(a)=\sqrt{2E}\left(1/(\beta\cdot f(a))+0.5\right)^{-1/2} \tag{8}$$

where a is the length-diameter ratio of a cylindrical casing.

According to equation (8), in contrast to long cylindrical casing (a >2), which the initial fragment velocity is equal to the velocity calculated by Gurney formula, the cylindrical casing with small length-diameter ratio ($a \leq 2$) has an initial velocity gain, which can be shown as:

$$g(a)=\frac{v(a)}{v(0)}=\frac{\left(1/(\beta\cdot f(a))+0.5\right)^{-1/2}}{(1/\beta+0.5)^{-1/2}} \tag{9}$$

Where $v(0)$ is the initial velocity calculated by Gurney formula, $v(a)$ is the initial velocity of a cylindrical casing with a specific length-diameter ratio.

3.2. Numerical simulation

The acknowledged model above was used to simulate the effect of the length-diameter ratio on the initial velocity (equation (6)). **Table** lists the eight cylindrical casings with different a, Figure shows the fragment velocity of the casing with different a, which means that the length-diameter ratio has a significant influence on the initial fragment velocity, especially when the length-diameter ratio is less than 2. The biggest fragment velocity of each cylindrical casing is treated as the initial velocity.

Table 2. Parameters and results of eight cylindrical casings modelled with various a.

No.	Length of Casing (L, mm)	Exterior diameter of charge (D, mm)	Interior diameter of casing (d, mm)	a	β	Fragment velocity $v(a)$ (m/s)	Velocity gain $g(a)$
1	7.5	30	24	0.3125	0.39	536.923	0.34822
2	15	30	24	0.625		881.017	0.57135
3	22.5	30	24	0.9375		1093.09	0.70888
4	30	30	24	1.25		1230.86	0.79822
5	37.5	30	24	1.5625		1306.12	0.84703
6	45	30	24	1.875		1392.52	0.90306
7	52.5	30	24	2.1875		1466.32	0.95092
8	60	30	24	2.5		1501.72	0.97388

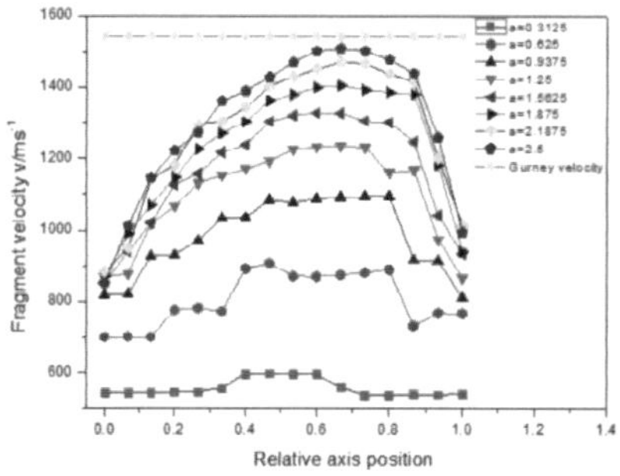

Figure 4. The velocity of gauge points with different a.

3.3. Simulation results

Figure is the front elevation of the outline distribution in the numerical simulations. The distributions of fragment show that when the casing expand to a certain extent, fragment velocity becomes stable [16].

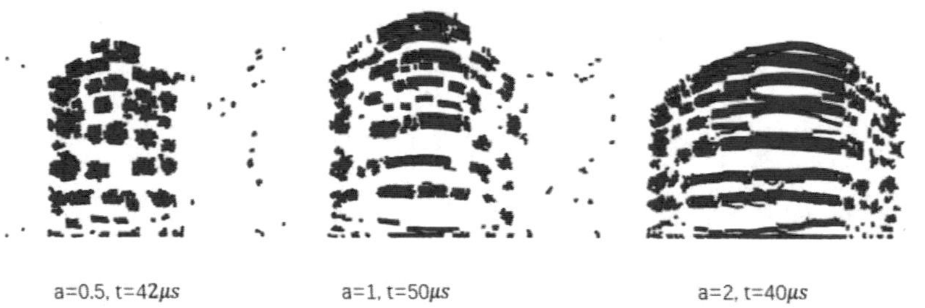

Figure 5. Front view of the outline distribution of cylindrical casings with various a.

Table shows the initial velocity gain of the eight cylindrical casing, which is drew against the length-diameter ratio in Figure . The initial velocity gain of the casing increases as the length-diameter ratio of the cylindrical casing gets bigger, and it gets close to 1 when a approaches 2. Therefore, the influence of the length-diameter ratio cannot be neglected.

It found that the modification function $f(a)$ which is an exponential form can accurately depict the influence of the length-diameter ratio. Therefore, considering that the initial velocity gain of the eight cylindrical casings in Figure , the modification function $f(a)$ that we can get by data fitting is shown as:

$$f(a) = ke^{-\frac{a}{t}} + b \tag{10}$$

where k, t and b are the constant. When subrogating equation (10) into equation (8), the initial velocity with different length-diameter ratio of the cylindrical casing is shown as:

$$v(a) = \sqrt{2E}\left(1/\left(\beta\cdot(ke^{-\frac{a}{t}}+b)\right)+0.5\right)^{-1/2} \tag{11}$$

The initial velocity gain could also be shown as equation (12) rather than equation (9).

$$g(a) = \frac{v(a)}{v(0)} = \frac{\left(1/(\beta\cdot(ke^{-\frac{a}{t}}+b))+0.5\right)^{-1/2}}{(1/\beta+0.5)^{-1/2}} \tag{12}$$

The results of simulation of the eight cylindrical casings (Figure) is fitted by using least square method, the fitting result gives k=-1.353, t=1.606 and b=1.212 (R^2=0.998).

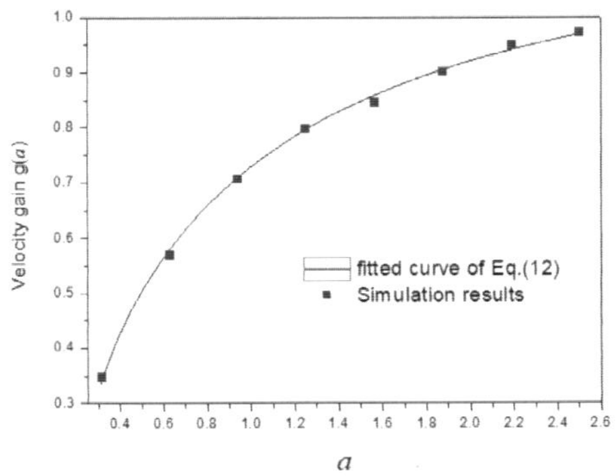

Figure 6. Simulated initial velocity gain of cylindrical casings with various a.

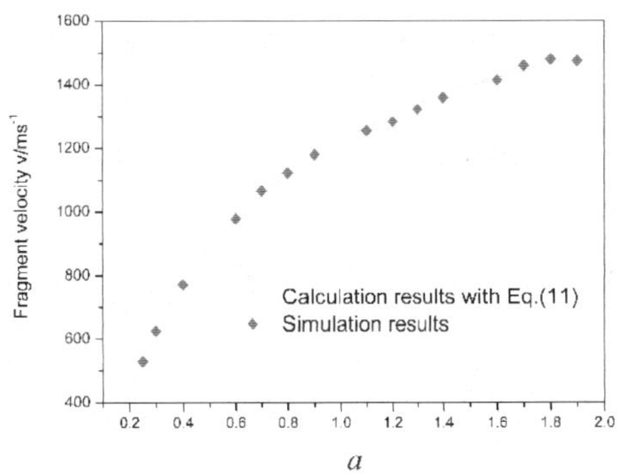

Figure 7. The initial fragment velocity of cylindrical casings calculated by simulation and equation (11) with another different a.

4. Validation of the proposed formula

The acknowledged numerical model needs to be employed to identify whether equation (11) can precisely predict the initial velocity with different length-diameter ratio. Cylindrical casings with another different a are calculated in both calculations with equation (11) and numerical simulation. The initial fragment velocities in numerical simulation and equation (11) are shown in Figure . It can be seen that the calculation results are almost the same with that calculated by equation (11).

5. Conclusion

According to the numerical simulation and calculation process above, we dedicate to the initial velocities of cylindrical casings with different length-diameter ratios. It obvious that the effect of the length-diameter ratio on the initial velocity should not be neglected. In our present work, there are following conclusions could be drawn:

- A simulation model is put forward, which is acknowledged by experimental data and X-ray images, and it can be applied to model fragmentation of cylindrical casing.
- A novel formula is proposed to describe the effect of the length-diameter ratio on the initial velocity of a cylindrical casing. The formula can predict the initial velocity initiated at the near end along the central of the casing with different length-diameter ratios ($a \leq 2$), and it is further verified by the validated simulation model, which shows satisfying results.

Acknowledgments
The authors wish to thank to Zhang Bo who provides simulation support, Feng Shun-shan offering the experimental data, and other seniors in our group.

References

[1] Gurney R 1943 The initial velocities of fragments from bombs, shells, and grenades *BRL Report 405*
[2] Charron Y J 1979 Estimation of velocity distribution of fragmenting warheads using a modified gurney metho *Air Force lnst of Tech* **113** 27-9
[3] Beijing Institute of Technology 1973 Explosion and its effects: compile group *Explosion and Its Effects* National Defence Industy Press
[4] Jaansalu K M and Dunning M R 2007 Fragment velocities from thermobaric explosives in metal cylinders *Propell Explos Pyrotech* **32** 80-6
[5] Kong X S, Wu W G, Li J, Liu F, Chen P and Li Y 2013 A numerical investigation on explosive fragmentation of metal casing using smoothed particle hydrodynamic method *Mater Des* **51** 729-41
[6] Feng S S 1987 An experimental investigation for the axial distribution of initial velocity of shells *Acta Armamentarii* **4** 60-3
[7] Huang G Y, Li W and Feng S S 2015 Axial distribution of fragment velocities from cylindrical casing under explosive loading *Int J Impact Eng* **76** 20-7
[8] Li W, Huang G Y and Feng S S 2015 Effect of eccentric edge initiation on the fragment velocity distribution of a cylindrical casing filled with charge *Int J Impact Eng* **80** 107-15
[9] Johnson R and Cook W H 1983 A constitutive model and data for metals subjected to large strains, high strain rates and high temperatures *The 7th international Symposium on Ballistic* (Hague, Netherlands) 541-7
[10] Chen G, Chen Z F, Tao J L, et al 2005 Investigation and validation on plastic constitutive parameters of 45 steel *Explos Shock Waves* **25** 451-6
[11] Mott N F 1943 *A theory of the Fragmentation of Shells and Bombs* (United Kingdom Ministry of Supply) A.C. 4035
[12] Mott N F 1943 *A theory of Fragmentation* (United Kingdom Ministry of Supply) A.C. 3348
[13] Mott N F 1947 Fragmentation of shell cases *Proc. Royal Society of London* **189** 300-8
[14] Glanville J P, Fairlie G and Hayhurst C 2003 Numerical simulation of fragmentation using AUTODYN-2D & 3D in explosive ordnance safety assessment *6th PARARI International Explosive Ordnance Symposium* (Canberra, Australia) 2-8
[15] Dynamics C 2011 *Release 14.0 Documentation for ANSYS AUTODYN* (USA: ANSYS Inc)
[16] Stronge W J, Ma X and Zhao L 1989 Fragmentation of explosively expanded steel cylinders *J Mech Sci* **31** 811-23

IOP Conf. Series: Materials Science and Engineering **629** (2019) 012021 doi:10.1088/1757-899X/629/1/012021

Water entry behaviours of hemisphere-head projectiles with high velocity

S Y Wu[1], Z Y Shao[1,2], S S Feng[1], T Zhou[1] and B Zhang[1]

[1]State Key Laboratory of Explosion Science and Technology, Beijing Institute of Technology, Beijing 100081, China

E-mail: shaozhiyu@bit.edu.cn

Abstract. The water-entry behaviours of projectiles with hemisphere-head were studied experimentally and theoretically, focusing on projectile dynamics and drag coefficient. Based on equivalent cone theory proposed by Baldwin, the equation of drag coefficient was developed to describe the resistance of hemispherical head projectiles from the time when the projectile contacts the water to the time the hemispherical head is submerged. A set of experimental apparatus was set up to record the water-entry process by high-speed photography, the variation of resistance during water-entry was obtained experimentally. Compared the acceleration calculated by conventional method, new method and experimental results, the new method is more consistent with the experimental value.

1. Introduction

As a kind of important underwater explosive weapon, air-launched water ammunition has always occupied an important position in the sea war because of its low cost, great power and flexible use. This kind of ammunition has the characteristics of high entry velocity and complex entry trajectory. In order to improve the reliability of the ammunition, the ballistic characteristics of projectile in the process of water-entry should be studied deeply. Water-entry of projectile is a complex physical process involving the coupling of solid, liquid and gas, and the projectile motion features in water are quite different from those in air. Due to the development of the weapons industry and the space industry, the issue of water-entry, especially in high velocity, has received significant attention. Early research dates back to 1929, when Von Karman [1] developed several mathematical methods to describe the impact of water upon rigid bodies. He was the first to use added mass instead of fluid action to analyze the impact of water-entry, and proposed the added mass method to calculate the impact load of water. Using the momentum conservation law to deduce the formula of impact load into water. Over the next hundred years, there was a lot of research undertaken by researchers, such as May, Schnitzer, Abelson, Waugh, and others [2-6]. Because of the complexity of water-entry, finding the motion law of projectiles during water-entry mainly rely on experimental study, and the velocity attenuation law is closely related to the drag coefficient. Therefore, to establish a simple and practical mathematical model to describe the motion characteristics of projectiles during water entry is particularly important. In recent studies, Guo *et al* [7–9] studied high-speed projectile striking into water horizontally and the effects of projectile nose shapes on velocity attenuations and drag coefficient. They considered the effect of initial velocity and head shape on the drag coefficient and established an empirical formula [9]. The premise of all these researches are assuming drag coefficient is a constant over the water-entry process. Indeed, the drag coefficient of the projectile is approximately a constant in the sailing process after it enters the water.

However, due to the change of the wetting surface during the process from touching the water surface to being submerged, the drag coefficient will change greatly. Actually in the early years, Baldwin [10] tested various nose shapes of projectiles, especially the cone body, and derived a set of equations to calculate the total resistance during the vertical water entry of projectiles with arch nose. In Baldwin's theory, drag coefficient at the initial stage of water entry is well considered.

In this paper, drag coefficient is analyzed during the water-entry process of hemisphere-head projectile and a new calculation model for drag coefficient was established theoretically. A set of experimental apparatus was set up to record the water-entry process by high-speed photography. The resistance of water during water-entry process was studied experimentally and theoretically.

2. Theoretical analysis

In the following analysis, the water-entry process means the process from projectiles touch the water surface to it is submerged. Previous report [11] showed that at an entry velocity of <8 *m/s*, the surface forces are important and the behavior of the missile depended greatly on its surface condition. Instead, the projectile velocity considered in this work is much higher (>50 *m/s*), which make the effect of surface condition negligible in the subsequent analyses. Considering an object with an initial velocity of v_0 penetrating into water along a straight trajectory in the $+x$ direction (figure 1). The motion equation of the object can be described by Newton's second law while ignoring gravity and buoyancy:

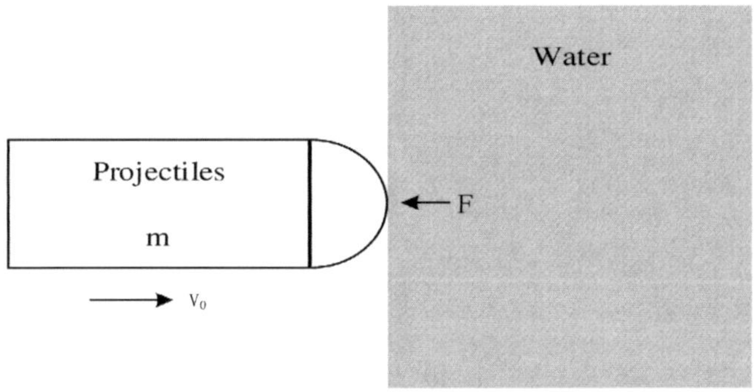

Figure 1. Horizontal penetration of water by projectiles with hemisphere-head.

$$m_b \frac{d^2x}{dt^2} = m_b \frac{dV_b}{dt} = \sum F_x = -\frac{1}{2}\rho_w A_0 C_d V_b^2 \tag{1}$$

where m_b and x denotes the object mass and penetration axis, respectively. V_b is the penetration velocity of the body, t is time, F_x is the drag force, and A_0 is defined as the projected frontal area of bodies, C_d represents the drag coefficient.

In this paper, the resistance to the head can be obtained from Newton's second law

$$F = -\frac{1}{2}\rho_w A_0 C_d V_b^2 \tag{2}$$

Obviously, the value of C_d is important to the analysis. In conventional resistance calculation, the drag coefficient is usually regarded as a constant, it is impossible to describe the resistance during water-entry process accurately, especially for the projectiles with arch nose-shape. In order to calculate the total resistance during the vertical water entry of projectiles with arch nose, Baldwin [12] developed the method of equivalent cone. In Baldwin's theory, the equivalent cone is composed of the cross section of the arch nose and water to a conical surface tangent to the surface of a pointed arch. This paper extends the approach in Baldwin to determine the value of C_d during the water entry of hemispherical head. The geometrical relationship of the system is showed in figure 2. Similar to arch nose, the hemispherical head will also make surface uplift of water, so there will form an equivalent water surface (*ews*) higher than the original water surface (*ows*). The first equivalent cone forms by *ows*, and the second equivalent cone forms by *ews*.

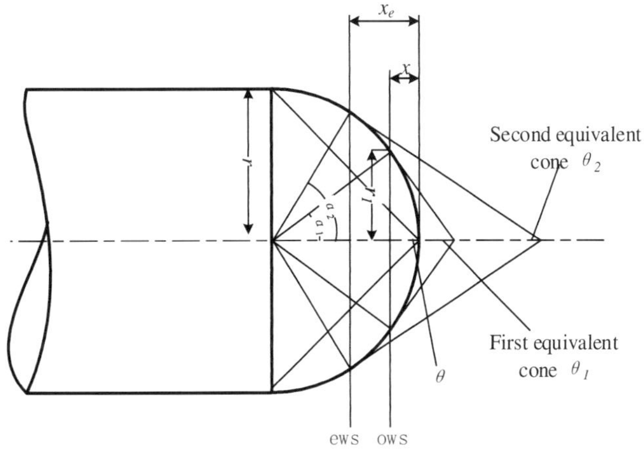

Figure 2. Geometrical relationship of hemisphere-head at water-entry process.

Based on equivalent cone theory proposed by Baldwin [12], C_d can be defined as

$$C_d = C_{dt} + C_{ds} + C_{df} \tag{3}$$

Where C_{dt} is transient drag coefficient, C_{ds} is quasi steady drag coefficient, and C_{df} is surface friction drag coefficient. C_{dt} is calculate by the first equivalent cone, C_{ds} and C_{df} are calculate by the second equivalent cone.

$$C_{dt} = \frac{2}{\rho_w A_0}\frac{dm}{dx_2} = 2.56(\frac{2r_1}{d})^2 tg\frac{\theta_1}{2} \tag{4}$$

Where m is added mass, x is the penetration distance, r_1 is radius of the first equivalent cone, and θ_1 is the aspect angel of the first equivalent cone.

$$C_{ds} = (\frac{2r_1}{d})^2 C_{ds}' \tag{5}$$

Where $C_{ds}' = 0.008088 + 0.122\theta_2 + 0.4364\theta_2^{\ 2} - 0.2811\theta_2^{\ 3} + 0.0587\theta_2^{\ 4}$, and θ_2 is the aspect angel of the second equivalent cone.

$$C_{df} = \frac{A_w}{A_0} C_f \tag{6}$$

Where unit surface friction coefficient C_f=0.0031, reference area $A_0=\pi r^2$, and the wetted cross area $A_w=\pi r^2$.

The water wetting factor can be determined as

$$\omega = \frac{x}{x_e} = 0.9985 - 0.1953\theta + 0.07047\theta^2 - 0.01528\theta^3 \tag{7}$$

Where θ is the vertex angle of water-entry body, for the hemispherical head $\theta=\pi/2$.

According to the geometrical relationship showed in figure 2, θ_1, θ_2, and r_1 can be obtained as a function of x

$$\theta_1 = \pi - 2\arccos(1 - \frac{x}{r}) \tag{8}$$

$$\theta_1 = \pi - 2\arccos(1 - \frac{x}{r\omega}) \tag{9}$$

$$r_1 = (2rx - x^2)^{\frac{1}{2}} \tag{10}$$

Substituting equations (4)-(10) into equation (3)

$$C_d = C_{dt}(x) + C_{ds}(x) + C_{df}(x) \quad (x \leqslant r) \tag{11}$$

Equation (11) shows that $C_d(x)$ is displacement-dependent. When x>r, C_{dt}, C_{ds}, C_{dt} will become constants, so will C_d.

3. Experiments

3.1. Experiment design

The schematic diagram of the test system is shown in figure 3. The experiment involved shooting a hemisphere head projectile into a water tank and recording the whole process with a high-speed camera. The required speed is 50 m/s-100 m/s, the front of the high-speed camera needs to be protected by bullet-proof glass.

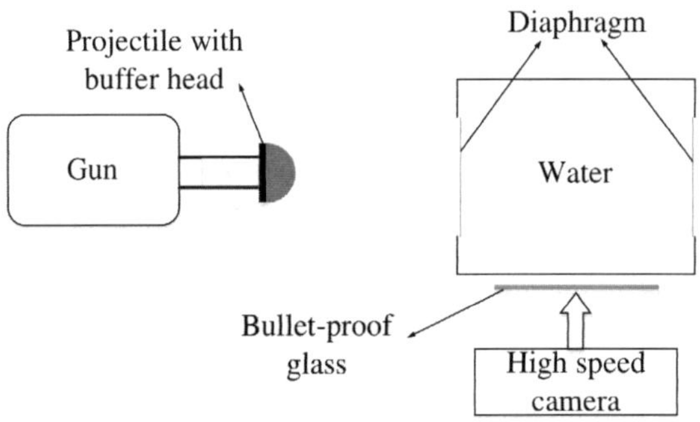

Figure 3. The schematic diagram of the test system.

3.2. Ballistic gun

Dynamic loading device used in this experiment is 40 mm smoothbore gun. Due to the larger specimen caliber and lower required speed compared with cannonball, the specimen is directly put at the muzzle, driven by explosive gas. The acceleration and initial velocity of the projectiles are adjusted by tuning the mass of the propellant in the cartridge.

3.3. High-speed camera and water tank

A Phantom digital high-speed camera was deployed to record the fluid, the dynamic of projectile upon the entry. The sampling rate was adjusted to 10000 FPS, one frame was taken every 100 μs, this sampling rate of exposure and resolution rate can satisfy experiment study. The transparent, impact-resistant water tank was made of polycarbonate composite material, and there was no cover on the tank. There were two holes in the front and the back pieces of the polycarbonate and sealing by two pieces of diaphragms. The projectile was fired into the front hole to penetrate the fluid horizontally and left the tank from the back hole either by itself or by manual unloading. Before each run, water was replenished and the tank was sealed with fresh diaphragms.

3.4. Projectile

The projectile with hemisphere head was made of 2A12 aluminum (150 mm in diameter, 3 kg mass). To reduce the mass, the hemispherical head is a shell structure. In order to facilitate acceleration, a nylon tail rod is connected to the projectile with a thread. There is a hole in the tail rod, high-density steel balls are stuffed into the holes to ensure the accuracy of the mass. Figure 4 is the sketch of the assembled projectile.

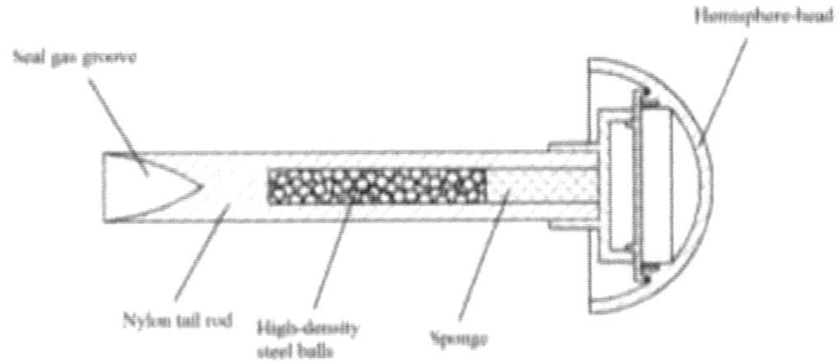

Figure 4. The sketch of the assembled projectile.

3.5. Water-entry process

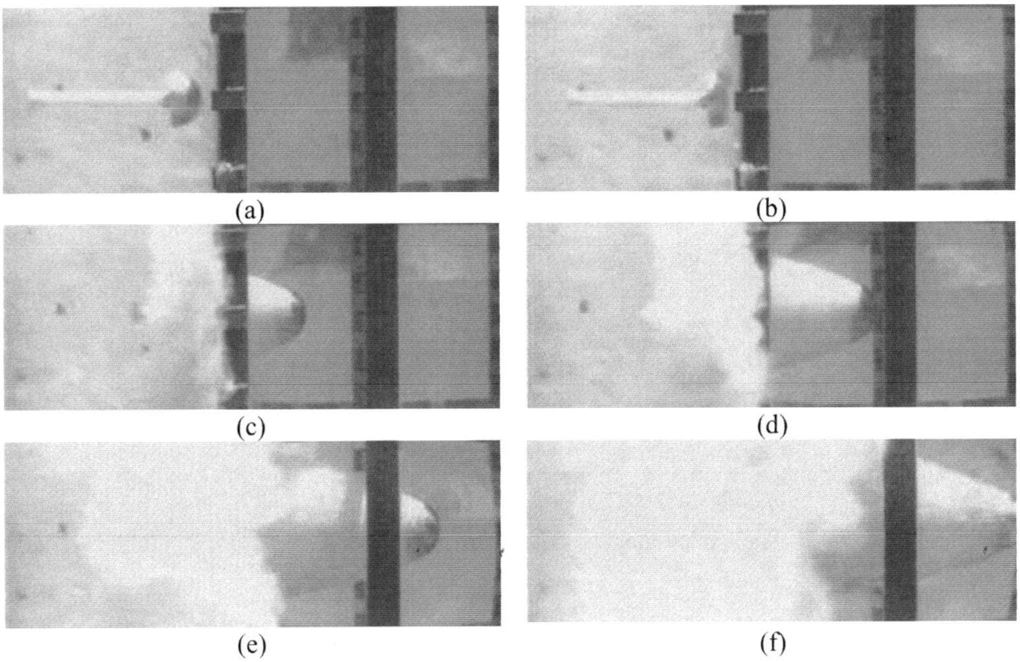

Figure 5. The water-entry process taken by high-speed camera. (a) -0.4 ms, (b) 0 ms, (c) 3.2 ms, (d) 5.5 ms, (e) 9.9 ms and (f) 14.6 ms.

Figure 5 shows the typical pictures taken by the high-speed camera. The projectile hits the water surface at a certain initial speed, and the zero point of time is when the projectile touches the water surface. In figure 5(c), it can be seen that the obvious tail emission and cavitation are formed, and then the cavitation is further expanded until the projectile penetrates out. Due to the limited volume of the water tank, the closure of the cavitation cannot be observed.

3.6. Experimental data analysis

Figure 6 demonstrates the velocity attenuation over time of the projectile, and figure 7 shows the experimental penetration distance with time. This paper is concerned about the movement at the initial stage of water entry. According to the observation, the projectile in this experiment has submerged the head into water within 1ms, so the collection points in the first 1 ms are denser. Integrating the curve shown in figure 6 to obtain the curve of acceleration over time, and then the resistance of water entry at the corresponding collection point can be calculated. The acceleration variations over time are shown in figure 8. As can be seen from the curve in figure 8, the impact at the initial stage of water entry is quite severe, and then the impact will decrease rapidly.

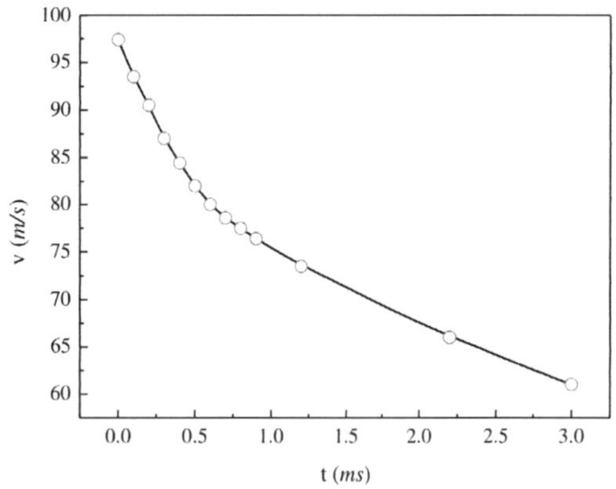

Figure 6. Velocity attenuation over time of the projectile in experiment.

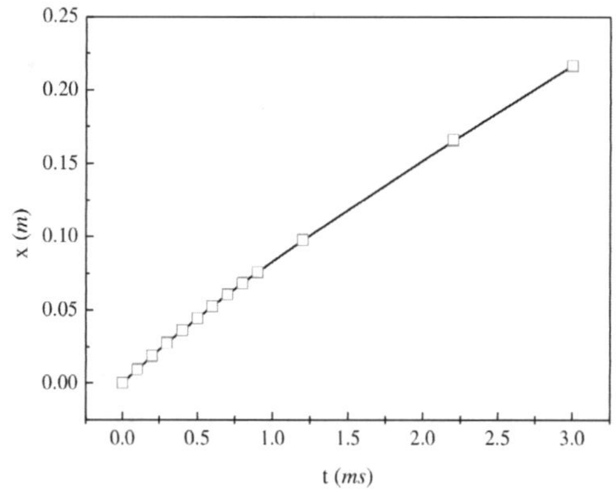

Figure 7. The experimental penetration distance with time in experiment.

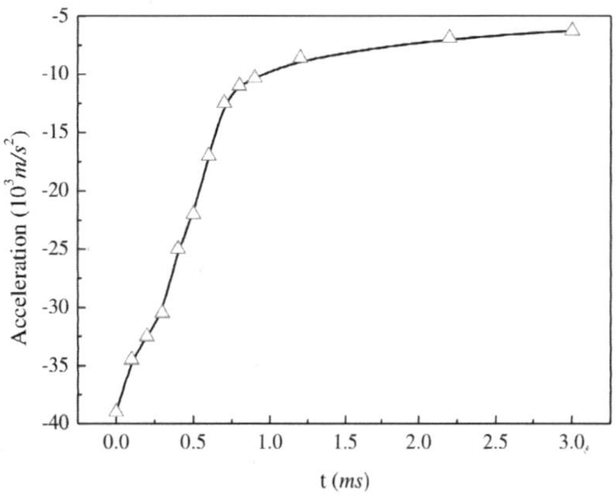

Figure 8. Acceleration variations over time in experiment.

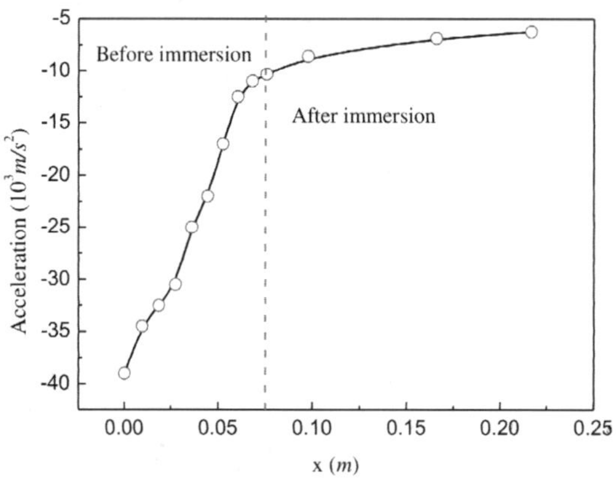

Figure 9. Acceleration variations over penetration distance in experiment.

Figure 9 shows the deceleration variations with penetration distance .The diameter of the hemispherical head used in the experiment was 0.15 m, and when the penetration depth reached 0.075 m, the hemispherical head was completely immersed in water. In figure 9, it can be seen that the curve starts to flatten after x=0.075 m. This shows that the water entry of projectiles with arch nose can be divided into two stages. The first stage is from the projectile contact with the water surface to the head fully immersed in the water, and the second stage is the projectile sailing in the water.

4. Discussion

4.1. Theoretical calculations

4.1.1. Conventional method. In conventional method, the drag coefficient is regarded as a constant during a water-entry process. This method is usually used in engineering calculation to calculate the resistance of a rigid object when it is running stably in the flow field. The value of the drag coefficient is obtained from a lot of engineering experience and is closely related to wetting surface. For hemisphere-head, the drag coefficient is 0.5 under subsonic velocity.

4.1.2. New method. As a matter of fact, to the hemispherical head projectile, at the initial stage of water entry, wetting surface is changeable. So the drag coefficient can't be regarded as a constant at this stage. According to the result of section 2, the model factors and calculated values are shown in table 1. The values in the table are based on the parameters of the projectile used in the experiments. The relationship between the drag coefficient and penetration distance is shown in figure 10. The drag coefficient calculated by the new method begins to approximate to a constant after the head is immersed, and is close to the empirical engineering value.

Table 1. Model calculation.

Factors	Calculated model
θ, rad	$\dfrac{\pi}{2}$
r, m	0.075
A_0, m^2	0.0177
m, kg	3
θ_1, rad	$\theta_1 = \pi - 2\arccos(1-\frac{x}{r}) = \pi - 2\arccos(1-\frac{x}{0.075})$
θ_2, rad	$\theta_2 = \pi - 2\arccos(1-\frac{x}{r\omega}) = \pi - 2\arccos(1-\frac{x}{0.06})$
ω	$\omega = \frac{x}{x_e} = 0.9985 - 0.1953\theta + 0.070047\theta^2 - 0.015280\theta^3 = 0.807$
r_1, m	$r_1 = \sqrt{2rx - x^2} = \sqrt{0.15x - x^2}$
C_{dt}	$C_{dt} = \frac{2}{\rho_w A_0}\frac{dm}{dx_2} = 2.56(\frac{2r_1}{d})^2 \text{tg}\frac{\theta_1}{2} = 455.1r_1^2 tg\frac{\theta_1}{2}$
C_{ds}'	$C_{ds}' = 0.008088 + 0.122\theta_2 + 0.4364\theta_2^2 - 0.2811\theta_2^3 + 0.0587\theta_2^4$
C_{ds}	$C_{ds} = (\frac{2r_1}{d})^2 C_{ds}' = (\frac{r_1}{0.075})^2 C_{ds}' = 177.8r_1^2 C_{ds}'$
C_{df}	$C_{df} = \frac{A_w}{A_0} C_f = 0.0031(\frac{r_1}{r})^2 = 0.551r_1^2$
C_d	$C_d = C_{dt} + C_{ds} + C_{df} = 455.1r_1^2 tg\frac{\theta_1}{2} + 177.8r_1^2 C_{ds}' + 0.551r_1^2$

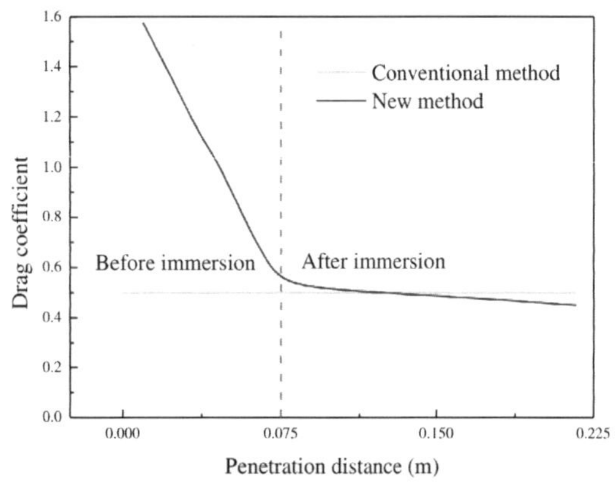

Figure 10. The relationship between the drag coefficient and penetration distance obtained by new and conventional method.

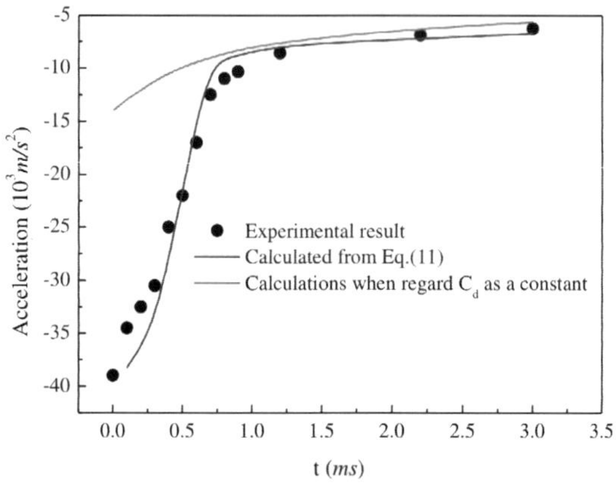

Figure 11. Comparison of experiment and theoretical calculations.

4.2. Compared with the experimental data

According to Newton's second law,

$$a = \frac{F}{M} = -\frac{\rho_w A_0 C_d V_b^2}{2m} \tag{12}$$

The theoretical acceleration can be calculated by substituting the experimental data into equation (12). The results calculated by traditional methods which regards drag coefficient as a constant are also turned out. The results of experiment and theoretical calculations are shown in figure 11. The new model shows good agreements with experimental result, in contrast, the conventional model is unable to describe the deceleration process of the projectile at the initial stage of water entry. Nevertheless, both the traditional model and the new model can well describe resistance of projectiles after the head immersed in water.

5. Conclusions

In this paper, the water entry process of projectile with hemisphere head has been studied experimentally and theoretically, especially the variation of drag coefficient during entry. The current work, within its scope, allows the following main conclusions to be drawn:

- By extent of the approach in Baldwin (1977), a new model of drag coefficient has been developed.
- It can be seen from the test data that when the projectile head is not fully immersed in water at the initial stage of entry, the drag coefficient will change over penetration depth.
- The comparison between experiments and theoretical analysis proves the deficiency of the conventional calculation method and the validity of the new model.

Acknowledgments

The authors would like to thank H T Qian and Y Q Ye for their kind help in experiments and would like to thank Bo Wang for the guidance on experiment design. All of them are the member of State Key Laboratory of Explosion Science and Technology, Beijing Institute of Technology, China.

References

[1] Von Karman T 1929 The impact of seaplane floats during landing *NACA TN 321* (Washington, USA: National Advisory Committee for Aeronautics) 2-8

[2] May A and Woodhull J C 1948 Drag coefficients of steel spheres entering the water vertically *J Appl Phys* **19** 1109-21

[3] May A 1970 Review of water-entry theory and data *J Hydronautics* **4** 140-2
[4] Schnitzer E 1953 Estimation of hydrodynamic impact loads and pressure distribution on bodies approximating elliptical cylinders with special reference to water landings of helicopters *NACA TN 2889* (Washington, USA: National Advisory Committee for Aeronautics)
[5] Abelson H I 1971 A prediction of water-entry cavity shape *Trans ASME J Basic Eng* 93 501-4
[6] Waugh G and Stubstad G W 1972 *Hydroballistics Modeling* (San Diego, CA: Naval Underwater Center)
[7] Guo Z, Zhang W and Wang C 2012 Experimental and theoretical study on the high-speed horizontal water entry behaviors of cylindrical projectiles *J Hydrodynamics Ser. B* **24** 217-25
[8] Guo Z T, Zhang W, Wei G and Ren P 2012 Numerical study on the high-speed water-entry of hemispherical and ogival projectile *2011: Proceedings of the Conference of the American Physical Society Topical Group on Shock Compression of Condensed Matter* (Chicago, Illinois) **1426** 64-7
[9] Guo Z, Zhang W, Xiao X *et al* 2012 An investigation into horizontal water entry behaviors of projectiles with different nose shapes *Int J Impact Eng* **49** 43-60
[10] Baldwin J L 1972 An experimental investigation of water entry (Corvallis, USA: Oregon State University, 1500 SW Jefferson Way Corvallis)
[11] May A 1952 Vertical entry of missiles into water *J APPL PHYS* **23** 1362-72
[12] Yang S X, Li N J and Xu X Z 2001 *Airborne Torpedo Technology* (Kunming: Yunnan Technical Press) (in Chinese)

IOP Conf. Series: Materials Science and Engineering **629** (2019) 012022 doi:10.1088/1757-899X/629/1/012022

Effect of groove structure on penetration performance of annular grooved projectiles impacting metal targets

Q Huang[1], S S Feng[1], X K Lan[1], C N Chen[1] and T Zhou[1,2]

[1]State Key Laboratory of Explosion Science and Technology, Beijing Institute of Technology, Beijing 100081, PR China

E-mail: zhoutong@bit.edu.cn

Abstract. With the continual development of advanced ammunition, a study on the effect of groove geometry on penetration performance is conducted. Based on ogive nose, an annular grooved projectile (AGP) was proposed and can generally be divided into two types (groove A and B) when considered the transition ways between groove and projectile nose. Difference in groove structures result in various penetration performance. According to dynamic cavity expansion theory and FEM simulation, the nose shape coefficient N^*, depth of penetration (DOP) and acceleration curves were analysed comparatively. The results also indicated that the firm embedded behaviour of groove B nose projectile is mainly due to the filling-material in the groove rather than the friction on contact surface, unlike traditional nose shape. Furthermore, a good agreement in calculation and simulation proved the analysis method was reasonable and feasible, and would have application potential in new concept warheads.

1. Introduction

In recent years, semi-armor-piercing high-explosive incendiaries (SAPHEIs) with delayed reactions have become important for reacting to various threats, including light armor aircraft, vehicles, and general support equipment [1]. The growing development increases the need for multipurpose SAPHEI concepts. Their penetration ability is mainly influenced by three approaches [2]: impact velocity, material property and projectile nose shape [3]. Under the bottleneck of existing technology and materials [4], geometrical optimization becomes an effective way on improvement [5].

Dynamic cavity expansion theory is verified by numerous experiments as a basis for evaluating the optimization [6]. Forrestal *et al* [7] established an empirical equation to estimate the penetration resistance. Jones *et al* [8] defined a related dimensionless constant names nose shape coefficient N^* and obtained an optimal single-curve nose shape with the minimum N^*. However, most efforts were mainly focused on single-curve geometry and its optimization while few work considered the rebounding phenomenon [9]. When impacting metal targets with low velocity, projectile with traditional nose shape is prone to rebound due to insufficient friction [10]. This phenomenon is unfavorable for explosive-filled projectile with delayed fuse and other applications.

Annular grooved projectile (AGP), refers to a projectile that contains one or more annular grooves in its nose. In this study, this projectile was proposed to solve the above problem, and was designed basing on a ogive shape and contained only one annular groove to simplify the analysis. Considered the different transition ways (whether smooth connections) between structures, AGP can be generally divided into two types, which are shown in Figure 1.

This paper describes two projectile types with different grooved structures comparatively. Numerical simulations were performed using LS-DYNA software. Then the effects due to different geometries of groove on penetration performance are discussed. Consequently, groove A nose shows a superiority of DOP while groove B nose exhibits an embedded behavior in penetration.

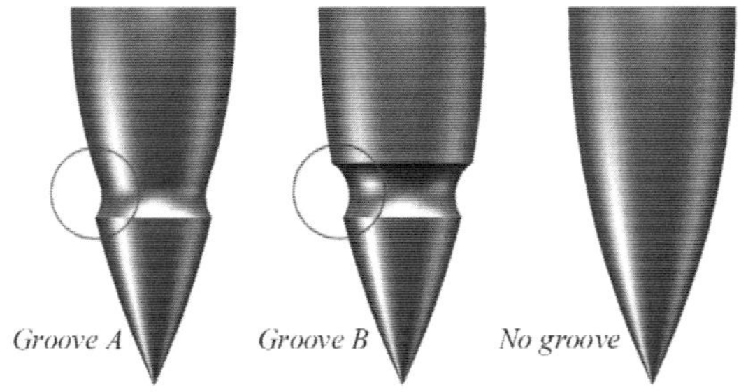

Figure 1. Shapes of two types annular grooved projectile: Groove A /B nose and corresponding ogive nose (from left to right).

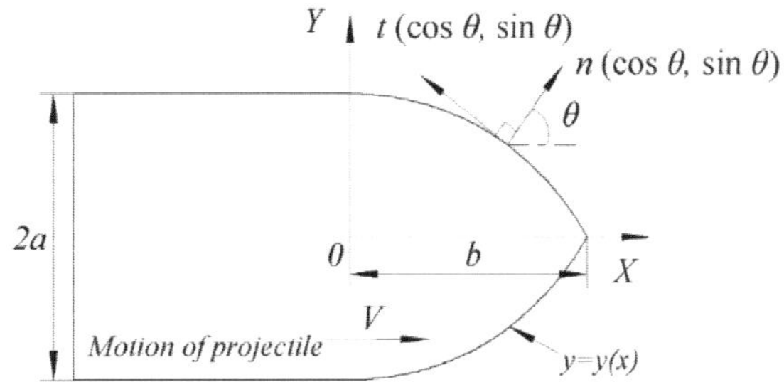

Figure 2. Side outline of an arbitrary nose.

2. Design of annular groove nose and related calculations

Assumed a non-deformable projectile having an arbitrary nose shape, as shown in Figure 2, impacts a target normally at velocity V_0 and proceeds to penetrate the target medium at a rigid-body velocity V. Dynamic cavity-expansion analysis yields the following relation between the normal compressive stress σ_n, and the projectile nose and normal expansion velocity v [11]:

$$\sigma_n = AY + B\rho v^2 \tag{1}$$

where Y and ρ are the yielding stress and density of the target material, respectively. A and B are dimensionless material constants. The particle velocity at the nose–target interface is $v = V\cos\theta$. The tangential stress on the nose was assumed to be determined by the frictional resistance on the interface: $\sigma_t = \mu_m \sigma_n$, where μ_m is the sliding friction coefficient upon impact. The resulting axial resistant force on the projectile nose can be integrated from the normal compressive and tangential stresses. Thus,

$$F = \pi a^2 \left(AYN_1 + B\rho V^2 N_2 \right) \tag{2}$$

where a is the radius of the projectile shank. N_1, N_2 are two dimensionless parameters relating to the nose shape and friction. If the nose shape can be represented by the function $y=y(x)$ for an arbitrary nose shape, as shown in Figure 2, then

$$N_2 = N^* + \frac{4\mu_m \iint_{A_n} \cos^2\theta \sin\theta \mathrm{d}A}{\pi d^2} \tag{3}$$

$$N^* = \frac{2}{a^2} \int_0^b \frac{yy'^3}{1+y'^2} \mathrm{d}x \tag{4}$$

The nose shape factor N^* is a dimensionless number and is only related to the geometrical nose shape [10]. It quantitatively characterizes the nose shape and can be used as a quantity for evaluating the inertial resistance acting on the projectile during penetration.

Ogive nose as a common nose shape, its calibre radius head (CRH) is generally between 3 and 5 in engineering practice [2]. In this case, the AGP in this study were optimally designed based on CRH 3 and 5 ogive nose. Assume that the nose shape can be represented by the function $y = y(x)$. This curve is usually monotonically increasing such as ogive or cone shape. However, when describing a segmented nose shape, such as annular grooved nose, there would be a decrease situation in $y(x)$. AGP can generally be divided into three parts and the details are exhibited in Figure 3.

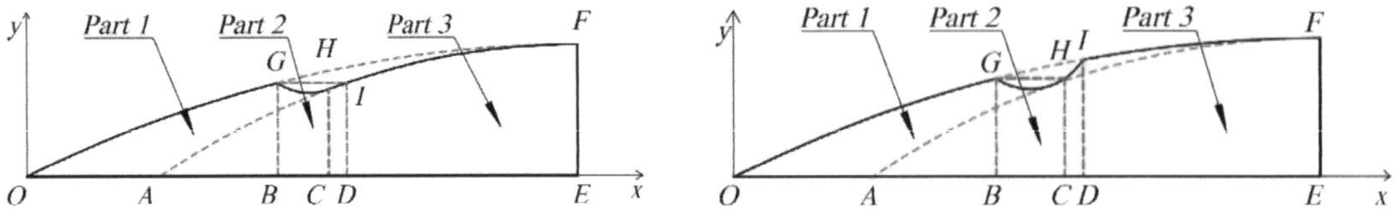

(a) (b)

Figure 3. Differences between modified and original projectiles. (a) Groove A nose and (b) Groove B nose.

Part 1(OBG) is similar to a CRH 5 ogive nose. This part need to have enough strength to keep better stability during penetration when considering non-ideal initial condition [12]. Therefore, the groove starts from the region where x/b = 0.3(OB). Part 2 (GBDI) is optimized by a circular arc and its length is controlled by the former and latter parts. The difference between groove A and B nose is also started from this part. Part 3(IDEF) of groove A is based on a CRH 3 ogive nose, while that of groove B nose is based on a CRH 5 ogive nose.

Take the groove A nose for example. Design of tangent in groove A nose is to avoid the stress concentration. OEFHG encloses the modified nose wherein OBG represents part 1, GBDI represents part 2, and IDEF represents part 3. AGF and OHF represents the shape of CRH 5 and CRH 3 respectively. GH shows the circular arc of the groove and GI is paralleled to x-coordinate. Dot H and F are points of tangency. For the modified shape, y_{OG}, y_{GI} and y_{IF} are the curves of segment OG, GI and IF respectively. According to the principle of subsection integral, the nose coefficients N* can be added together form these three parts, as given by equation (5):

$$\begin{aligned} N^* &= N_1^* + N_2^* + N_3^* \\ &= \frac{2}{a^2}\left(\int_0^{x_B} \frac{y_{OG} y_{OG}'^3}{1+y_{OG}'^2}\mathrm{dx} + \int_{x_B}^{x_D} \frac{y_{GI} y_{GI}'^3}{1+y_{GI}'^2}\mathrm{dx} + \int_{x_D}^{x_E} \frac{y_{IF} y_{IF}'^3}{1+y_{IF}'^2}\mathrm{dx} \right) \end{aligned} \quad (5)$$

As a groove structure, the curve of GI segment contains a decreasing function segment, which means that the derivative of the y_{GI} could be negative. Considering the limitations of classical cavity expansion theory, corresponding simplified outlines instead of actual outlines were used in calculation. Figure 4 presents the relationship of N* about four types of noses. The x-coordinate is based on CRH 3 ogive nose.

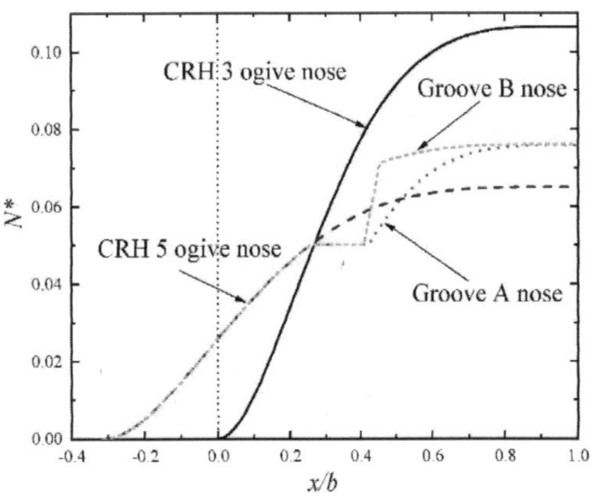

Figure 4. The relationship between N* of the four projectiles types and nose position.

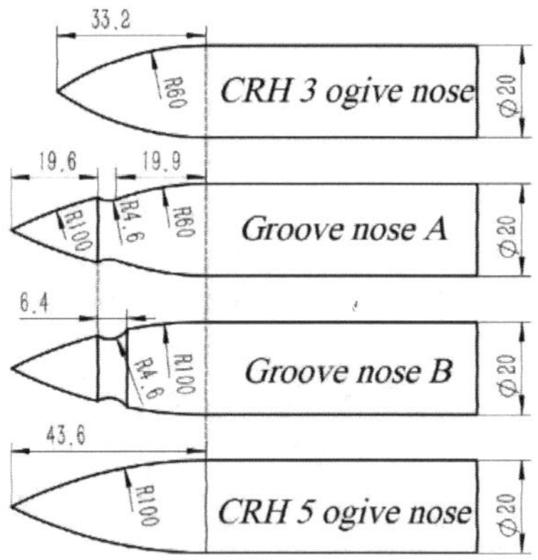

Figure 5. Dimensions of four types of projectiles (mm).

3. Materials and methods

IOP Conf. Series: Materials Science and Engineering **629** (2019) 012022 doi:10.1088/1757-899X/629/1/012022

The annular grooved projectiles (AGPs) used in this study are designed based on ogive shape. The significance of the groove structure is to obtain a relatively superior performance in penetration. Dimensions of four kinds of projectiles are shown in Figure 5 and the length of projectile shank was 3 times its diameter. Rounded structures at both ends of the groove and chamfer of the tip were designed to avoid stress concentration.

In this study, all of the targets were 30 mm thick and 100 mm in radius. They were made of the 45 steel. As a ductile metal, this material has an ultimate tensile strength of 496 MPa [13]. Furthermore, the projectile was fabricated from a high-strength material, i.e. the 30CrMnSiNi2A steel having the ultimate tensile strength of 1767 MPa [14]. These hardened steel projectiles experienced negligible inelastic deformation during the impact process and consequently can be considered as 'rigid body' [15]. To simplify the analysis, the coefficient of dynamic friction between the projectile and the target was also assumed to be a constant.

LS-DYNA was used for the simulations. The symmetry in the problem was exploited by modelling only half the projectile and the plate. To reduce the computation time, the refined shell elements were used, as shown in Figure 6. The mesh comprised 8-node brick elements with reduced integration and stiffness-based hourglass control with exact volume integration. Element size of the projectile was 0.5×0.5×0.5 mm^3 while that of target (the impact region) was 0.2×0.2×0.2 mm^3. Outside of the impact region, the mesh became bigger radially towards target edge. Contact between different parts of the model was modeled using an eroding single surface segment-based formulation [16]. The target was modeled using a modified version of Johnson-Cook constitutive material model [17]. The main material parameters used in the simulation are listed in Table 1. Moreover, the target was fully clamped at the boundary.

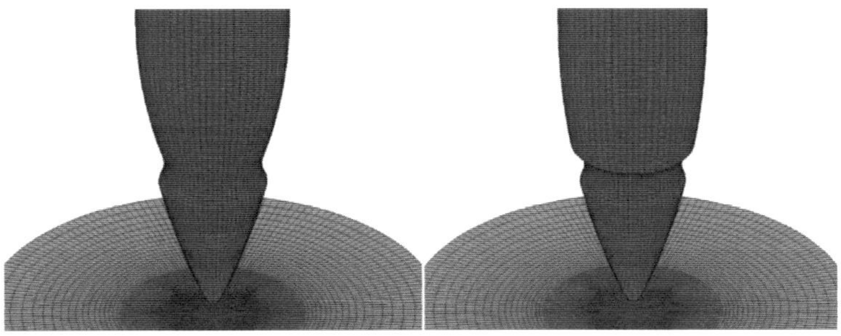

Figure 6. Finite element mesh of AGP impacting a metallic target.

Table 1. Main parameters for J-C constitutive relations of target material [13].

A/(MPa)	B/(MPa)	C	n	m	$T_r/(\mathrm{K})$	$T_m/(\mathrm{K})$	$\rho_t/(\mathrm{kg/m^3})$	E/(GPa)	D_1	D_2
496	434	0.07	0.307	0.804	293	1763	7800	210	0.636	1.936

4. Numerical simulation results and discussion

Numerical simulations of four types projectiles were carried out under the same initial conditions. All of the impact velocities were under the ballistic limits to prevent perforating. In this case the penetration process can generally be divided into two phases: forward phase and rebound phase. The former indicates the process before the projectile reaches the maximum DOP, while the latter represents the remaining penetration process until the projectile is no longer moving. As a result, the performance of projectiles in these two phases could exhibit different penetrating performance.

4.1. Superiority of DOP in forward phase

Three velocities were selected for the simulation and the variation of deceleration, velocity and DOP of those projectiles were analysed. Figure 7 shows the deceleration time-history curves of four types of

IOP Conf. Series: Materials Science and Engineering **629** (2019) 012022 doi:10.1088/1757-899X/629/1/012022

projectiles. In general, the groove A nose projectile has a smaller peak deceleration value and a wider pulse, whereas the CRH 3 ogive nose projectile has the largest peak. It indicates that the groove A has small penetration resistance and long penetration time which benefits the DOP. The vertical dashed line in this figure represents the moment when groove structure contacted target. Before this moment, three types projectiles except the CRH 3 one had the same deceleration value due to the same outline in Part 1 (Figure 3). However, only the curve of groove B nose projectiles increases sharply after this moment and becomes negative after reaching zero.

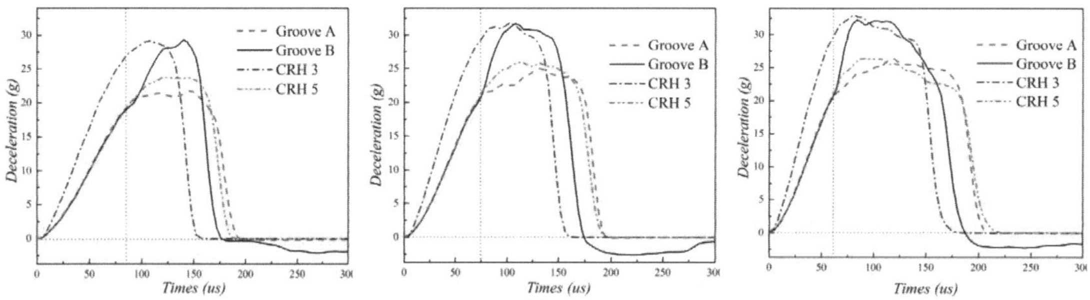

(a) (b) (c)

Figure 7. The deceleration time-history curves at different impact velocities. (a) 250 m/s, (b) 300 m/s and (c) 350 m/s.

Since the process of AGP vibrates in the target is much longer than that of penetrate, only the first 300 μs data was displayed and ignored the subsequent lengthy vibration curve segments. Figure 8 exhibits the corresponding velocity time-history curves of those projectiles. The vertical dotted line and the curve tendency are similar to that of deceleration curve. Different from other nose shapes, only groove B nose projectiles obtain a rebound velocity that approaching zero obviously.

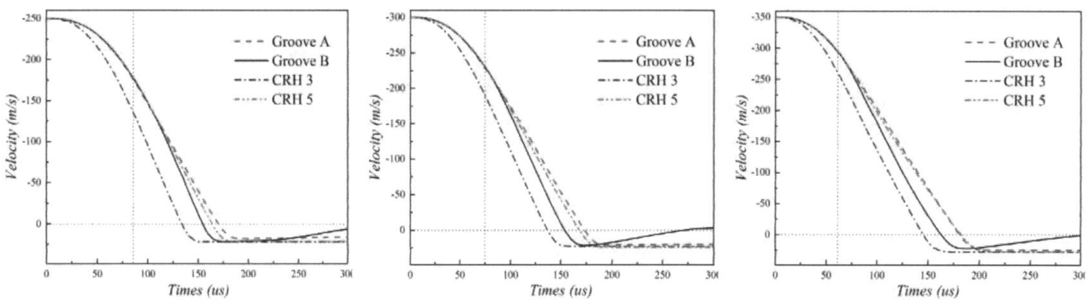

(a) (b) (c)

Figure 8. The velocity time-history curves at different impact velocities. (a) 250 m/s, (b) 300 m/s and (c) 350 m/s.

The curves shown in Figure 9 clearly reflect the effect of groove geometry on penetration performance. Considered the coefficient N^* and k, the DOP of CRH 3 projectiles generally 20% ~ 30% lower than that of CRH 5 projectiles. Furthermore, two groove geometries show visible difference with an increasing impact velocity: the groove A projectile exhibits a superiority of DOP under low-velocity impact, especially when the dimensionless DOP is less than 2. This phenomenon could be explained though the coefficient N^*.

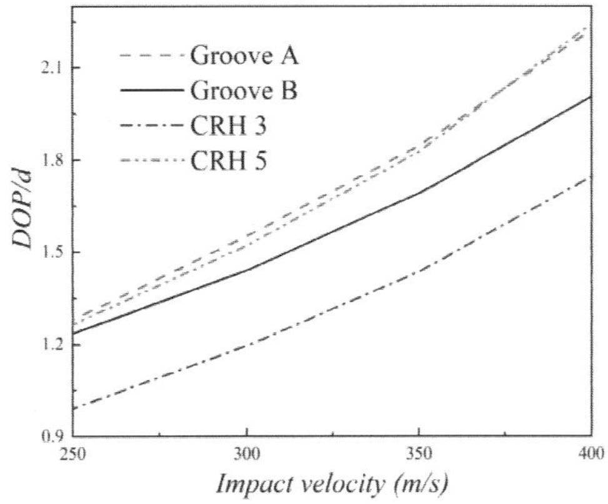

Figure 9. Dimensionless penetration depth history of four types projectiles.

4.2. Superiority of embedment in rebound phase

The embedding phenomenon of projectile was studied in rebound phase. Due to elastic recovery of target material [15], projectile has a rebound tendency after reaching maximum DOP. This phenomenon would cause projectile rebounded from the target, which is unfavourable for explosive-filled projectile with delayed fuse.

Figure 10 shows the penetration process of three types projectiles impacting 30 mm steel plates at 300 m/s respectively. Some typical physical behaviour of targets were well-captured in the numerical simulation, such as petals and denting on the impact surface. Three moments, which represented three typical relative positions between groove and target were selected. The CRH 5 ogive projectile as a comparison, exhibits that the target material is flowing along the projectile contour. This phenomenon is different from annular groove projectile. Regardless of groove A or B noses, the target material does not flow along the contour of groove in the initial stage which caused a cavity in the groove.

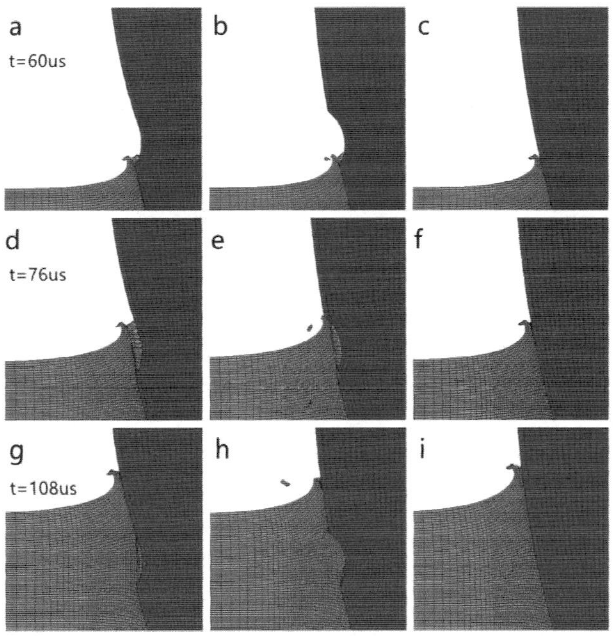

Figure 10. Penetration process of the groove A/B and CRH 5 projectiles respectively.

With DOP increased, target material is squeezed and filled into the groove structure contributed by region HI (in Figure 3(b)) of groove B nose, while the cavity in groove A nose has not been filled until the end.

The groove-filling processes of groove B nose are shown in Figure 11 to obtain more detailed deformation in penetration. After reaching the maximum DOP, projectile has a rebound tendency and a long-term vibration due to the elastic recovery of target [17]. In the case that projectile successfully prevented rebounding in the first vibration, it would embed into target. For ogive projectile, the resistance only comes from friction between projectile and target is generally unstable and may not be sufficient to limit the rebounding.

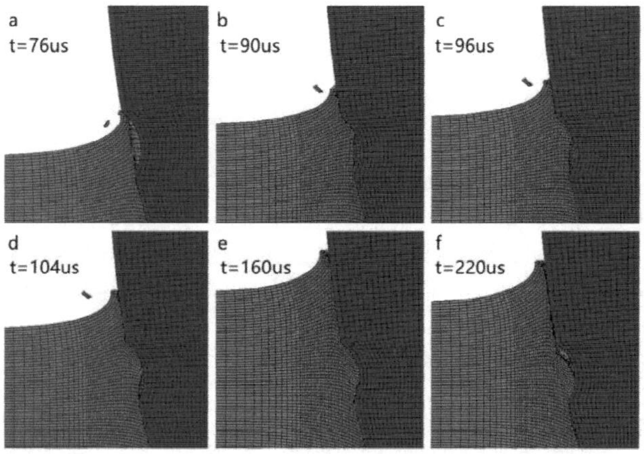

Figure 11. Penetration process of the groove B nose projectile.

However, groove B provides another relatively stable embedded method based on the plastic deformation of filling-material in the groove. According to simulation results, groove B projectile has a embedding superiority in rebound phase. Furthermore, compared to other nose shapes, this superiority of groove B indicated that the firmly embedding behaviour was mainly due to plastic deformation of the filling-material rather than friction between the projectile and target.

5. Conclusions

The effect of groove geometries on impacting metal targets was studied. Based on the experimental and numerical observations, the conclusions are written as follows:

- compared with traditional ogive projectile, the annular grooved projectiles were proposed and exhibited DOP superiority or embedded behaviour with different groove structure.
- influence of the nose shape coefficient N^* of annular grooved projectile on penetration performance was discussed based on calculation and simulation results.
- groove A projectile has a deeper DOP than corresponding ogive projectile during low speed impact. As impact velocity decreases, this superiority caused by nose shape becomes more significant.
- groove B projectile has a firm embedded behaviour during low-velocity impact. This behaviour is mainly due to the plastic deformation of filling-material rather than friction between the projectile and target.

However, the effect of groove geometry is complex and is influenced by the target material, groove size, target thickness, and impact velocity. Therefore, further theoretical and experimental researches have to be carried out to support the findings in this study. It is hoped that this framework will effectively optimize the design of complex nose details to achieve a variety of penetration performance.

Acknowledgments
This study was financially supported by the National Natural Science Foundation of China [grant number 11472053].

References

[1] 2014 *Nammo Ammunition Handbook* Edi. 2
[2] Frew D J *et al* 1998 Penetration of concrete targets with ogive-nose steel rods *International Journal of Impact Engineering* **21** 489-97
[3] Forrestal M J *et al* 1994 An empirical-equation for penetration depth of ogive-nose projectiles into concrete targets *International Journal of Impact Engineering* **15** 395-405
[4] Frew D J, Forrestal M J and Hanchak S J 2000 Penetration experiments with limestone targets and ogive-nose steel projectiles *Journal of Applied Mechanics-Transactions of the Asme* **67** 841-5
[5] Teng X *et al* 2007 Protection performance of double-layered metal shields against projectile impact *Journal of Mechanics of Materials and Structures* **2** 1309-30
[6] Li Q M and Chen X W 2003 Dimensionless formulae for penetration depth of concrete target impacted by a non-deformable projectile *International Journal of Impact Engineering* **28** 93-116
[7] Forrestal M J, Borvik T and Warren T L 2010 Perforation of 7075-T651 aluminum armor plates with 7.62 Mm Apm2 bullets *Experimental Mechanics* **50** 1245-51
[8] Jones S E *et al* 1998 On the optimal nose geometry for a rigid penetrator *Computational Mechanics* **22** 413-7
[9] Zukas J A and Scheffler D R 2001 Impact effects in multilayered plates *International Journal of Solids and Structures* **38** 3321-8
[10] Dey S *et al* 2007 On the ballistic resistance of double-layered steel plates: An experimental and numerical investigation *International Journal of Solids and Structures* **44** 6701-23
[11] Jones S E and Rule W K 2000 On the optimal nose geometry for a rigid penetrator, including the effects of pressure-dependent friction *International Journal of Impact Engineering* **24** 403-15
[12] Liu J, Pi A and Huang F 2015 Penetration performance of double-ogive-nose projectiles *International Journal of Impact Engineering* **84** 13-23
[13] Zhang L *et al* 2003 Dynamic fracture and mechanical property of D6a, 921 and 45 steels under low shock pressure *Chinese Journal of High Pressure Physics* **17** 305-10
[14] Niu Q *et al* 2017 Analysis of chip serration phenomenon in side milling of ultrahigh-strength steel 30crmnsini2a *International Journal of Advanced Manufacturing Technology* **88** 985-93
[15] Song Q *et al* 2017 A similarity method for predicting the residual velocity and deceleration of projectiles during impact with dissimilar materials *Advances in Mechanical Engineering* **9** 1-14
[16] Johnson G R and Cook W H 1985 Fracture characteristics of 3 metals subjected to various strains, strain rates, temperatures and pressures *Engineering Fracture Mechanics* **21** 31-48
[17] Dong Y *et al* 2011 Study on characteristics of partial penetration of projectile into steel panel *Chinese Journal of High Pressure Physics* **25** 565-71

Study on the protective properties of polyurethane foam for water-entry projectile

Y Q Ye[1], Z Y Shao[1,2] and S Y Wu[1]

[1]State Key Laboratory of Explosion Science and Technology, Beijing Institute of Technology, Beijing 100081, China

E-mail: shaozhiyu@bit.edu.cn

Abstract. When the projectile enters into the water at high velocity, the impact force will be huge, therefore, it is necessary to protect the projectile. In this paper, polyurethane foam is selected to study its protective properties for water-entry projectile. The water-entry test was designed to compare the numerical simulation results, in order to verify the feasibility and the validity of the model. In particular, the numerical simulation method was used to compare the motion and response characteristics of unprotected and protected projectile. The results demonstrated that the polyurethane foam has good protective properties and comparing the protective effect of different density polyurethane foam by numerical simulation methods, it can be seen that the better protection is achieved at a density of 0.17 g/cm^3 and 0.40 g/cm^3.

1. Introduction

In the naval blockade, it is a common blockade method to attack the target ship by airdropping blocking ammunition. When the entry velocity is too high, the water-entry impact will have an adverse effect on the projectile, and even make the projectile invalid: therefore, it is necessary to protect the water-entry projectile [1]. Under this background, in this work, a kind of nose cap with polyurethane foam as protective material is designed. The protective properties of polyurethane foam by means of test and numerical simulation method was analyzed.

2. Numerical check

In order to study the protective properties, the impact response of the polyurethane foam, and the water-entry projectile kinematics problems, the finite element program LS-DYNA was selected to simulate the projectile water-entry process. The water-entry test was designed to prove the validity of the LS-DYNA calculation results.

2.1. Numerical model

In order to realize a numerical simulation, a 3D model, identical to the test conditions, was implemented. Since the air has negligible effect on the calculation results, the air domain was ignored [2]. The structure was symmetrical, therefore used a quarter model. The system of elements was cm-s-g. Simulating the projectile with a quarter of an aluminum cylinder, its quality was comparable to the test piece with a thickness of 2.5 cm, whereas the radius was slightly smaller than the protective material. The protective material was a quarter sphere with a radius equal to 7.5 cm, the sides of the cube water was 60 cm. The asymmetric surfaces of the waters were set to non-reflective surface to simulating the infinite waters situation. The diagram and the grid of the model are shown in figure 1.

IOP Conf. Series: Materials Science and Engineering **629** (2019) 012023 doi:10.1088/1757-899X/629/1/012023

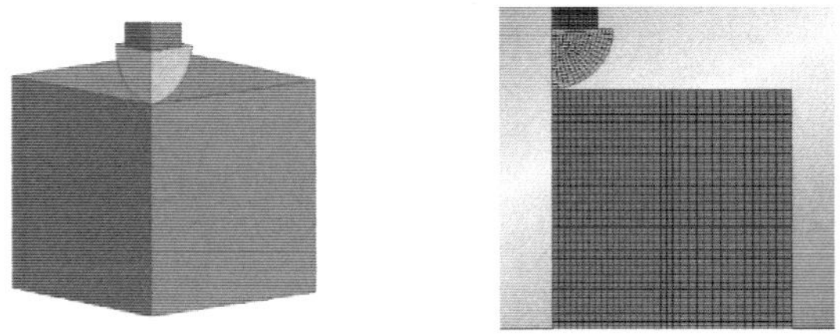

Figure 1. The diagram and the grid of the model.

The material model of the foam was selected as *MAT_CLOSED_CELL_FOAM, which is well suited for low density closed cell polyurethane foams; the material parameters are shown in table 1.

Table 1. The material parameters of polyurethane foam.

$P(g/cm^3)$	E(MPa)	A(MPa)	B(MPa)	C	P_0(MPa)	Ø	γ_0
0.25	9.93×10^{-5}	1.71×10^{-5}	1.45×10^{-5}	-4.27	1.0×10^{-6}	0.208	0

The constitutive relationship is shown in equation (1).

$$\delta y=A+B(1+C\gamma) \tag{1}$$

A, B, C are constants given by equations (2), (3), (4). γ is volumetric strain given by equation (5) [3].

$$A=23.7*\emptyset^{1.676} \tag{2}$$

$$B=19.2*\emptyset^{1.645} \tag{3}$$

$$C=2.21\text{-}31.1\emptyset \tag{4}$$

$$\gamma =V\text{-}1+\gamma_0 \tag{5}$$

Where ρ is the material density, E is the Young's modulus, γ_0 is the initial foam pressure, γ_0 is the initial body strain, φ is the ratio of the density of the foam to the polymer.

Set the initial velocity of the projectile to 100 m/s, the obtained water-entry process is shown in figure 2. The polyurethane foam was deformed, its deformation process experiences a linear elastic section, a yield section and a compact section [4].

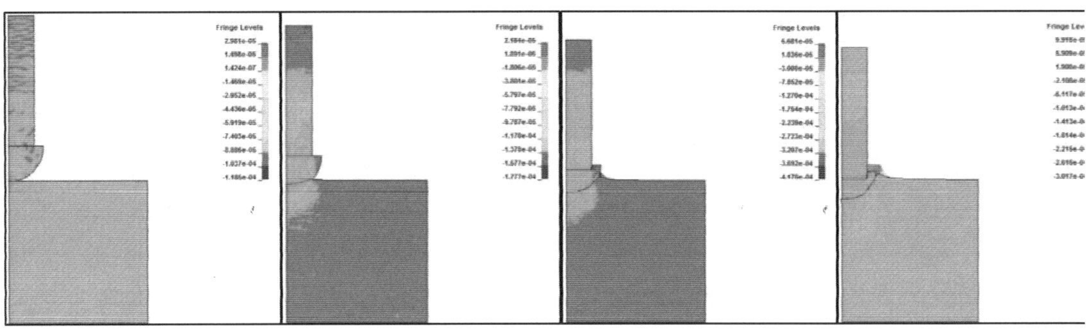

(a) (b) (c) (d)

Figure 2. Water-entry diagram (hit the water-deformation-compaction). (a) t=3.99 μs, (b) t=48.0 μs, (c) t=105 μs and (d) t=148 μs.

2.2. Water-entry test

The schematic diagram of the test system is shown in figure 3. It consists of a ballistic gun, a test piece, a high-velocity camera and a tank.

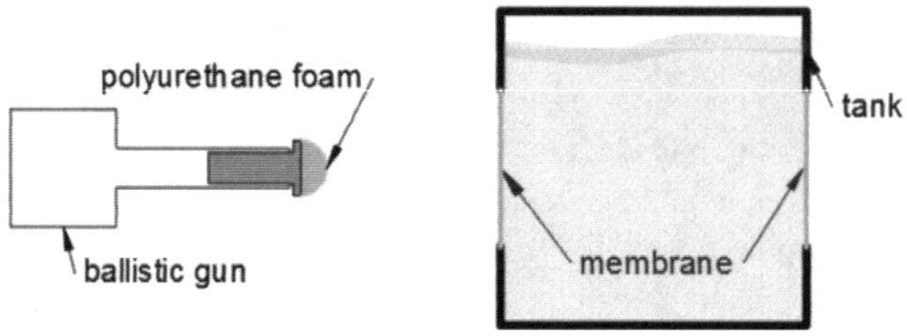

Figure 3. The schematic diagram of the test system.

The distance between the tank and the gun muzzle was 2 m. The water tank was 1 m long, 0.8 m wide and 0.6 m high; the center opening diameter was 0.4 m. The test required a velocity of 50 m/s-100 m/s. The high-velocity camera was placed at a distance of 3 m, in front of it was a bullet-proof glass.

The dynamic loading device used in this test was a 40 mm diameter smoothbore. Because the test piece was super-caliber and the required velocity was not very fast, this test piece was directly loaded from the muzzle, and the power burning gas pushed the sabot.

A Phantom color high-velocity camera was used, the sampling rate was adjusted to 10000 fps, the screen exposure and resolution at this sampling rate can meet the needs of test. The test piece was mainly divided into three parts: sabot, collet, and protective head. The sabot was 0.3 m long, its tail was concave and radius was 0.04 m. The sabot and the protective head were connected by the collet. The protective head of the test piece was a hemisphere with a radius of 0.15 m. It is made of polyurethane foam with a density of $0.25\ \mathrm{g/cm^3}$ and loaded by the ballistic gun to reach about 100 m/s velocity. The schematic diagram and the physical picture of the test are shown in figure 4.

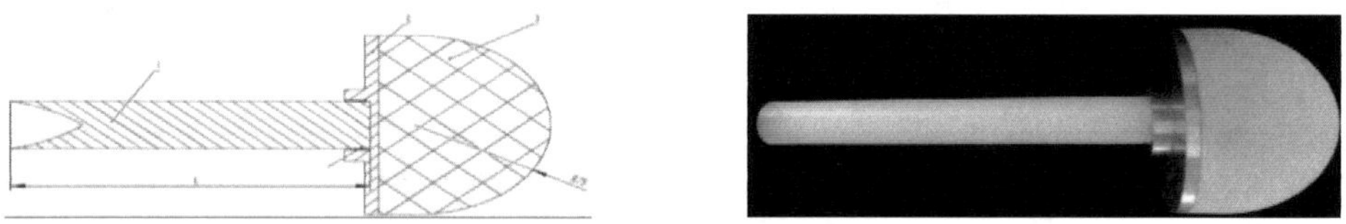

Figure 4. The schematic diagram and the physical picture of the test.

2.3. Comparison of test and numerical results

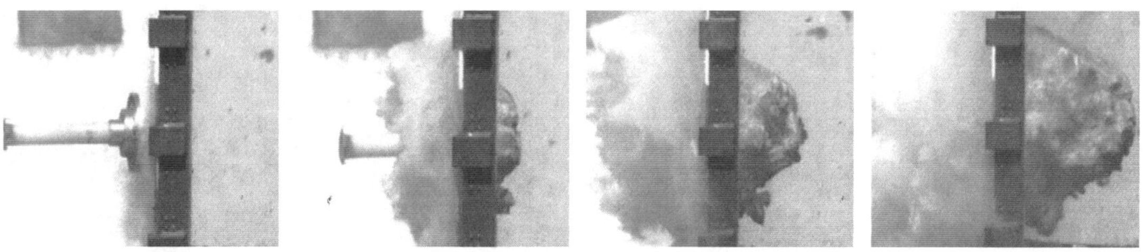

(a) (b) (c) (d)

Figure 5. Water-entry process of polyurethane head. (a) t=0.6 ms, (b) t=6.0 ms, (c) t=13 ms and (d) t=21 ms.

Figure 5 shows the water-entry process of the test piece; figure 6 shows the recovered protective head. It can be seen that the polyurethane head was quickly compacted and, then, largely crushed, which facilitates the detachment of the nose cap in water [5].

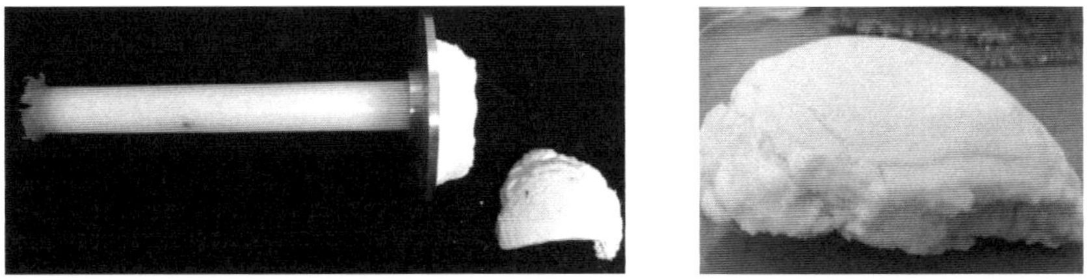

Figure 6. The recovered protective head.

Comparing the test results with the numerical results, figures 7 and 8 show the penetration distance-time and velocity-time curve. As it can be observed from these figures, the test lines conformed closely with the numerical lines and the test results verified the feasibility and validity of the model, indicating that the water entry process can be simulated by this model.

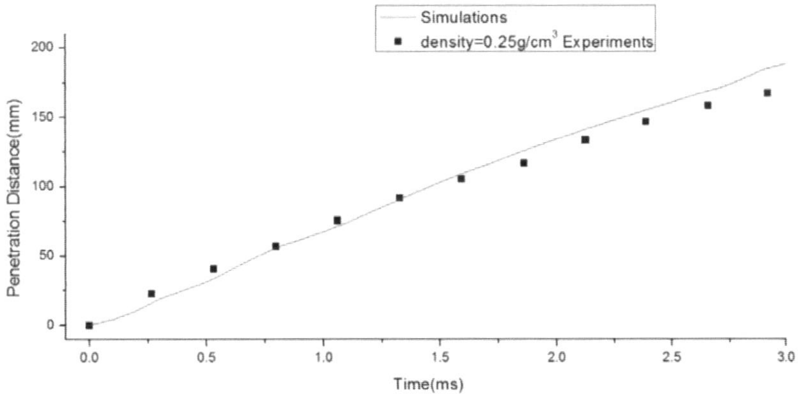

Figure 7. The penetration distance-time curve.

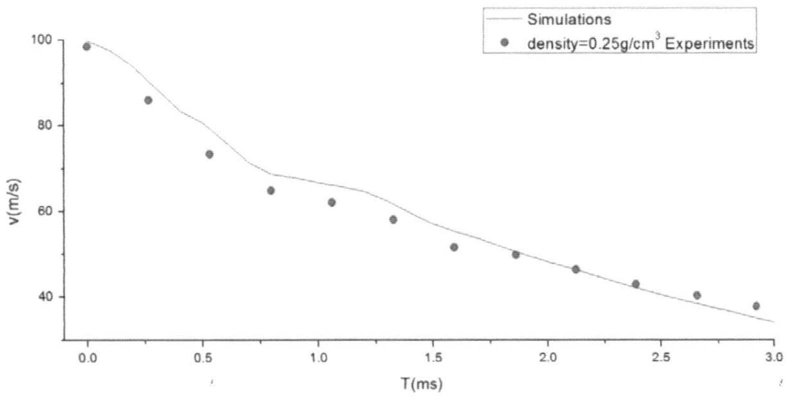

Figure 8. The velocity-time curve.

3. The protective properties of polyurethane foam

3.1. Comparison between protected and unprotected

Modeling the water-entry process of the single projectile and the projectile with polyurethane foam, the models of the two cases are shown in figure 9.

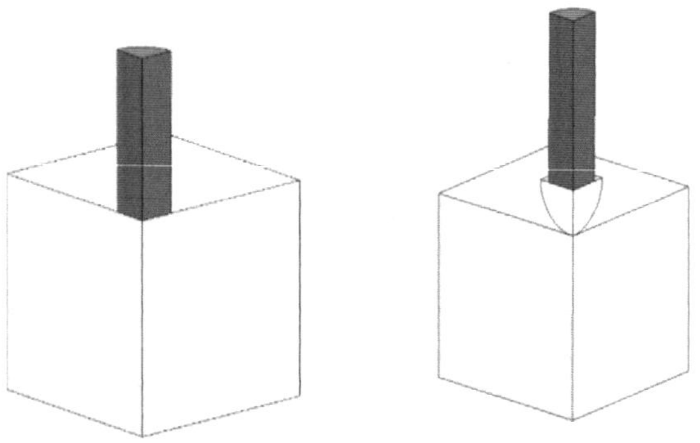

Figure 9. Model diagram.

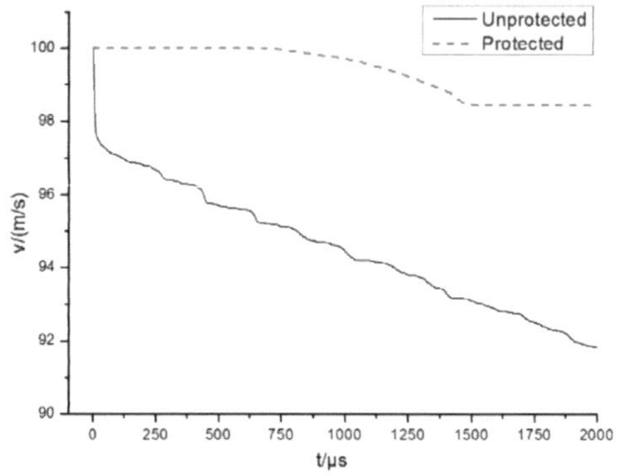

Figure 10. Protected and unprotected velocity-time curve.

The velocity of the projectile was displayed in two cases, and the display time was 2000 μs. As shown in figure 10, the velocity of both decreased, but the velocity reduction was different. In the case of unprotected, the velocity dropped instantaneously at the beginning; this behavior was not observed in the case of polyurethane foam.

Observing the acceleration, it changed in two cases. As shown in figure 11, related to the case of unprotected, the acceleration-time curve showed a step, the acceleration rose in an instant, because there was a large impact forces when the projectile entered into the water unprotected. This stage was the key stage of protection; when protected, this stage disappeared instead.

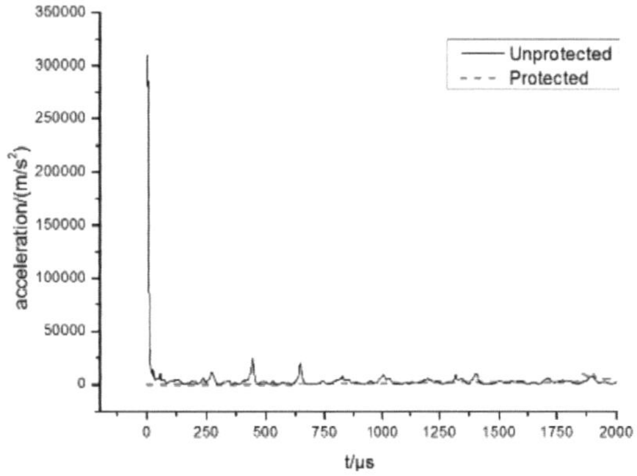

Figure 11. Protected and unprotected acceleration-time curve.

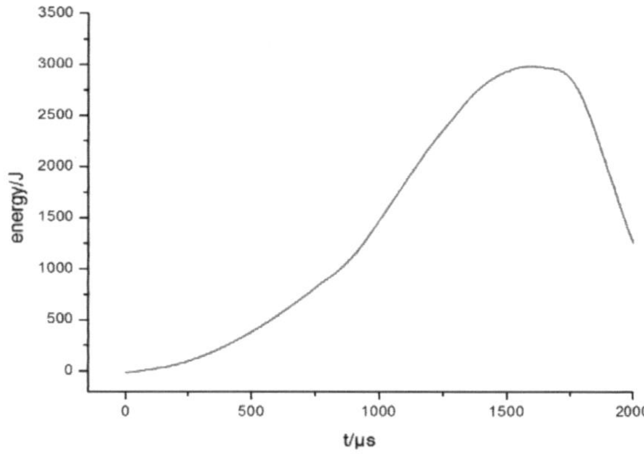

Figure 12. Energy absorption.

The protective mechanism of the protective material consisted of the absorption of a large amount of energy when subjected to impact [6]. Part of the polyurethane foam was selected to observe its energy changes, as shown in figure 12.

The yield strength is the yield limit when the metal material yields. The external force greater than the yield strength will permanently invalidate the part and cannot be recovered. Setting the yield stress of the projectile material to 355 Mbar, the ls-dyna divided the projectile into a number of units, then detected the force of each projectile unit during the water-entry process, the yield units fraction was shown in two cases, as shown in figure 13, in the case of unprotected, the yield volume fraction appeared to be a peak at about 17.2%. After adding the protective head, the maximum yield volume fraction was reduced to 0.4%.

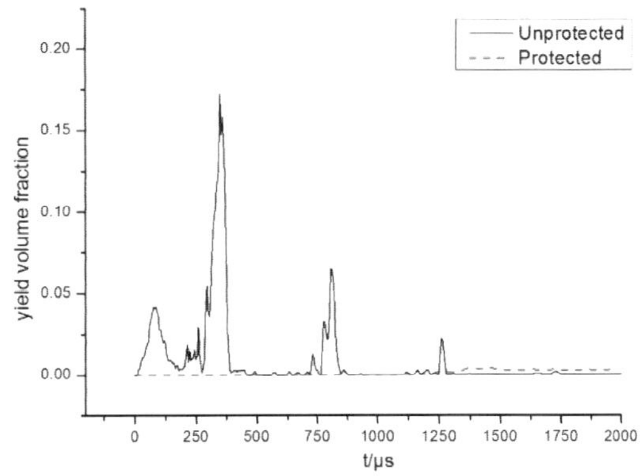

Figure 13. Yield volume fraction.

Combined with the above analysis, protective effect was evaluated by four standards, as follows:

- Velocity: small velocity change means the motion was stable, achieving protective effect.
- Energy absorption: the protective material absorbed energy to achieve protective effect.
- Maximum acceleration: the smaller the acceleration, the smaller the impact force, achieving protective effect.
- Yield stress: the smaller the stress on the projectile the smaller the damage to the projectile, protective effect is achieved.

Using the above four standards to illustrate the role of protective materials, the comparison results of unprotected and protected are shown in table 2.

Table 2. Comparison of unprotected and protected.

Standard		Protection situation	
		unprotected	protected
Standard 1	velocity change (m/s)	8.2	1.6
Standard 2	energy absorption (J)	None	2985.2
Standard 3	maximum acceleration (m/s^2)	310057	13379
Standard 4	yield volume fraction	17.2%	0.4%

It can be seen that using protective material, it can reduce the change of the body velocity by 80% and the acceleration peak by 96%. Operating in this manner, the yield volume fraction was reduced by 98%, the protective nose cap can ensure the placidity move velocity, the projectile was effectively protected and the protective effect was obvious.

3.2. The protective performance of different density

The density of materials has a great influence on the protective effect, therefore the effect was studied. The chosen densities are: 0.17 g/cm^3, 0.20 g/cm^3, 0.34 g/cm^3, 0.40 g/cm^3.

Figure 14 shows the mechanism of polyurethane foam (the density was 0.20 g/cm^3); in particular, the polyurethane foam was compressed, compacted, and crushed. At the beginning, the bottom of the foam was first compressed, the projectile propelled in it, and then the top of the foam was compressed. After that, the foam was compacted, and finally crushed. When the density was bigger, the sooner was compacted. However, the smaller the density, the easier crushed. The smaller the density, the larger the pores of the polyurethane foam are more likely to be destroyed.

The velocity of the projectile before it crushed was displayed in four cases, the display time is 2000 μs. As shown in figure 15, the velocity of all decreased with time, but the velocity reduction was different.

The velocity of the projectile before it crushed was displayed in four cases, the display time is 2000 μs. As shown in figure 15, the velocity of all decreased with time, but the velocity reduction was different.

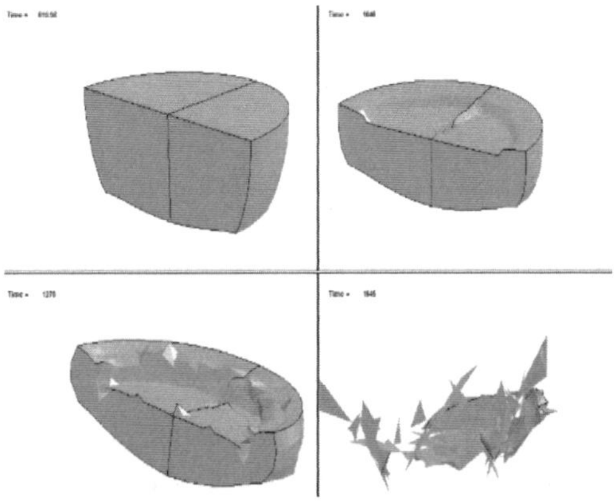

Figure 14. The mechanism of polyurethane foam.

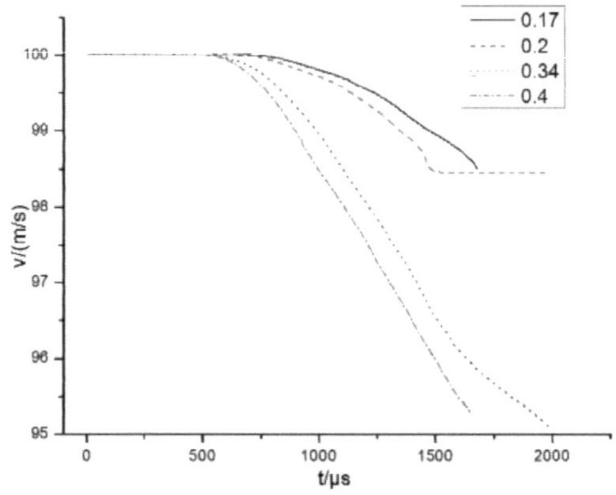

Figure 15. Velocity-time curve comparison.

Further the acceleration change was observed. As shown in figure 16, the acceleration had a peak moment. At this time, the projectile was subjected to a very strong impact. The paper thinks, at this time, the foam material is compacted. The four cases were obviously different, the higher the density, the less time it took to be compacted. The time it took to be compacted were 1.66 s, 1.45 s, 1.47 s, 930 s respectively. As the density increases, the pores of the polyurethane foam became smaller and it was compacted faster.

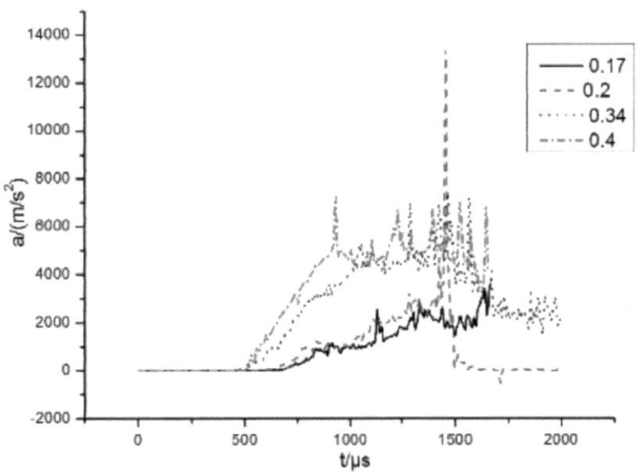

Figure 16. Acceleration-time curve comparison.

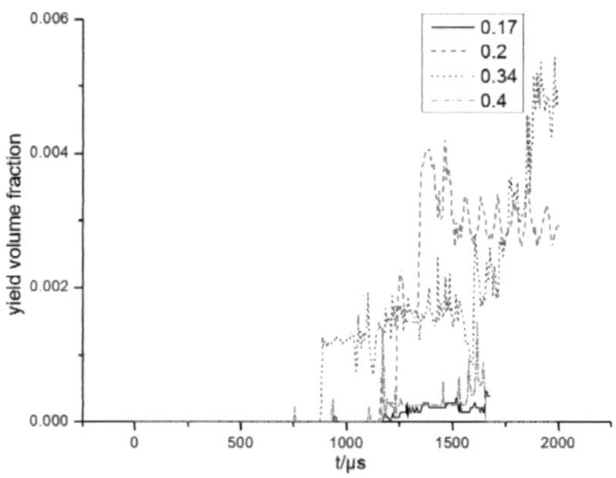

Figure 17. Yield volume fraction.

As above, setting the yield stress of the projectile material to 355 Mbar, the yield fraction in four cases was shown in figure 17.

Combined with the above analysis, protective effect was evaluated by four standards: (1) velocity change; (2) maximum acceleration; (3) yield volume fraction; (4) compaction time, some are the same as above. The fourth standard was compaction time. The longer the time used for compaction, the longer the action time, and the better effect of protection. Using the four standards to illustrate the role of density of polyurethane foam, the comparison results are shown in table 3.

In the table, the density of the two groups with better effect is marked under each standard. It can be seen that the better protection was achieved at a density of 0.17 g/cm^3 and 0.40 g/cm^3.

When the density was small, the action time was longer, the energy absorption was more, and the velocity change was more gently, which will better protect the projectile. However, when the density was very low, it may appear that it was crushed before compacted which will affect its protective effect, so the protective material should choose a moderate density.

Table 3. Comparison of different density.

Standard		Density (g/cm^3)			
		0.17	0.20	0.34	0.40
Standard 1	velocity change (m/s)	1.5	1.6	4.1	4.7
Standard 2	maximum acceleration (m/s^2)	3618.7	13379	7398.9	7265.5
Standard 3	yield volume fraction	0.00046	0.0042	0.0054	0.0015
Standard 4	compaction time (s)	1.7	1.5	1. 5	0.93

4. Conclusion

In this work, the design of a water-entry impact test, useful to compare with the numerical results was performed, to verify the feasibility and validity of the model.

Using four standards to evaluate the protective properties of polyurethane foam, it can be seen that polyurethane foam had good protection properties.

The effect of the density of the polyurethane foam was studied. Using four standards to compare the protective properties of polyurethane foams with different densities, it can be seen that the better protection was achieved at a density of 0.17 g/cm^3 and 0.40 g/cm^3.

Reference

[1] Howard E A 1959 *P. Protective Nose Cap for Torpedoes* US2889772
[2] Shi D Y 2005 *Based on Dynamic Analysis ANSYS/LS-DYNA 8. 1* (China: Tsinghua University Press) pp 96-178
[3] N-eilsen, Morgan M S and Krieg H D 1987 *SANDIA REPORT* **225** 86-2927
[4] Qian L X, Liu F and Qu M 2015 *J. Torpedo Technology* **23** 257-61
[5] Xuan J M, Song Z P and Yan Z H 1999 *J. Torpedo Technology* **2**45-50
[6] Qian L P 1993 *Design and Manufacture of Plastic Products* (China: Tongji University Press)

IOP Conf. Series: Materials Science and Engineering **629** (2019) 012024 doi:10.1088/1757-899X/629/1/012024

Mechanical characterization of mortars used in the restoration of historical buildings: an operative atlas for maintenance and conservation

A Grazzini[1,2], M Zerbinatti[1] and S Fasana[1]

[1]Department of Structural Geotechnical Building Engineering, Politecnico di Torino, Corso Duca degli Abruzzi 24 - 10129, Turin, Italy

E-mail: alessandro.grazzini@polito.it

Abstract. The complexity of the architectural and structural restoration of historical buildings requires a careful choice in the use of repair mortars compatible with the original materials of the existing building. It is important to set up an atlas of reference, compared to many mixtures of mortars and conglomerates, useful to support the operational choices in the project of restoration, recovery, and maintenance of historical buildings. The aim is also to encourage the use of local materials with a "short supply chain", compatible with buildings and applied by traditional techniques. The research makes use of experimental bending, compression and elastic modulus tests on different mixtures of mortars. This paper describes some of the first results obtained so far, because the research is still underway for many different mixtures to test and catalogue. The comparison between the data and the reference values that will be identified, will lead to set up a reference atlas of mechanical behaviour of repair mortars in order to direct the design and operational choices in function of requirements and performance (of the wall systems) of each specific cases in the field of the restoration of historical buildings.

1. Introduction

The choice of materials represents a fundamental part to correctly direct a repair work of historical buildings, ensuring their conservation over time. The choice is therefore subject to the verification of the compatibility from the aesthetic / historical point of view (figure 1a) and the compatibility from the physical-mechanical and chemical point of view (figure 1b). In the specific case, the mortars used in the historical buildings (but also in the new buildings), often need of replenishment or a total replacement of materials (of wall joints, or external plasters, e.g.), due to the natural degradation during the time or to an extraordinary event (such as an earthquake). Often many restoration work proved to be ineffective shortly (detachment of repair mortars) due to evident chemical-physical-mechanical incompatibility in comparison to the masonry support (figure 1b) [1, 2]. The interface surface between plaster and masonry is the site of mechanical or thermo-hygrometric strains, due to thermal variations, rising damp [3, 4] or seismic forces. To mitigate these strains it is essential that the repair mortar has similar physical-mechanical characteristics to those of the historical repaired plaster and of the historical masonry. At the Department of Structural Geotechnical Building Engineering (DISEG) of the Politecnico di Torino, several experiments were carried out by fatigue and thermo-hygrometric stresses tests for validating the most durable repair mortar compatible with a specific historical masonry [1-5]. The experimental prequalification phase proved to be very useful for the choice of the most compatible material in important restoration sites [2, 6].

Figure 1. (a) Patches of cement/lime based plaster on historic façade; (b) deterioration (with detachment) of cement plaster applied for the restoration of a masonry subject to moisture.

In this research, the mechanical characteristics are evaluated (dynamic elastic modulus, bending and compressive strength) of certain formulations of mortars produced and tested at the Mortars Laboratory of the DISEG Department (Politecnico di Torino), thus trying to investigate the choice from the side of mechanical compatibility with existing walls. The research is in progress [7] and in this paper the first results will be presented. The goal is indeed to provide useful data to help draft an atlas as a tool for assessing the behaviour and compatibility of repair mortar for the building heritage and therefore can be a useful tool for the operators (designers, Superintendences and workforce) in the decision-making process for the planned intervention. The research is part of an international scientific debate sensitive to the characterization of mortars within restoration building sites [8-10].

2. Types and composition of mortars

The mortar is a conglomerate composed of a mixture of water, sand, binder and any additives. Dosing appropriately, a mixture is formed with behaviour plastic, which through irreversible chemical reactions hardens over time. It is also important to highlight the differences between mortars, pastes, plasters and concretes, which although similar in various aspects differ in the composition (table 1).

Table 1. Types and composition of mortars.

Pastes	Plasters	Mortars	Concretes
Water	Water	Water	Water
Binders	Binders	Binders	Binder (cement or *pozzolana*)
Any additives	Fine aggregates (max 0.5 mm)	Aggregates (max 4 mm)	Sand (max 4 mm)
	Any additives	Any additives	Gravel (max 8 mm)
	Adjuvants	Adjuvants	Medium gravel (max 65 mm)
			Any additives

Thanks to technological progress it was possible to produce mortars with specific compositions, expanding the field of application of the mortars:

– *Mortars for restoration*: of monuments, are generally mortars special with *ad hoc* formulations, which vary case by case. Once hardened, the mortar must have similar characteristics compatible with those of the material on which is applied. These mortars must be carefully evaluated for the following properties (after diagnostic campaign): elastic modulus, compressive strength, flexural strength, expansion coefficient, porosity and permeability.

– *Mortars for renovation*: to deny water penetration through underground and above-ground walls, used in tanks, ducts, sewers, foundations, canals, tunnels, ducts, dams, for leaks and for filling cavities. Characterised by a rapid hardening and a high and immediate mechanical strength.

– *Strengthening mortars:* as part of the repair mortars for strengthening work are used to fill failures or macro-porosity and to perform anchors by fixing. Among these mortars is the so-called strengthening grout, which does not it is nothing but an expansive mortar. The grout is used to fill the cavities of the masonry by pouring or even by injection.

– *Mortars for recovery:* the recovery work consists in repairing a structure. High-strength chemical and mechanical synthetic mortars are used, because must be able to withstand stress challenging. In these particular mortars it is common to find thin fibers added, for example glass fibers or polyamide fibers, which are used to create a widespread "reinforce". This category also includes mortars used to reconstruct the cover of a work in reinforced concrete.

– *Mortars for masonry:* this type of mortar is linked to the construction of brick or stone walls, and is used to ensure the correct installation and binding of the masonry elements. For this use almost all types of mortar can be used, however the compatibility with the mechanical characteristics of the support is fundamental (enough, with diagnostic support).

– *Mortars for finishing work:* all those mortars that are used for various uses, as the creation of a shaving layer filler on a surface, so to allow the application of coatings and painting, can therefore be considered as an adhesive bridge for subsequent plastering. They are also used for gluing non-structural elements such as insulating tiles or panels or for sealing joints or joints.

Since the mortars of the conglomerates consist of a mixture of various elements, depending on the composition of this mixture we can distinguish various types of mortars:

– *Aerial mortars:* formed by water, sand and an aerial binder as lime aerial (powder or putty lime). The main feature of these mortars is that they can only harden on contact with the air. They therefore serve for plasters and walls in elevation stressed in a manner reduced and exposed to the air.

– *Hydraulic mortars*: with hydraulic properties, or able to harden even in contact with water, therefore suitable for work in damp and cold environments. They are composed of water, sand and natural hydraulic lime or of hydrated lime in powder (aerial) with the addition of aggregates with pozzolanic properties of natural origin (e.g. *pozzolana*) or artificial (e.g. *cocciopesto* or kaolin).

– *Cement mortars*: with cement as a binder. Strength development (mechanical and chemical) is rapid, waterproof and they have a remarkable duration.

– *Compound or bastard mortars:* consist of a mixture of two or more different types of binders. E.g. cement is added to an air lime mortar to give superior mechanical strength, or a cement mortar is added with lime aerial to improve its workability. Lime combined with gypsum guarantees a quick setting.

3. Mortar production in laboratory test and experimental results

The mortars made belong to different types: aerial, hydraulic, cement and bastard mortars. The cement used is of the type 32,5 and 42,5 R. The aerial lime corresponded to *"Calce del Brenta"*, a special magnesic putty lime more than three years aged. Siliceous sand was extracted from the seabed of the tertiary basin of the Po Valley, wet washed in a cyclone at the quarry. The first type of *cocciopesto* corresponded to a Tra̕ss of natural origin, ground into very fine powder, without chemical additives, free of cement. The second type was a reactive aggregate obtained from crushed tiles and bricks without clinkerization, only aggregate less than 150 μm were used. The natural raw pearlite, a variety of volcanic rock effusive, was with medium grain size (0.3-1.2 mm) and coarse grain size (0.8-2.5 mm), non-toxic, inert, non-combustible. The mortar specimens were prepared in a non-conditioned laboratory environment with a temperature of approximately 20 +-2 °C and U.R. 60:70%. For each of the formulations made, 9 specimens were produced standardized (40x40x160) mm, of which only 3 will be broken. The remaining 6 specimens will become part of the DISEG laboratory archive in order to allow possible future comparisons or checks. The mortars were coded by the letters G, H, I , L, M, N, O and P (table 2).

Table 2. Mortars compositions.

		MORTAR COMPOSITIONS								
		AGGREGATES				BINDERS				
						AERIAL LIME		HYDRAULIC LIME		
		SILICEOUS SAND OF QUARRY	COCCIOPESTO	POZZOLANA	PERLITE	LIME PUTTY	HYDRATED LIME IN POWDER	HYDRAULIC LIME (NHL 3,5)	CEMENT PORTLAND (32,5)	CEMENT PORTLAND (42,5)
SAND MORTAR AND HYDRAULIC LIME	G1	3						1		
	G2	2.5						1		
	G3	12				1		3		
SAND MORTAR AND CEMENT PORTLAND (32,5)	H1	3							1	
	H2	12				2			2	
	H3	12				1			3	
	H4	12				3			1	
SAND MORTAR AND CEMENT PORTLAND (42,5)	H5	3								1
	H6	12				2				2
	H7	12				1				3
	H8	12				3				1
SAND MORTAR AND COCCIOPESTO	I1	3	100% fine			1				
SAND MORTAR AND PERLITE	L1	3			100% fine	1				
	L2	2			100% fine	1				
SAND MORTAR AND HYDRATED LIME IN POWDER	M1	2.5					1			
	M2	3					1			
	M3	3.5					1			
SAND MORTAR AND LIME PUTTY	N1	2.5				1				
	N2	3				1				
SAND MORTAR + COCCIOPESTO + HYDRAULIC LIME IN POWDER	O1	2.5	60%				1			
	O2	3	60%				1			
SAND MORTAR + POZZOLANA + HYDRAULIC LIME IN POWDER	P1	2.5		30%			1			
	P2	3		30%			1			

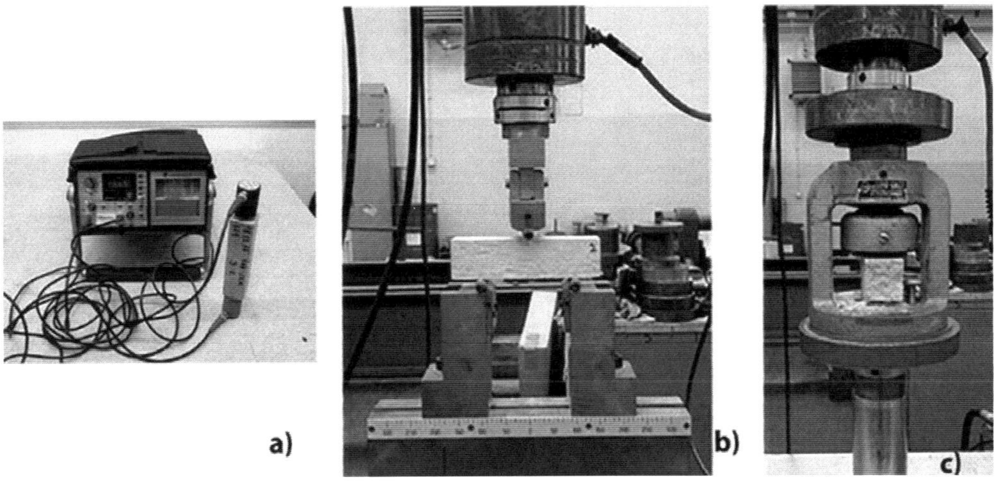

Figure 2. (a) ultrasonic test; (b) bending test; (c) compression test on mortar specimens.

The main measurement performed in the experimentation concerned the determination of the elastic modulus of different mortar formulations by means of the ultrasonic test (figure 2a). The purpose of this non-destructive survey was to detect anomalies in the ultrasonic waves materials (cracks, porosity or cavities). Sound is an elastic mechanical vibration, produced by expansion and compression of the particles that make up a propagation medium. The ultrasonic are therefore sound waves (longitudinal) with a high frequency, beyond the audible threshold. Conventionally the ultrasounds range from the frequency of 20 kHz to 1 GHz. The time required for a wave to pass through a material was measured. If an ultrasonic wave encounters a vacuum layer or a material during its journey different, undergoes alterations. The ultrasonic test is described by the UNI EN 12504-4: 2005 standard [11]. The purpose of the test was to determine the propagation speed of the ultrasonic wave. This value will then be used to calculate the dynamic elastic modulus E_d by (1):

$$E_d = v_m^2 \cdot \rho_m \tag{1}$$

where v_m is the average propagation speed (m/s), ρ_m is the average density of the formulation (kg/m^3).

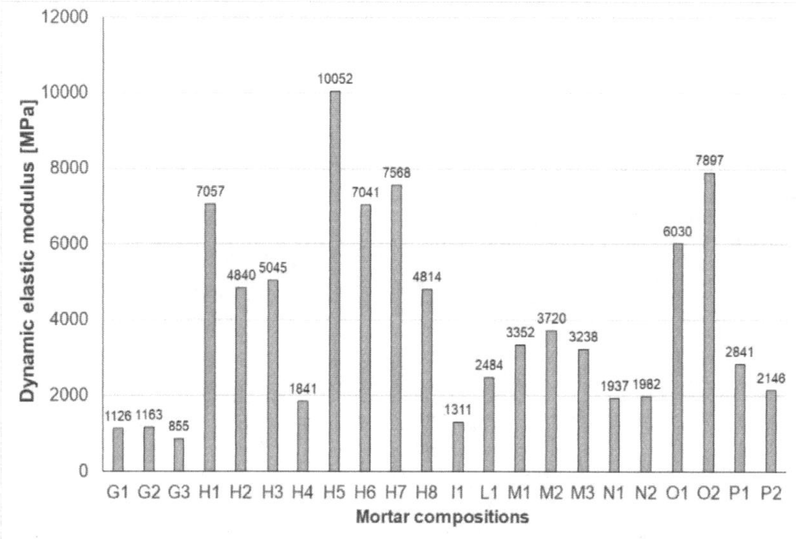

Figure 3. Medium values of dynamic elastic modulus for each mortar composition.

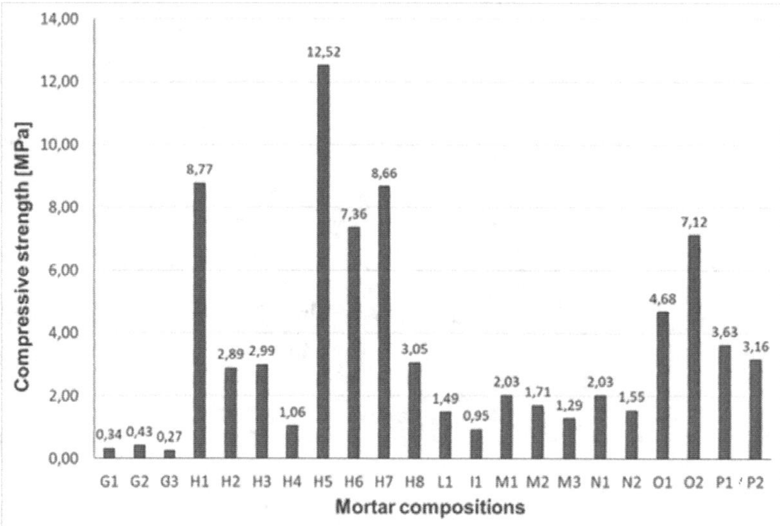

Figure 4. Medium values of compressive strength for each mortar composition.

Since the elastic modulus is directly proportional to the speed of transit of the pulses ultrasonic, higher speed formulations also have higher values of Ed. As can be seen in figure 3, the set of specimens that presented the average Ed value greater was made by cement or cement and lime putty (category H), with the exception of the specimens with a preponderant amount of lime putty compared to cement (H4 and H8). Among the category H, the greater Ed was obviously recorded by the specimens composed of 42.5 cement (H5, H6, H7, H8). Within category H cement specimens have returned very high values: it can be seen that as the lime / cement putty ratio increased, the value of Ed decreased. Also the category

O2, which was made with sand, *cocciopesto* and hydrated lime powder with 3:1 aggregate/binder ratio, shown a high Ed value. The high speed of this formulations corresponded to a reduced porosity of the specimens. While the absolute lowest values of Ed were recorded by the specimens made with hydraulic lime (category G). Also among the hydraulic lime mortars, the G3 mortar (which presented lime putty in its composition) had a lower elastic modulus compared to the other two mortars (G1 and G2) which had hydraulic lime as the binder. Comparing the values of the G1 and G2 mortars it was possible to see how the quantity of hydraulic lime increased the elastic modulus. The aggregate / binder ratio in fact affected the density of the specimens, which was directly proportional to Ed. The tests were performed with two different types of sonic probes (50 kHz and 120 kHz), finding results comparable.
For a comparison with the experimental results of the elastic modulus, the bending (figure 2b) and compressive (figure 2c) tests described by the UNI EN 1015-11:2007 standard [12] were also performed. The results of the compressive strength shown in figure 4 confirmed the similar trend already found by the measurement of the dynamic elastic modulus. A more rigid mortar corresponds to higher compressive strength values, and vice versa. The bending test was performed with 3 supports. For technical reasons the data of flexural tests on G-H-I-L mortars were not available, however the trend detected by the results of the remaining formulations confirmed the mechanical characteristics previously obtained (figure 5).

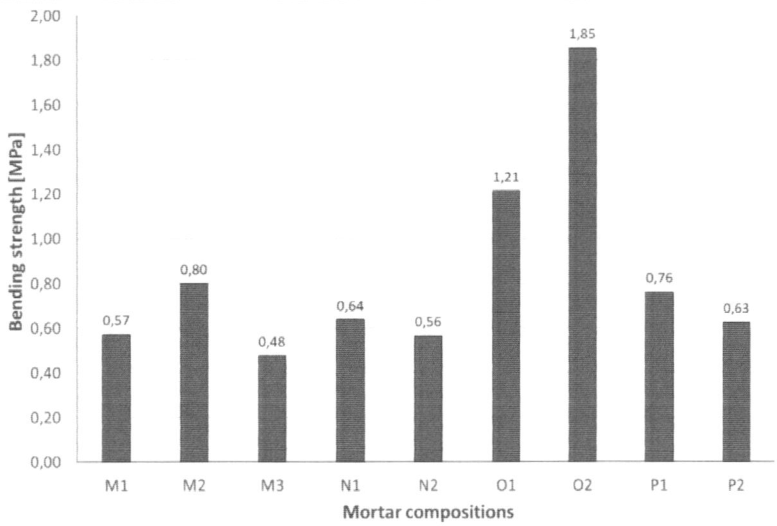

Figure 5. Medium values of bending strength for each mortar composition.

4. Discussion

The elastic modulus of the materials has a high importance especially in the presence of coupled materials such as e.g. in the case of a plaster applied to masonry support. The mortar, once applied, begins the phases of setting and hardening which inevitably lead to the emergence of controlled deformations and strains, which are only partially transmitted to the masonry. Thermal cycles, shrinkage due to plaster curing, etc., would not compromise the durability of the mortar if it were not bound to the rigid support. The case of plaster applied to the wall support, however, corresponds of a constraint, the displacements due to expansion or compression phenomena are prevented. As a result, strains arise in the material applied: the strains created in the material must balance that due to the constraint. The value is proportional to the elastic modulus of the material; therefore, the tensile or compressive stresses will be greater the greater the rigidity of the material itself. These are the reasons why the elastic modulus of materials is extremely important in the case of coupled materials. To avoid the degradation phenomena just described it is necessary that between the two materials (the new and the pre-existing) there is a certain mechanical compatibility and that the values of the elastic modules are compatible. This means that the elastic modulus of the plaster applied on an old masonry must be similar, better if inferior, to that of the support [2, 5]. In this way the stresses are borne by the wall, having greater stiffness. Mechanical compatibility is closely linked to the value of the elastic modulus as it is the parameter that best represents the strength and deformability of the material. The atlas under construction will contain the elastic modulus of the different mortar mixtures (the aim is carry out data for 100 mortar mixes) so that the designer can compare the values with those of the historical masonry on which will be applied.

5. Conclusions

The technical operators which take an interest in maintenance, conservation and restoration of the heritage architectural may need to analyse data that helps correct it formulation of a mortar. The premixed solutions on the market no longer contain the preparation methods and formulations that were once handed down from master to master. The formulations made for this paper seek to regain part of the knowledge that has been partially lost due to industrialization. Industrial mixtures will never have a perfect coincidence with mortars in place, perhaps not even full physical-mechanical compatibility. Therefore, those who work in the sector, following surveys on historic mortars in operation will have to formulate ad hoc compatible mixtures. These analyses (sometimes very expensive) are indicated for protected monuments and buildings. For widespread construction, clients and designers do not usually carry out diagnostic investigations. Have a reference text containing formulations and data to support the project and construction site can help technical operators.

References

[1] Bocca P, Grazzini A and Masera D 2011 Fatigue behaviour analysis for the durability prequalification of strengthening mortars *Journal of physics. conference series* **305** 012041

[2] Bocca P and Grazzini A 2010 Durability evaluation of strengthening mortars applied to historical masonry structures *Proc. 2th Int. Conf. on Sustainable Construction Materials and Technologies (Ancona)* **3** 1419-29

[3] Bocca P, Valente S, Grazzini A and Alberto A 2014 Detachment analysis of dehumidified repair mortars applied to historical masonry walls *Internationa Journal of Architectural Heritage* **8** 336-48

[4] Grazzini A, Lacidogna G, Valente S and Accornero F 2019 Detachment monitoring of repair mortar applied to historical masonry stone by acoustic emission technique *Proc. Int. Conf. on Structural Analysis of Historical Constructions (Cusco)* **18** (Cham: Springer) 2197-205

[5] Bocca P and Grazzini A 2013 Mechanical properties and freeze-thaw durability of strengthening mortars *Journal of Materials in Civil Engineering* **25** 274-80

[6] Formia A, Serra C L, Zerbinatti M and Tulliani J M 2014 The plasters of the Sacro Monte of Varallo Sesia. From the characterization to the proposition of a restorative mix *Case Studies in Construction Materials* **1** 46-52

[7] Fasana S, Zerbinatti M and Grazzini A 2018 Mortars and plasters mixes for ancient buildings and their mechanical characteristics. First results of a research project (in progress) *Proc. In. Conf. on Science and Cultural Heritage (Bressanone)* **2018** 271-81

[8] Freire T, Silva A S, Veiga R and De Brito J 2010 Characterization of Portuguese historical gypsum mortars: a comparison between two case studies *Materials Science Forum* **636-637** 1258-65

[9] Veiga R, Fragata A, Velosa A L, Magalhaes A C and Goreti M 2010 Lime-based mortars: viability for use as substitution renders in historical buildings *International Journal of Architectural Heritage* **4** 177-95

[10] Veiga R, Silva A S, Tavares M, Santos A R and Ampreia N 2013 Characterization of renders and plasters from a 16th Century Portuguese Military Structure: chronology and durability *Restoration of Buildings and Monuments* **19** 223–8

[11] UNI EN 12504-4: 2005 Prove sul calcestruzzo nelle strutture - Parte 4: Determinazione della velocità di propagazione degli impulsi ultrasonici

[12] UNI EN 1015-11:2007 Metodi di prova per malte per opere murarie - Parte 11: Determinazione della resistenza a flessione e a compressione della malta indurita

IOP Conf. Series: Materials Science and Engineering **629** (2019) 012025 doi:10.1088/1757-899X/629/1/012025

An analytical method for J-resistance curve based on a new fracture criteria and finite element aided testing method

D Yao[1,3], L X Cai[2], J C Luo[1], L Sun[1] and P Z Li[1]

[1]Nuclear Power Institution of China, Chengdu, China
[2]Applied Mechanics and Structure Safety Key Laboratory of Sichuan Province, School of Mechanics and Engineering, Southwest Jiaotong University, Chengdu, China

E-mail: di_yaodic@163.com

Abstract. *J*-resistance curve of ductile material is important to describe the ability of a crack structure or material to resist fracture. Experimental and theoretical methods are developed to obtain *J*-resistance curve of metallic material. However, there are rare reports to predict *J*-resistance curve directly by tensile test on ductile materials. In this paper, an analytical method is proposed to predict the *J*-resistance curve for metallic material. Based on a new method, true stress-strain curve up to failure of ductile material is obtained. Thus, fracture threshold in different constrains are calculated using different kinds of specimens. A simple logarithmic-relationship between σ_{1f} and σ^*_f (Yao-Cai criteria) is proposed. The new fracture criterion and full-range constitutive relationship are then used to describe the fracture behaviour near the crack tip. Stress triaxiality and critical fracture first-principle stress around crack tip are obtained. Therefore, fracture procedure is divided into two parts: crack tip blunting and steady crack growth. The relationship of *J*-integral and crack growth amount *R* is established respectively according to the blunting line equation and HRR stress field. The predicted results for test specimens (CT, SENT, CIET) with different initial crack length of the A508-3 steel coincide well with the test results. The results show that stress triaxiality at crack tip is a typical characteristic feature following the constraint effect and the Yao-Cai criteria is recommended for the fracture threshold of ductile material.

1. Introduction

The J-integral proposed by Rice [1] is one of the most widely used fracture parameters. Research in ductile fracture and acquisition of ductile fracture toughness JIC is important to structural integrity assessment issues, forming analysis of cutting, punching problems and damage accident analysis.

The experimental methodfor J-R curve of ductile material has been established over the years. Such as unloading compliance method, electric potential drop method, and load separation theory [2]. However, the test procedure needs relatively long time and a great deal of material is required to obtain the J-R curve to characterize the in-plane effect of crack tip constraint. To solve the problem, some theoretical methods were put forward, such as finite element analysis (FEA) method, J-Q and J-A2 parameter prediction method, cohesive method and micromechanical method.

FEA method for fracture toughness of ductile materials is the most common manner which has been conducted by many researchers. Acharyya. S [3], Benseddiq N [4], Xue Z [5] and Hutchinson [6] calculated the fracture toughness of ductile by GTN constitutive model, but the mesh size varies. There are no standard size for meshing in FEA to calculate the J-⊿a curve at present. The J-Q and J−A2 approaches have been frequently applied to describe the constraint effect of the crack tip [7-10]. By the method above, a set of J-⊿a curve is obtained. However, this type of test needed to determine the fracture toughness. Cohesive and micromechanical methods [11-12] are new approaches developed in recent years. Worth to emphasis that the accuracy of these methods needs verification.

Analytical approach to crack growth resistance was initiated by McClintock [13] first proposed an analytical method to obtain fracture toughness of a mode III crack.w This approach culminated in the work of Rice and coworkers [14] continued McClintock's work for mode I crack. In this mode, solutions for the crack opening profile behind the advancing tip were obtained. The refined numerical calculations of SHAM [15] were used to establish the relation of the opening profile on K and ⊿a. A propagation criterion was imposed on the solution requiring that the near-tip opening profile be invariant once propagation is under way. The outcome of the approach was the prediction of the crack growth resistance curve, K (⊿a). However, the J-⊿a prediction method is rare reported.

In this study, a new method [16-19] is proposed to acquire the true stress-strain curve of the A508-3 steel up to failure using a notched round bar specimen. According to the analysis of failure mechanism of ductile materials, the relationship between fracture threshold and stress triaxiality at the fracture is studied by the tensile tests applying different kinds of specimens. Based on the ftrue stress-strain curve for ductile materials, the Yao-Cai fracture criterion [20] of the A508-3 steel is obtained. By dividing the crack propagation procedure into two parts, the relationship between *J*-integral and blunting propagation value $⊿a_B$, steady propagation value $⊿a_f$ was obtained, respectively. In plane-strain condition, stress distribution in front of crack tip is assumed to be satisfying the HRR stress field. Thus, the quadratic function relationship *J*-integral and $⊿a$ is obtained.

2. Research conditions

2.1. *Material and test system*

The servo-hydraulic universal testing machine 809 MTS, the static/dynamic 632.12C-21 MTS extensometer (25 mm gauge length, 50% measuring range), the centering grips system and the Crack Opening Displacement 632.02F-20 MTS extensometer (5 mm gauge length, 4mm measuring range) were employed for the experiment. Tensile test was conducted under displacement control mode at velocity of 0.02 mm/s. An analysis of the Young's modulus determined in four directions of the material showed that the test error did not exceed 0.5%. It was possible by means of centering grips (the alignment system).

Table 1. Chemical composition of the A508-3 steel.

Material chemical	C /%	Mn /%	Si /%	P /%	S /%	Ni /%	Cr /%	Mo /%
A508-3 steel	0.25	1.12-1.58	0.15-0.37	≤0.018	≤0.005	0.57-1.03	0.1-0.25	0.4-0.6
Material chemical	V /%	Cu /%	Al /%	Ti /%	Co /%	B /%	Nb /%	Ca /%
A508-3 steel	≤0.02	≤0.15	≤0.025	≤0.015	≤0.25	≤0.003	≤0.01	≤0.015

The A508-3 steel employed in reactor pressure vessel was investigated. Tab. 1 shows the chemical composition of the A508-3 steel. Uniaxial true stress-true strain curve before necking of the steel examined was captured in a tensile test on a smooth round bar specimen. This characteristic can be described by Ramberg-Osgood relationship model,

$$\begin{cases} \varepsilon = \varepsilon_e + \varepsilon_p \\ \varepsilon_e = E\sigma \\ \dfrac{\varepsilon}{\varepsilon_0} = \dfrac{\sigma}{\sigma_0} + \alpha\left(\dfrac{\sigma}{\sigma_0}\right)^n \end{cases} \tag{1}$$

where, $\sigma_0=(R_{p0.2}+R_m)/2$ is flow stress, $R_{p0.2}$ is yield stress, R_m represents ultimate tensile strength,

$\varepsilon_0 = \sigma_0/E$ is flow strain, and E is Young's modulus. All mechanical parameters from equation (1) are listed in table 2. Figure 1 shows the tensile curves of the A508-3 steel before necking. The material is assumed to be isotropic, and it satisfies the basic assumption of elastic-plastic mechanical theory.

Table 2. Mechanical parameters of the A508-3 steel.

Materials and specimens	E /MPa	$R_{p0.2}$ /MPa	R_m /MPa	σ_0 /MPa	ε_0 /%	α	n
A508-3 steel-1#	195300	505	634	570	0.292	2.47	8.46
A508-3 steel-2#	195200	485	643	564	0.289	2.68	8.40

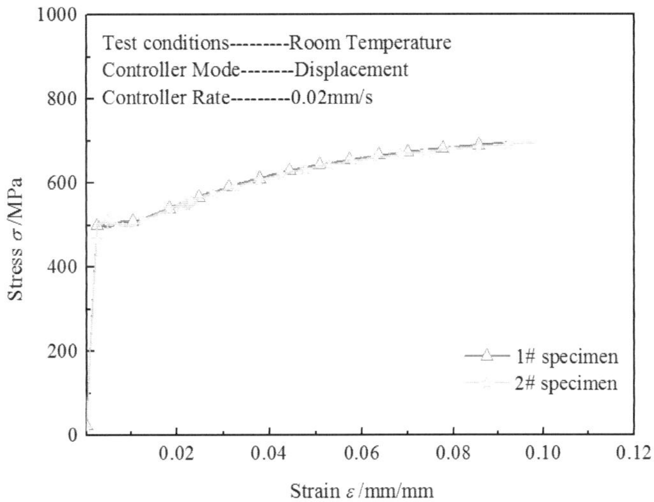

Figure 1. The uniaxial tensile test results for the A508-3 steel.

2.2. *Test specimens*

Fracture criterion for the A508-3 steel was obtained applying the specimens shown in figure 2. The true stress-strain curve was elaborated by notched bar specimen (NRB). Figure 3 shows the specimens for fracture toughness and J-R curve representing different constraint. Compact tension (CT) specimen contained a/W=0.5 and a/W=0.7, while a/W=0.5 expressed the single edge-notched bending (SENB) and small-sized C-shaped inside edge-notched tension (CIET) specimens.

As shown in the figure 3, compared with CT and SENB specimen, CIET specimen is a kind of small sized components. Less material consumption is needed and CIET specimen satisfies loading separation assumption in a wide range of crack size [21]. Thus, loading separation method was used to obtain the fracture toughness of the A508-3 steel by CT, SENB and CIET specimen.

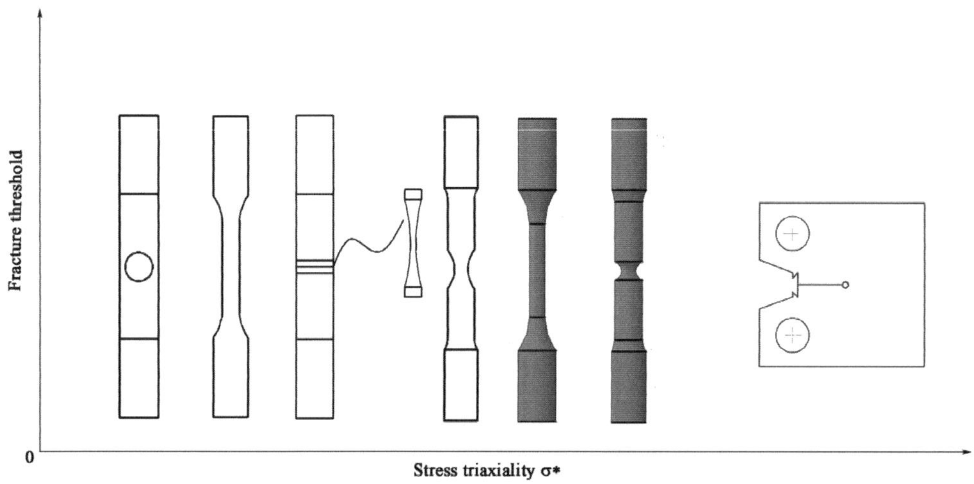

Figure 2. Testing specimens for fracture criterion of the A508-3 steel: a) Circular hole sheet (CHS) specimen, b) Straight sheet (SS) specimen c)Sheet specimen with notch in the direction of thickness (SNT), d) Sheet specimen with notch in the direction of width (SNW), e) Straight round bar (SRB) specimen f) Notched round bar (NRB) specimen, g) Blunt cracked compact tensile.

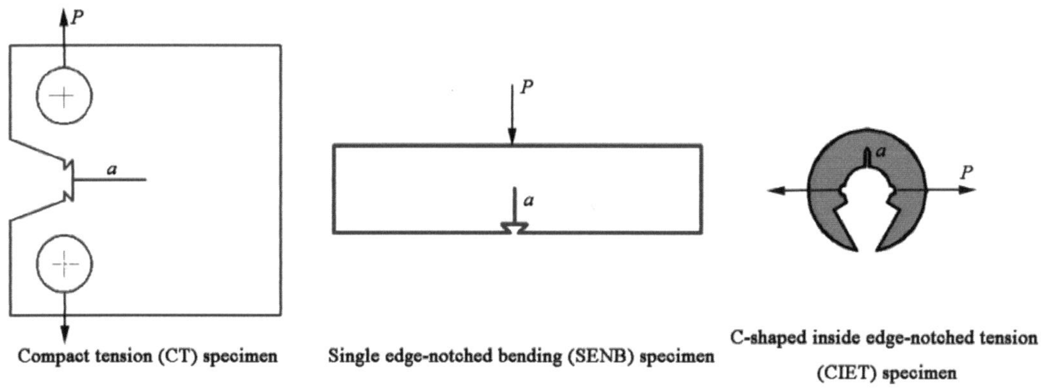

Figure 3. Test specimens for *J*-R curve of the A508-3 steel.

2.3. FAT method fortrue stress-strain curve up to failure of the A5085-3 steel (Please renumber the figures submitted in this section)

Figure 4 shows the basic procedure of FAT method. The true stress-strain curve and its Ramberg-Osgood model are shown in figure 5, where the strain hardening exponent is equal to n=8.40 and the yield offset is equal to α=2.68.

As demonstrated in the figure 4, the true stress-strain curve coincides well with the uniaxial tensile results, but when after necking, the results differ. Thus, the true stress-strain curve was used to simulate the stress field at crack tip of CT specimen, as shown in figure 6.

Compared with the HRR stress field,

$$\sigma_{ij}(r,\theta) = \sigma_0 [\frac{J}{\alpha\sigma_0\varepsilon_0 I_n r}]^{\frac{1}{n+1}} \tilde{\sigma}_{\theta\theta} \bigg|_{n}^{\theta=0^{\circ}} \quad (2)$$

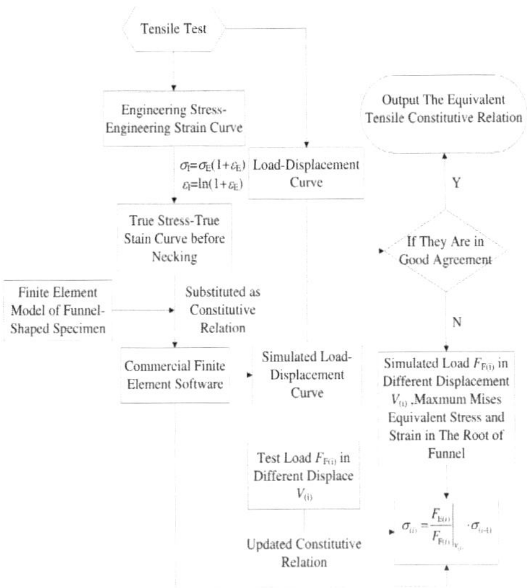

Figure 4. The basic procedure of FAT method.

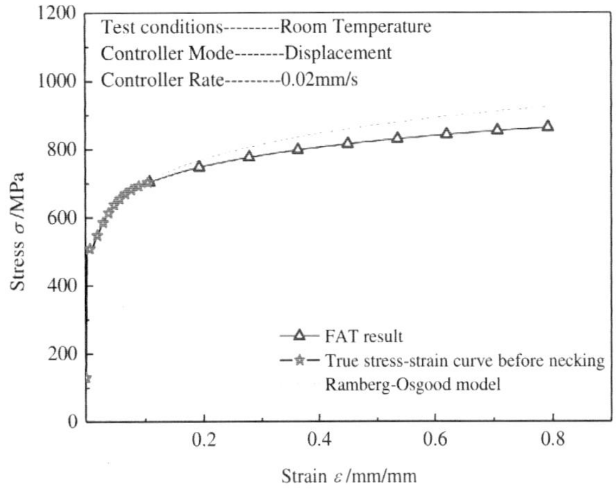

Figure 5. The true stress-strain curve of the A508-3 steel.

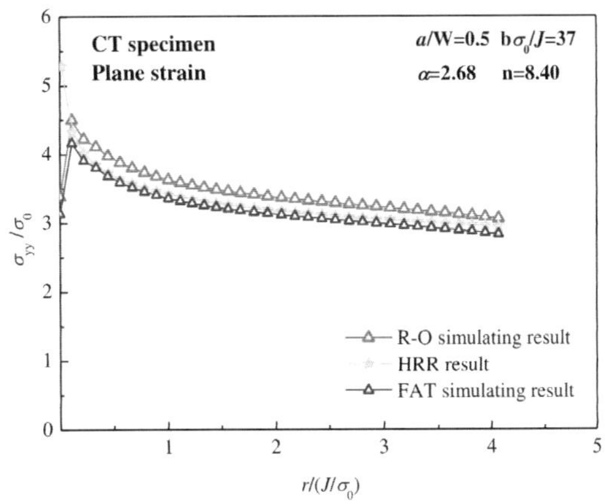

Figure 6. Results of stress field simulation, obtained by the use of R-O model and full-range constitutive relationship.

The full-range constitutive relationship simulating result coincide well, which verifies the validity of FAT method.

2.4. Stress mode fracture criterion for the A5085-3 steel on the Yao-Cai model

A stress mode fracture criterion based on FAT method has been put forward [20] to obtained the stress mode fracture criterion of ductile materials. As shown in figure 7, the critical fracture first-stress in a large range of stress triaxiality is obtained. Yao-Cai criterion (equation 3) describing the relationship between the fracture first-principle stress and stress triaxiality is then established.

$$\begin{cases} \bar{\sigma}_{1f} = \ln(\sigma^*_f) + \bar{S} \\ \bar{S} = \dfrac{\sigma_{1f}\big|_{\sigma^*_f=1}}{\sigma_{1f}\big|_{Uniaxial\ plane\ stress}} \\ \bar{\sigma}_{1f} = \dfrac{\sigma_{1f}}{\sigma_{1f}\big|_{Uniaxial\ plane\ stress}} \end{cases} \quad (3)$$

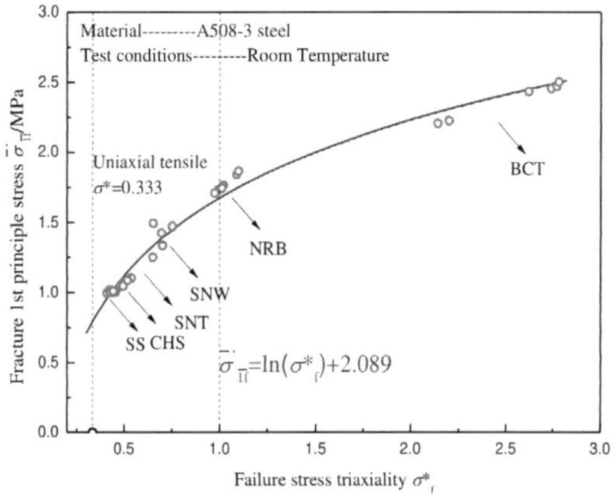

Figure 7. Yao-Cai criterion.

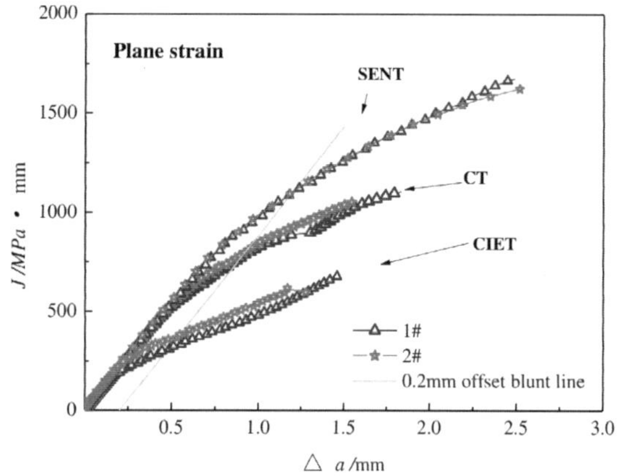

Figure 8. *J*-R curve of three types of specimens.

As shown in figure 6, Yao-Cai criterion for the A508-3 steel has a simple logarithmic law, Where, $\sigma_{1f}\big|_{Uniaxial\ plane\ stress}$ =655MPa, $\sigma_{1f}\big|_{\sigma^*_f=1}$ =1368MPa. Therefore, critical fracture first-stress in any stress triaxiality can be obtained. For example, for CT specimen with a/W=0.5, the stress triaxiality value σ_f* at crack tip equals to 3.0. The critical fracture first-stress σ_f=2097MPa, which is much larger than fracture first-stress in notched round bar specimen.

2.5. Experiments of J-R curve for the A508-3 steel

According to the ASTM E1820-15a, fracture tests were finished. Loading-separation method was used to obtain the fracture toughness and *J-R* curve by a use CT, SENB and CIET specimens. Figure 8 shows the test results of different kinds of specimens.

As we can see in the figure 7, the J-R curve of SENB specimens with a/W=0.5 assume to be the highest fracture toughness. The lowest fracture toughness is obtained from CIET specimens with a/W=0.64. The difference is because the constraint at crack tip of the specimens is various. The *J*-Q and J–A_2 approaches are generally used to characterize crack tip constraint, but in this study, stress triaxiality is applied to describe the constraint effect.

3. Analytical Method for *J*-Resistance curve

The crack propagation procedure is divided into two parts: crack tip blunting and steady crack growth. The relationship of *J*-integral and crack growth amount *R* is established respectively.

3.1. Crack tip blunting

Crack tip blunting is described completely by Landes.et.al [22]. A linear relationship exist between crack stretch zone width Δa_B and *J*-integral,

$$J = \frac{2.5\sigma_0}{d_n}.\Delta a_B = \frac{2.5E}{d_n^*}.\Delta a_B \tag{4}$$

where, d_n* is material related parameter, which can be referred from literature [22],

$$\begin{cases} d_n^* = D_n\left(\sigma_0 / E\right)^{n-1} \\ D_n = 0.787 + 1.554n - 2.45n^2 + 16.952n^3 - 38.206n^4 + 33.13n^5 \\ \sigma_0 = R_{p0.2}10^t \\ t = \left[n/(n+1)\right]\log(E\varepsilon_y / R_{p0.2}) \\ R_{p0.2}/R_m = \left[1/(1+\varepsilon_s)\right]\left[(2.718/n)\ln(1+\varepsilon_s)\right]^n \\ \varepsilon_y = R_{p0.2}/E + 0.002 \end{cases} \tag{5}$$

The equation is then simplified as follows,

$$J = 3.75R_m\Delta a_B \tag{6}$$

When necking occurs,

$$\varepsilon_n = \frac{1}{n} \tag{7}$$

Therefore, the ultimate tensile strength R_m can be obtained,

$$R_m = \sigma_0\left[\frac{1}{\alpha}\left(\frac{1}{n\varepsilon_0} - \frac{\sigma}{\sigma_0}\right)\right]^{\frac{1}{n}} \tag{8}$$

The blunting line equation is improved,

$$J = 3.75\sigma_0\left[\frac{1}{\alpha}\left(\frac{1}{n\varepsilon_0} - \frac{\sigma}{\sigma_0}\right)\right]^{\frac{1}{n}}\Delta a_B \tag{9}$$

3.2. Steady crack growth mechanism

The steady crack growth procedure for mode I crack is shown in figure 9. The stress state at crack tip is equivalent to tensile processing of a series of notched-round bar specimens. Therefore, the step-by-step steady crack growth occurs in front of the crack tip. The propagation amount has gradually increasing trend. So we assume the *J*-integral as the driving energy for crack growth.

CT specimen with *a*/*W*=0.5 was taken as an example to show the prediction method in detail.

Figure 10 shows the evolution law of stress triaxiality when the loading increases. As we can see, the stress triaxiality at a certain region in front of crack tip is almost constant.

We take the distribution law of stress triaxiality when b_0/J=63 as the representative law of the whole loading procedure. The critical fracture stress envelope is then obtained. As shown in figure 11, the Y-stress increases as *J*-integral increasing. Thus, a critical *J*-integral J_0 is existing, under which the maximum Y-stress reaches critical value:

$$\sigma_{yy\max} = \sigma_f = \tilde{\sigma}_{\theta\theta}\Big|_{n}^{\theta = 0^\circ} = \ln(\sigma^*_{yy\max}) + \bar{S} \tag{10}$$

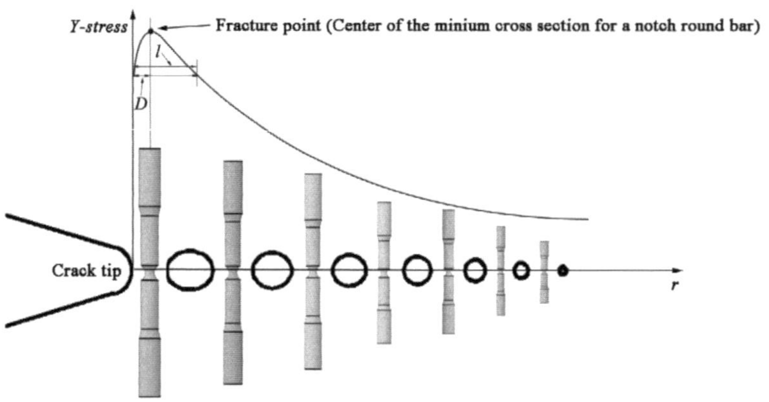

Figure 9. Equivalent step-by-step steady crack growth.

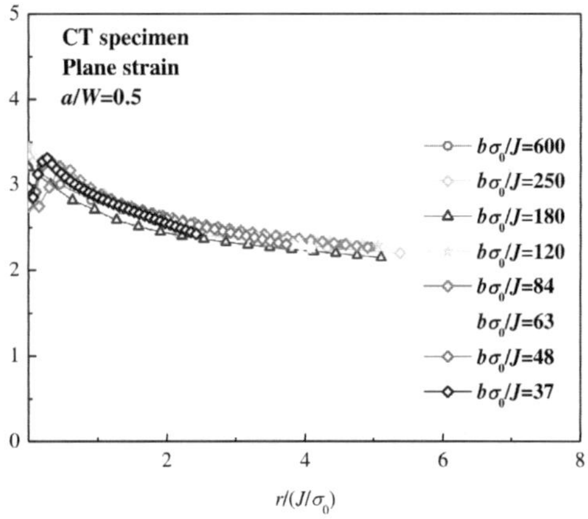

Figure 10. Distribution of stress triaxiality in front of crack tip as loading increase.

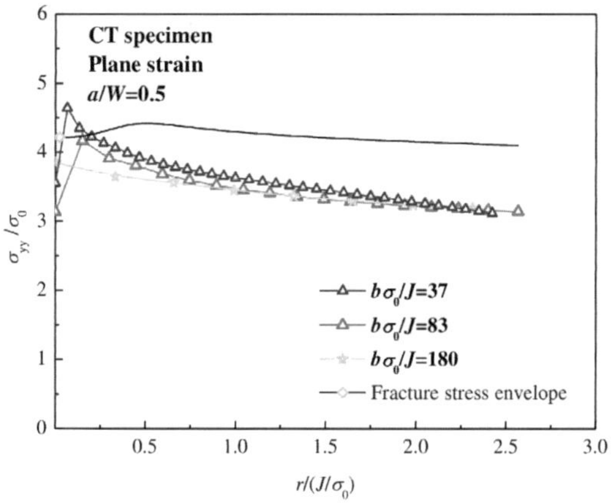

Figure 11. Acquisition process of J_0

As shown in the figure, the critical J_0=222 MPa·mm, the maximum stress triaxiality σ^*_{max}=3.1 and the maximum Y-stress σ_{yymax}= 2121MPa. Therefore, the propagation value D_0 is obtained, according to the HRR field:

$$\sigma_{yy\max f} = \sigma_0 [\frac{J_0}{\alpha\sigma_0\varepsilon_0 I_n D_0}]^{\frac{1}{n+1}} \tilde{\sigma}_{\theta\theta} \Big|_{n}^{\theta = 0^{\circ}} \tag{11}$$

$$D_0 = \frac{J_0}{\alpha\sigma_0\varepsilon_0 I_n} \left(\frac{\sigma_{yy\max}}{\sigma_0 \tilde{\sigma}_{\theta\theta} \Big|_{n}^{\theta = 0^{\circ}}} \right)^{-(n+1)} \tag{12}$$

For CT specimen with a/W=0.5 of A508-3 steel, D_0=0.0577mm. After the first-step propagation, we assume J-integral as the driving energy of crack propagation. The increment value of J-integral remains unchanged every step:

$$J = iJ_0 \tag{13}$$

When the crack propagation reaches the i step, the stress distribution still satisfies the HRR field after the maximum value, although the stress redistribute. According to the HRR field, the crack propagation value D_i is obtained:

$$D_i = \frac{J_i}{\alpha\sigma_0\varepsilon_0 I_n}\left(\frac{\sigma_{yy\max nominal}}{\sigma_0 \tilde{\sigma}_{\theta\theta}\Big|_{n}^{\theta=0^\circ}}\right)^{-(n+1)} \tag{14}$$

where, $\sigma_{yymaxnominal}$ represents the average critical stress in [0~l] region, as shown in figure 8. For CT specimen with a/W=0.5 of A508-3 steel, the average stress triaxiality $\sigma^*_{fnominal}$=2.9, as shown in figure 11. Therefore, the fracture critical Y-stress is obtained as follows:

$$\sigma_{yynominal} = \begin{cases} \varphi(\sigma^*_{fnominal}) & \sigma_{yynominal} < \sigma_{yy\max nominal} \\ \ln(\sigma^*_{fnominal}) + \bar{S} & \sigma_{yynominal} = \sigma_{yy\max nominal} \end{cases} = 2096MPa \tag{15}$$

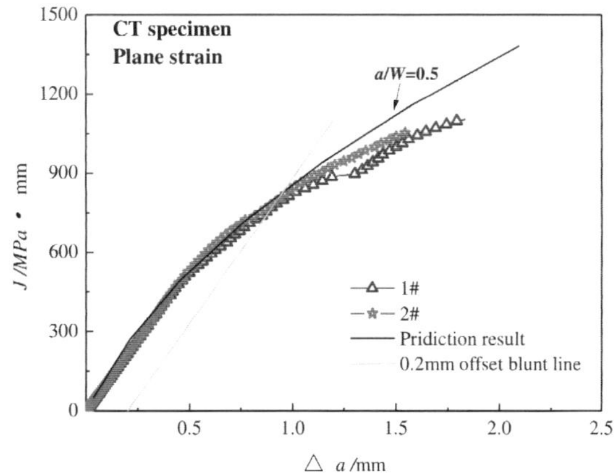

Figure 12. Prediction and experimental J- Δa of CT specimen with a/W=0.5.

The total steady crack propagation value Δa_f is calculated as below,

$$\Delta a_f = \sum_0^i D_i(i) = \sum_0^i \frac{J_i}{\alpha\sigma_0\varepsilon_0 I_n}\left(\frac{\sigma_{yy\max nominal}}{\sigma_0 \tilde{\sigma}_{\theta\theta}\Big|_{n}^{\theta=0^\circ}}\right)^{-(n+1)} \tag{16}$$

3.3. *Theoretical equation for J- ⊿a*

When $J<J_0$, only blunting propagation occurs, but when J reaches J_0, blunting propagation and steady propagation occur at the same time. Therefore, the total crack growth value can be obtained,

$$\Delta a = \begin{cases} \Delta a_B = \dfrac{J_i}{3.75\sigma_0\left[\dfrac{1}{\alpha}\left(\dfrac{1}{n\varepsilon_0} - \dfrac{\sigma}{\sigma_0}\right)\right]^{\frac{1}{n}}} & J \le J_0 \\ \Delta a_f + \Delta a_B = \sum_0^i \dfrac{J_i}{\alpha\sigma_0\varepsilon_0 I_n}\left(\dfrac{\sigma_{yy\max nominal}}{\sigma_0 \tilde{\sigma}_{\theta\theta}\Big|_{n}^{\theta=0^\circ}}\right)^{-(n+1)} + \dfrac{J_i}{3.75\sigma_0\left[\dfrac{1}{\alpha}\left(\dfrac{1}{n\varepsilon_0} - \dfrac{\sigma}{\sigma_0}\right)\right]^{\frac{1}{n}}} & J > J_0 \end{cases} \tag{17}$$

Combined with equation (12), $i=J/J_0$, the relationship between J-integral and Δa is obtained as follows.

$$J=\begin{cases}\begin{cases}3.75\sigma_0\left[\frac{1}{\alpha}\left(\frac{1}{n\varepsilon_0}-\frac{\sigma}{\sigma_0}\right)\right]^{\frac{1}{n}}\Delta a_B & \Delta a\le\dfrac{J_0}{3.75\sigma_0\left[\frac{1}{\alpha}\left(\frac{1}{n\varepsilon_0}-\frac{\sigma}{\sigma_0}\right)\right]^{\frac{1}{n}}}\\ \dfrac{\sqrt{\beta^2+4\alpha\Delta a}-\beta}{2\alpha} & \Delta a>\dfrac{J_0}{3.75\sigma_0\left[\frac{1}{\alpha}\left(\frac{1}{n\varepsilon_0}-\frac{\sigma}{\sigma_0}\right)\right]^{\frac{1}{n}}}\end{cases}\\ \begin{cases}\alpha=\dfrac{A}{2J_0^2}\\ \beta=\dfrac{A+2B}{2J_0}\\ A=D_0\left(\dfrac{\sigma_{yy\max f}}{\sigma_{yy\max\text{nominal}}}\right)^{n+1}\\ B=\dfrac{\alpha\sigma_0\varepsilon_0 I_n D_0\sigma_{yy\max f}^{n+1}3.75\sigma_0\left[\frac{1}{\alpha}\left(\frac{1}{n\varepsilon_0}-\frac{\sigma}{\sigma_0}\right)\right]^{\frac{1}{n}}}{\left(\sigma_0\tilde{\sigma}_{\theta\theta}\Big|_{n}^{\theta=0^\circ}\right)^{n+1}}\end{cases}\end{cases} \tag{18}$$

As we can see in the equation, J-integral and Δa have a simple quadratic function relationship. All the parameters are obtained by tensile tests on smooth round bar and notched round bar specimens. The Yao-Cai criteria is obtained in tensile tests using notched round bar and center circle sheet specimens. J_0 is calculated by FEA simulation based on the full-range constitutive relationship and Yao-Cai criteria. Therefore, an analytical method for J-Resistance curve of ductile materials is established.

4. Prediction results of the analytical method for J-resistance curve

4.1. Prediction results of CT specimen with a/W=0.5

Figure 12 shows the prediction result of CT specimen with a/W=0.5. As shown in the figure, the prediction result coincides well with the experimental results.

4.2. Prediction results of CT specimen with a/W=0.5 and a/W=0.7

Figure 13 shows the stress triaxiality distribution in front of the crack tip of CT specimen with a/W=0.5 and a/W=0.7. Table 3 shows the parameters for J-Δa curve of the two types of specimens. The stress triaxiality value of CT specimen with a/W=0.5 is larger than the specimen with a/W=0.7. Therefore, the constraint for CT specimen with a/W=0.5 is larger than in the case of the specimen with a/W=0.7. The critical Y-stress is also different. Stress triaxiality represents the in-plane constraint effect for fracture toughness. Figure 14 shows the prediction results and experimental results, both coincide well.

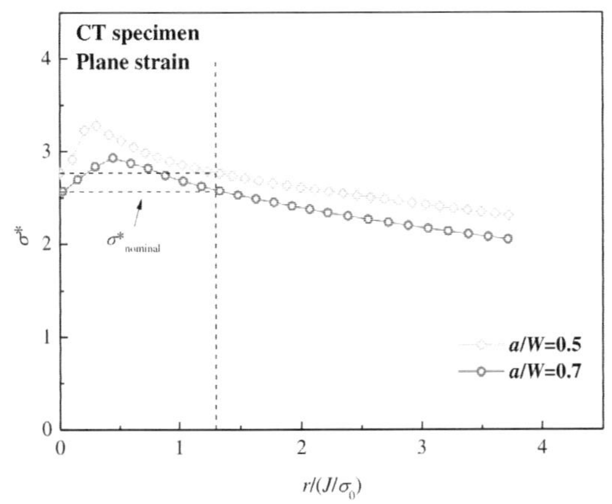

Figure 13. Stress triaxiality distribution in front of crack tip of the two types of specimens.

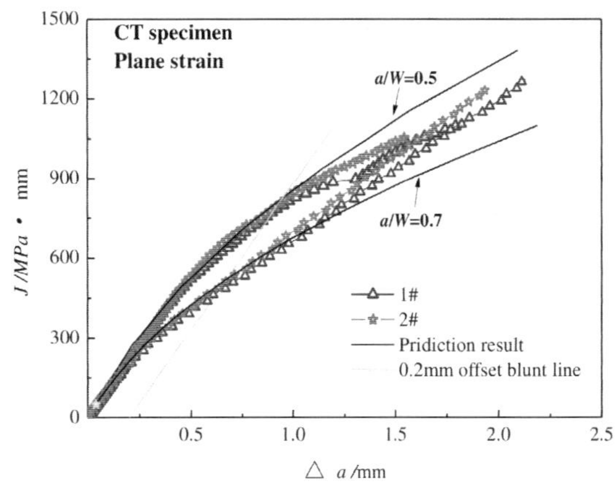

Figure 14. Prediction and experimental J-⊿a of CT specimen with a/W=0.5 and a/W=0.7.

Table 3. parameters for J-⊿a curve of the two types of specimens.

CT specimen	D_0 /mm	$\sigma^*_{nominal}$ /MPa	$\sigma_{yymaxnominal}$ /MPa	J_0 /MPa • mm
a/W=0.5	0.057	2.97	2096	222
a/W=0.7	0.055	2.48	1971	156

4.3. Prediction results of CT, SENB and CIET specimens with a/W=0.5

The prediction J-⊿a curve for three types of specimens are shown in figure 15. As we can see, the analytical method can preferable predict the J-⊿a curve of different constraint types of specimens.

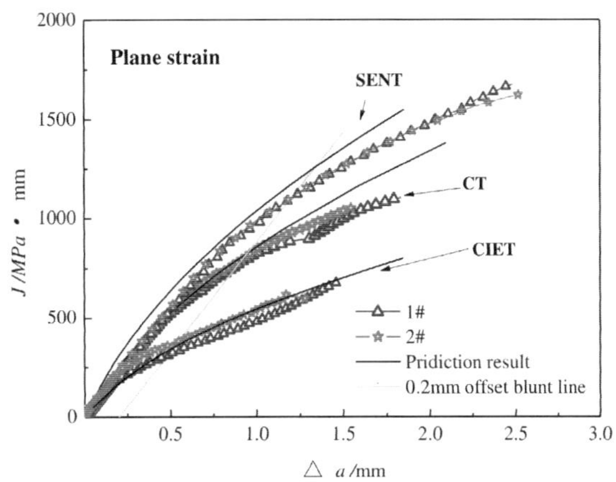

Figure 15. Prediction and experimental J- ⊿a of CT, SENT and CIET specimen with a/W=0.5.

5. Discussions and conclusion

Based on the true stress-strain curve of ductile materials, the Yao-Cai fracture criterion for the A508-3 steel is obtained. By dividing the crack propagation procedure into two parts, the relationship between J-integral and blunting propagation value ⊿a_B, steady propagation value ⊿a_f was obtained, respectively. In plane-strain condition, the stress distribution satisfies the HRR stress field. Thus, the quadratic function relationship J-integral and ⊿a is obtained. The analytical method can be concluded as follows,

- Finish the tensile experiment on various types of specimens;
- Calculate the full-range constitutive relationship basing on tensile test on notched-round specimen;
- Calculate Yao-Cai criteria employing data from tensile experiment on notched-round and flat with a hole specimens;
- Establish the basic parameters for FEA from the full-range constitutive relationship;
- Predict the J-Δa curve of ductile material by equation (17), where J_0 is obtained by FEA.

Therefore, by finish a small number of tensile tests on ductile material, the J-Δa curve for different constraint can be obtained. This can reduce material consumption.

Acknowledgments
This study was co-supported by the National Natural Science Foundation of China (No. 11472228) and the Sichuan Youth Science and Technology Innovation Team Projects (No. 2013TD0004).

References

[1] Rice J R 1968 A path independent integral and the approximate analysis of strain concentration by notches and cracks *ASME J. Appl. Mech.* **35** 379-86
[2] Begley J A and Landes J D 1972 The J-integral as a fracture criterion *Fracture Mechanics ASTM STP* **514** 1-23
[3] Xue Z, Pontin M G, Zok F W and Hutchinson J W 2010 *Engineering Fracture Mechanics* **77** 492-509
[4] Benseddiq N and Imad A 2008 *International Journal of Pressure Vessels & Piping* **85** 219-27
[5] Acharyya S and Dhar S 2008 *Journal of Materials Science* **43** 1897-909
[6] Tekoğlu C, Hutchinson J W and Pardoen T 2015 On localization and void coalescence as a precursor to ductile fracture *Philosophical Transactions of the Royal Society of London A: Mathematical, Physical and Engineering Sciences* **373** 20140121
[7] Cravero S, Bravo R and Ernst H 2008 Constraint evaluation and effects on JR resistance curves for pipes under combined load conditions *The Eighteenth International Offshore and Polar Engineering Conference* (Vancouver, Canada)
[8] Ding P and Wang X 2012 An estimation method for the determination of the second elastic–plastic fracture mechanics parameters *Engineering Fracture Mechanics* **79** 295-311
[9] Wang Z X and Liu R 2012 Analysis and simulation of elastic-plastic fracture parameters at crack tip under biaxial loading and uniaxial loading *Applied Mechanics and Materials. Trans Tech Publications* **166** 2883-6
[10] Zhu X-K, Kim Y, Chao Y-J and Lam P 2003 Constraint effect on 3D crack-front stress fields in elastic-plastic thin plates *ASTM Special Technical Publication* **1417** 270-87
[11] Woelke P B, Shields M D and Hutchinson J W 2015 Cohesive zone modeling and calibration for mode I tearing of large ductile plates *Engineering Fracture Mechanics* **147** 293-305
[12] Liao F, Wang W and Chen Y 2015 Ductile fracture prediction for welded steel connections under monotonic loading based on micromechanical fracture criteria *Engineering Structures* **94** 16-28
[13] McClintock F A, Kaplan S M and Berg C A 1966 Ductile fracture by hole growth in shear bands *International Journal of Fracture* **2** 614-27
[14] Rice J R and Sorensen E P 1978 Continuing crack-tip deformation and fracture for plane-strain crack growth in elastic-plastic solids *Journal of the Mechanics and Physics of Solids* **26** 163-86
[15] Tvergaard V and Hutchinson J W 1992 The relation between crack growth resistance and fracture process parameters in elastic-plastic solids *Journal of the Mechanics and Physics of Solids* **40** 1377-97
[16] Cai L X, Yao D and Bao C 2011 Testing method to obtain the full-range true strain-stress curve *The Chinese Congress of Theoretical and Applied Mechanics (Zhengzhou, China)* (in Chinese)

[17] Yao D and Cai L X 2012 An approach for full-range uniaxial constitutive relationship of 304 stainless steel *Journal of Sichuan University (Engineering Science Edition)* **2012(S2)** 163-6
[18] Yao D, Cai L X and Bao C 2014 Approach for full-range uniaxial constitutive relationships of ductile materials *China Measurement & Test* **40** 5-13 (in Chinese)
[19] Yao D, Cai L X, Bao C and Ji C B 2014 Determination of stress-strain curve of ductile materials by testing and finite element coupling method *Chinese Journal of Solid Mechanics* **35** 226-40 (In Chinese).
[20] Yao D, Cai L and Bao C 2016 A new fracture criterion for ductile materials based on a finite element aided testing method *Materials Science and Engineering: A* **673** 633-47
[21] Bao C, Cai L X, He G W *et al* 2016 Normalization method for evaluating J-resistance curves of small-sized CIET specimen and crack front constraints *International Journal of Solids and Structures* **94** 60-75
[22] Landes J D 1995 The blunting line in elastic - plastic fracture *Fatigue & Fracture of Engineering Materials & Structures* **18** 1289-97

An active actuator based on giant magnetostrictive composite pendulum for vibration isolation

J M Chen[1,2], A Jin[1,3], L Wang[1,2,4] and Y X Tang[1,2]

[1]Center of Ultra-precision Optoelectronic Instrument Engineering, Harbin Institute of Technology, Harbin 150080, China
[2]Key Lab of Ultra-precision Intelligent Instrumentation (Harbin Institute of Technology), Ministry of Industry and Information Technology, Harbin 150080, China
[3]Key Laboratory of Micro-systems and Micro-structures Manufacturing, Ministry of Education, Harbin Institute of Technology, Harbin 150080, China

E-mail: hit_wanglei@hit.edu.cn

Abstract. This paper proposes an active actuator based on giant magnetostrictive composite pendulum for vibration isolation, laying the foundation for vertical micro-vibration isolation. The performance of a single-stage pendulum vibration isolation system can be improved by increasing the pendulum length, along with the influence of the environment on the system increases. In precise experiments, effective isolation of vertical micro-vibration improves the accuracy of the experiments. Interference caused by micro-vibration can be attenuated by designed magnetostrictive actuators. Then, a suspension actuator consists of a composite pendulum and a driving device is designed. The composite pendulum is composed of a quartz fiber coated with a giant magnetostrictive material (GMM). The GMM can achieve the desired small strain in a designed magnetic field. GMM is applied on a quartz fiber to form a composite pendulum. Therefore, low-frequency active vibration isolation of single pendulum is realized. Compared with the passive vertical vibration isolation system, the system gets better vibration isolation and anti-interference ability. Compared with the traditional active vertical vibration isolation system, it has the advantages of system simplification, high integration and high sensitivity.

1. Introduction

There is a large demand for precise instruments with high measurement sensitivity, repeatability and measurement uncertainty in ultra-precision measuring, such as scanning probe microscopy (SPM), atomic force microscopy (AFM), holographic interferometer and confocal microscope. However, even the small vibration of amplitude or frequency may interfere the measurement result. That means, as the increasing accuracy of the instruments, the vibration of the external has an increasing influence on the accuracy of the instrument. The vibration isolation technology has become a bottleneck that limits the improvement of instrument accuracy, especially in suppressing the vibration interference signal of ultra-low frequency. Several systems, such as gravitational wave detection, dark matter detection, strong magnetic fields, and pulsar formation studies, all require a low-vibration environment. Compared to the external vibration, the measured signal is particularly weak, making it important to isolate the external vibration, especially the impact of ground vibration. In the absence of an ultra-low frequency vibration isolation system, the desired signals received by the gravitational wave detector may be annihilated in the noise. Ultra-low frequency (0.1 Hz or less) micro-vibration isolation technology has become a hotspot in the field.

Low-frequency vibration isolation system is mainly divided into two types, passive vibration isolation system [1,2] and active vibration isolation system [3-6]. The passive vibration isolation system based on single pendulum is a mature one [1,2,7-11]. Increasing the swing length can improve the vibration isolation performance. The suspension of single pendulum has a certain weight. Therefore, a string mode resonance will occur if it is subjected to lateral disturbance under a certain tension. String mode resonance limits the length of the pendulum. In order to make the pendulum swing at a low frequency and with better vibration isolation performance, Kim K S *et al* combined the single pendulum and the spring to form an inverted pendulum [12], which is symmetrically distributed on both sides of the square vibration isolation platform. The inverted pendulum has a simple structure and reliable operation. VIRGO, a large interferometer designed to detect gravitational waves, proposed a 9-stage composite pendulum vibration isolation system SA (Super-Attenuator) [13,14]. It combines multi-stage single pendulum series and active vibration isolation, leading the improvement of the low frequency vibration isolation performance. However, the method has three defects. Firstly, the actuator and sensor are fixed on the first and second stage pendulums, making it difficult to isolate the vibration on the test mass. Secondly, the structure of the system is complex because of the installation of multiple sensors and actuators, making it heavily depend on initial installation. Thirdly, fixing the sensor on the test mass introduces low-frequency noise, such as thermal noise. Therefore, an active actuator based on giant magnetostrictive composite pendulum is proposed for vibration isolation.

2. Design of the active actuator

The suspension actuator, which utilizes the hysteresis effect of the GMM, is composed of a drive system and a composite pendulum. The drive system generates the desired magnetic field for driving the composite pendulum, including the variable magnetic field and the bias magnetic field. The variable magnetic field is generated by energizing the drive coil, and the bias magnetic field is provided by the permanent magnet. The composite pendulum is formed by coating the surface of a quartz fiber with GMM.

2.1. Design of the overall structure

The suspended active actuator consists of a composite pendulum and a drive unit. The composite pendulum comprises a quartz fiber and a coating of GMM coated on the quartz fiber. The composite pendulum senses external vibration and transmits vibration signal to the actuator. The driving device consists of an upper guiding magnet, a permanent magnet, a driving coil, a skeleton and a lower guiding magnet. The skeleton is a structure with a through hole in the middle, and its cross section is "H" shaped. The upper guiding magnet is fixed on the top of the skeleton, the lower one is fixed at the bottom, and the coil is nested round of the skeleton. The driving current of the driving coil is adjusted according to the vibration signal obtained by the system, in order to provide a driving magnetic field for the composite pendulum. The permanent magnet is nested round the upper guiding magnet, the driving coil, the skeleton and the lower guiding magnet, for providing a bias magnetic field for the composite pendulum. The composite pendulum passes through the center of the driver. The test mass is suspended by the vibration isolating element through the quartz fiber below the outer support structure.

Based on the designed structure, the suspension active actuator, wherein the composite pendulum senses external vibration, transmits the vibration signal to the actuator. The actuator actively controls the driving current of the driving coil according to the collected vibration signal to change the driving magnetic field of the composite pendulum. The magnetic field, with the help of the giant magnetostrictive effect of the GMM, makes the composite pendulum to produce a strain opposite to the direction of the external vibration, thereby reducing the interference of external vibrations on the system. The schematic of the active actuator is shown in figure 1.

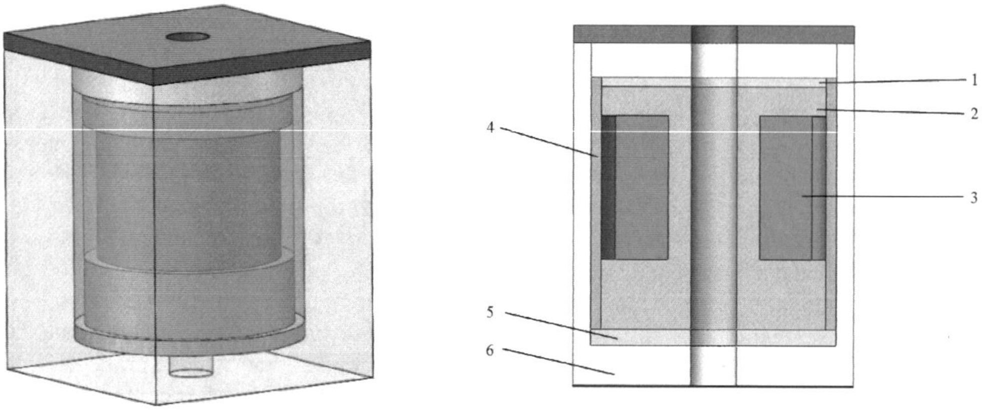

Figure 1. Schematic of the active actuator, in which, 1) and 5) are magnetizers, 2) is coil frame, 3) is coil, 4) is Permanent Magnet and 6) is shell.

2.2. Design of the composite pendulum

The composite pendulum is a quartz fiber coated with giant magnetostrictive powder on the surface. In this paper, the quartz fiber is single mode fiber. Firstly, the GMM and the glue are mixed at a mass ratio of 5:1 and stirred uniformly. Secondly, the fiber optic is completely immersed in the mixed solution, then slowly taken out after being sufficiently contacted. The process of taking out the fiber optic should be steady to ensure uniform adhesion of the mixed solution. Thirdly, the removed quartz fiber is placed on the mould to heat for drying, at 150°C for 3 minutes. After that, a composite pendulum is made. The schematic of the composite pendulum is shown in the figure 2. As shown, its size matches the size of the drive system to ensure that the amount of elongation is as large as possible.

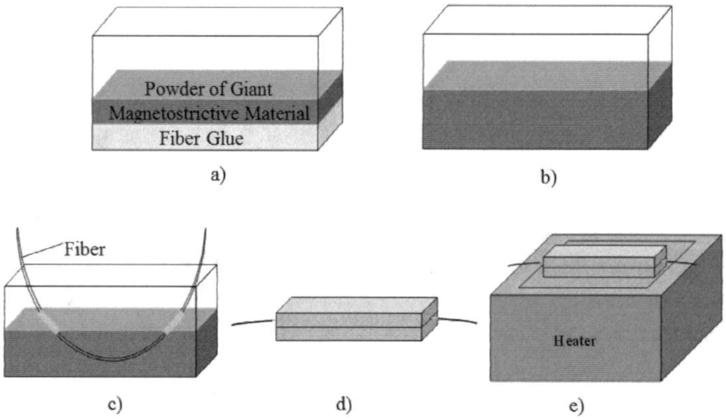

Figure 2. Schematic of the process of making a composite , in which, a) the ratio of weight between powder of GMM and fiber glue is 5:1; b) stir the powder of GMM and fiber glue uniformly; c) immerse the quartz fiber; d) take out the quartz fiber steady and place it in the mould; e) Thermoset the composite pendulum at 150°C.

In figure 3(a), the diameter of polyimide-layer, cladding, and core is 250 μm, 125 μm and 9 μm respectively. To improve the sensitivity of composite pendulum, polyimide-layer of area to be coated on the fiber is peeled off. So, the GMM is coated on the cladding of fiber optic directly and the schematic of composite pendulum is shown in figure 3(b). The coated length is 40 mm, the size of the drive system, and the image of composite pendulum is shown in figure 3(c). The diameter of composite pendulum is 203.15 μm. The thickness of magnetostrictive film is about 39 μm.

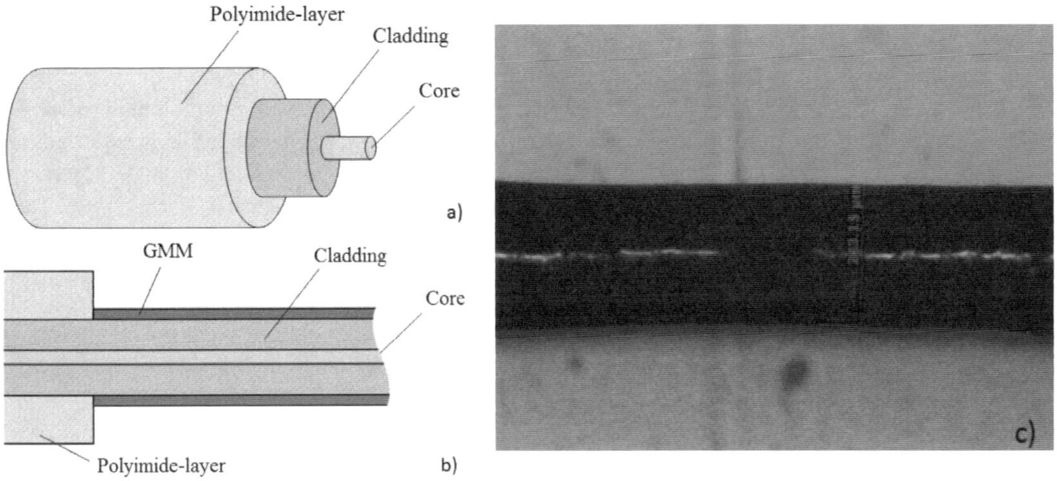

Figure 3. (a) is the schematic of fiber optic, (b) is the schematic of composite pendulum and (c) is the image of composite pendulum.

3. Experiments and results

The magnetic field is co-generated by the main coil in the giant magnetostrictive actuator (GMA), the Helmholtz compensation coil and the permanent magnet. The GMA is placed vertically. One end of the quartz fiber coated with the GMM is fixed on the bracket which is on the top of GMA. A certain weight is suspended at the other end. The composite pendulum is located on the central axis of the coil. The intensity of the central magnetic field is controlled by the magnitude of the input direct current. The coating layer of the GMM varies in a varying magnetic field, leading the change in the length of the cycloid. The schematic of the system is shown in figure 4.

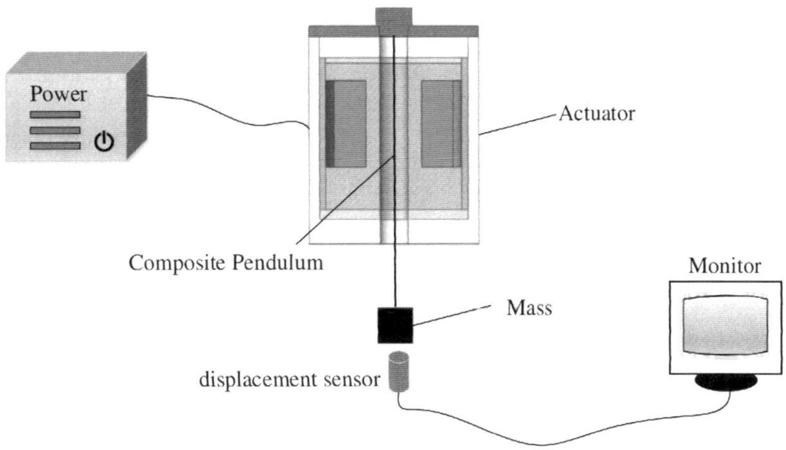

Figure 4. The schematic of the overall system.

The displacement sensor employed for the experiment is eddyNCDT 3010 displacement measuring system. The eddyNCDT 3010 operates without contact using eddy current technology. It can be used for making measurements on targets made of either ferromagnetic or non-ferromagnetic electrically conductive materials without need to contact. 1) In order to ensure the measurement sensitivity of the sensor, if the thickness of the test mass is greater than 0.6 mm, the diameter of the measured surface should be greater than 1.5 times the diameter of the probe; 2) Irregular surface of the measured object will cause additional error to the actual measured value, especially for vibration measurement. The additional error signal is superimposed with the actual vibration signal, which is electrically difficult to separate. So, the measured surface should be smooth, there should be no defects such as nicks, holes, bosses, grooves, etc. 3) the characteristics of the sensor are related to the conductivity and permeability of the measured object. If the measured object is a magnetic conductive material, magnetic effect and eddy current effect exist simultaneously. The magnetic effect and eddy current effect on the contrary, it will cancel part of the eddy current effect, resulting in the low sensitivity of sensor. If the object is non-magnetic or weak magnetically conductive material (such as aluminum, material of the test mass), the eddy current effect is relatively strong due to weak magnetic effect. So, the sensor sensitivity is higher.

The relationship between the pendulum strain and the driving current was measured. The driving current increases at the step of 0.4 A. The result is shown in figure 5.

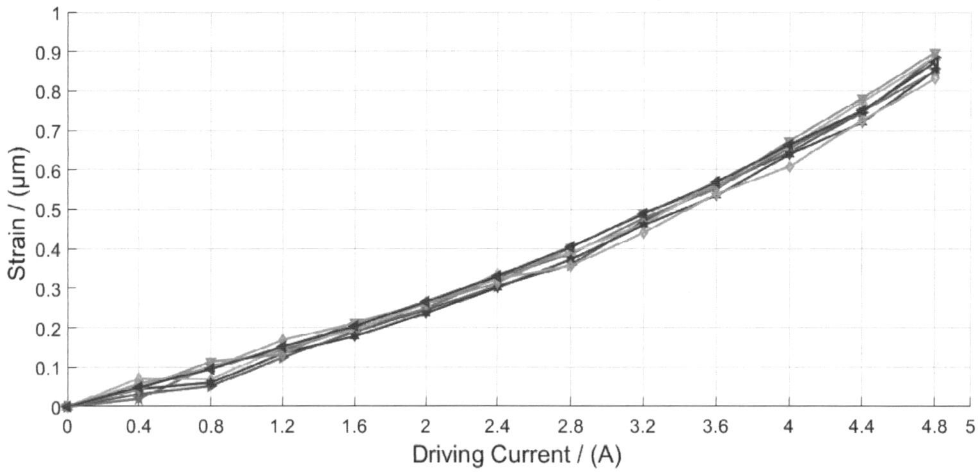

Figure 5. The measured data of the relationship between pendulum strain and driving current.

The strain of the composite pendulum in the driving magnetic field has a good linear relationship with the driving current. The repeatability of the pendulum strain calculated is ±2.38%.

4. Conclusions

In the composite pendulum, the GMM is coated on the surface of the quartz fiber. Therefore, the giant magnetostrictive effect of the GMM, the effect of the dimensional change under the action of the magnetic field, makes the composite pendulum a unity of the sensor and the compensation for vibration. When a vibration interference occurs, the current of the driving coil changes, leading the change of the magnetic field of the coating of the GMM. As a result, the coating of the GMM produces an external strain with the opposite vibration direction to compensate the external vibration, thereby reducing the interference of external vibration to the system and achieving the purpose of active vibration isolation.

The composite pendulum, which combines the perceptual vibration and the compensating vibration, only includes the quartz fiber and the giant magnetostrictive coating, thus realizing system simplification, self-sensing and active actuation.

The sensor and the actuator are not directly mounted on the test mass, making it possible to solve the problem of introducing low frequency noise in the conventional structure, in which the sensor and the actuator are mounted in the test mass. By reducing the noise source, the actuator gets the better vibration isolation of the low frequency vibration.

Acknowledgments
This study was financially supported by National Natural Science Foundation of China (51675137; 51375121) and National key R&D Program of China (2016YFF0200700).

References

[1] Shoemaker D *et al* 1988 Noise behavior of the garching 30-meter prototype gravitational wave detector *Phy Rev D* **38** 423-38
[2] Bradasehia C *et al* 1989 First results on the electronic cooling of the Pisa seismic noise super-attenuator for gravitational wave detection *Phys Lett A* **137** 329-33
[3] Wen S, Mittleman R, Mason K *et al* 2014 Hydraulic external pre-isolator system for LIGO *Class Quantum Gravity* **31** 235001
[4] Blom M R, Beker M G, Bertolini A *et al* 2015 Vertical and horizontal seismic isolation performance of the advanced virgo external injection bench seismic attenuation system *Phys Proc* **61** 641-7
[5] Matichard F, Lantz B, Mittleman R *et al* 2015 Seismic isolation of Advanced LIGO: Review of strategy, instrumentation and performance *Class Quantum Gravity* **32** 185003
[6] Jeon B G, Chang S J, Kim S W *et al* 2015 Base isolation performance of a cone-type friction pendulum bearing system *Struct Eng Mech* **53** 227-48
[7] Bertolini A, DeSalvo R, Galli C *et al* 2006 Design and prototype tests of a seismic attenuation system for the advanced-LIGO output mode cleaner *Class Quantum Gravity* **23** S111
[8] Saulson P R 1994 *Fundamentals of Interferometric Gravitational Wave Detectors* (World Scientific, Singapore)
[9] Fabbro R *et al* 1987 Three-dimensional seismic super-attenuator for low frequency gravitational wave detection *Phys Lett A* **124** 253-7
[10] Fabbro R Del *et al* 1988 Performance of a gas spring harmonic oscillator *Rev Sci Instrumy* **59** 292-7
[11] Kim K S and Rydberg A 2009 Tunable inverse pendulum vibration isolation system *U.S. Patent 7,543,791*
[12] Bradaschia C, Del Fabbro R, Di Fiore L *et al* 1989 First results on the electronic cooling of the Pisa seismic noise super-attenuator for gravitational wave detection *Phys Lett A* **137** 329-33
[13] Del Fabbro R, Di Virgilio A, Giazotto A *et al* 1988 First results from the Pisa seismic noise super-attenuator for low frequency gravitational wave detection *Phys Lett A* **132** 237-40
[14] Aasi J, Abbott B P, Abbott R *et al* 2015 Advanced ligo *Class Quantum Gravity* **32** 074001

Optimization for cantilever piezoelectric energy harvester with a cavity

L H Chen[1,2], J T Xue[1,2], L Q Chang[1,2] and F H Yang[3,4]

[1]College of Mechanical Engineering, Beijing University of Technology, Beijing 100124, China
[2]Key Laboratory of Nonlinear Vibration and Strength of Mechanical Structures, Beijing 100124, China
[3]School of Statistics and Mathematics, Central University of Finance Economics, Beijing 100081, China

E-mail: yangfh_cufe@163.com

Abstract. The structure optimization of piezoelectric energy harvester aims at frequency reduction, broadening bandwidth and increasing output mainly. This paper employs a cantilever with cavity modelled by the finite element software COMSOL Multiphysics and the effect of the cavity on transducer is studied. The first order characteristic frequency is obtained by modal analysis module and the frequency response curve is solved by frequency domain analysis module. Comparison results show that the introducing of cavity improves output voltage of energy harvester and reducing device nature frequency. Moreover, effect of the cavity size parameters such as the thickness, length and position on frequency, output and conversion rate of the harvester is studied, after that the energy harvester structure is optimized.

1. Introduction

With the continuous development of wireless sensor, the requirements for energy supply are increasing. Vibration energy harvester can absorb ambient vibration and convert into electric energy which has become an aroused general interest in recent years as its long-life, low-cost and environment friendly compared with traditional chemical batteries [1-5]. PEH (piezoelectric energy harvester) has high conversion rate and good compatibility with MEMS technology, making them more suitable for practical engineering applications [6,7]. The cantilever structure has a better characteristic of energy capture as it can generate large strain in piezoelectric layer when it vibrates [8,9], and most of current energy harvesters adopt cantilever structure.

The main goals of PEH optimization are frequency reduction, broadening bandwidth and increasing output. The voltage across the resistance reflects the amount of electrical energy converted per unit time and performance of the converter depends on output voltage. Introducing a cavity in the cantilever structure not only reduce the system nature frequency but also improve the output voltage, which is a simple and effective optimization method. Wang *et al* [10] designed a PZT layer-air-PZT layer composite cantilever structure, and found that the output voltage is significantly improved. Reddy *et al* [11] designed a PEH with a rectangular cavity and showed that an improved output voltage can be obtained compared to the beam without a cavity. Raju *et al* [12] developed and demonstrated a PEH with multiple rectangle cavities to harvest high power. It was found that the device with two cavities captured the highest voltage compared with the others. Ramalingam *et al* [13] designed a variable-section beam with a trapezoidal cavity and then added a low-frequency beam as an optimization, reducing the nature frequency and widening the working band.

Raju *et al* [14] introduced cavities in two different variable width cantilever PEH and compared their conversion efficiency.

Most of the existing researches studied the influence of the presence or absence of cavity on device output from perspective of theoretical, there is few researches are about effect of cavity size parameters on output of the PEH. In this paper, a rectangle cavity is constructed in the cantilever beam modelled by finite element software COMOL Multiphysics. On the premise of ensuring the stability of the structure, the influence of cavity size parameters on output voltage, power and conversion rate is studied. This work finds that not all cases with cavity play a role in improving PEH performance and the size of cavity has a great impact on the results, and the optimal parameters are finally given.

2. Modeling

The structure of the PEH is shown in figure 1, which consists of base beam, piezoelectric layer and circuit part. The piezoelectric layer is attached to upper surface of the cantilever beam and a cavity is introduced in the base beam, where l_b, l_p, l_c are length of beam, piezoelectric layer and cavity respectively; w is width of structure; h_b, h_p, h_c are thickness of beam, piezoelectric layer and cavity respectively; t is the distance between the upper surface of the cavity and base beam; R is the resistance of the external circuit; V is the voltage across the resistor.

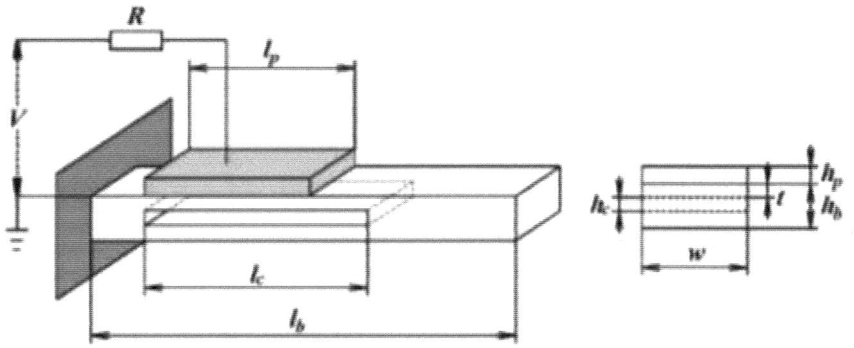

Figure 1. Schematic for cantilever-type PEH with cavity.

Table 1. Dimension and properties of device.

Description	Symbol	Value	Units
Length	l_b	150	mm
	l_p	75	mm
Width	w	20	mm
Thickness	h_b	6	mm
	h_p	0.5	mm
Young's modulus	E_b	70	Gpa
	E_p	56	Gpa
Density	ρ_b	2700	$\text{Kg} \cdot \text{m}^{-3}$
	ρ_p	7500	$\text{Kg} \cdot \text{m}^{-3}$
Piezoelectric stress constant	e_{31}	-6.62	$\text{C} \cdot \text{m}^{-2}$
Dielectric constant	ε_{33}^S	1470	
Load	F_A	100	$\text{N} \cdot \text{m}^{-2}$
Resistance	R	5e4	Ω
Damping coefficient	η_b	0.015	
	η_p	0.025	

IOP Conf. Series: Materials Science and Engineering **629** (2019) 012027 doi:10.1088/1757-899X/629/1/012027

The PEH is modelled by the finite element software COMSOL Multiphysics as shown in figure 2. The material of the base beam is selected as aluminium and the material of the piezoelectric layer is selected as PZT-5H. The parameters and geometric dimensions are shown in table 1. The cantilever beam structure is modelled by two-dimensional solid module and tetrahedral element is used for discretization. The complete grid contains about 5000 domain cells and 832 boundary elements and the freedom degree of solution is 10366. Considering the piezoelectric effect, the force field is coupled with the electric field and the finite element model is established, modal analysis and domain analysis are carried out then. In order to study the effect of the cavity size parameters, parametric scanning for h_c, t and l_c is performed.

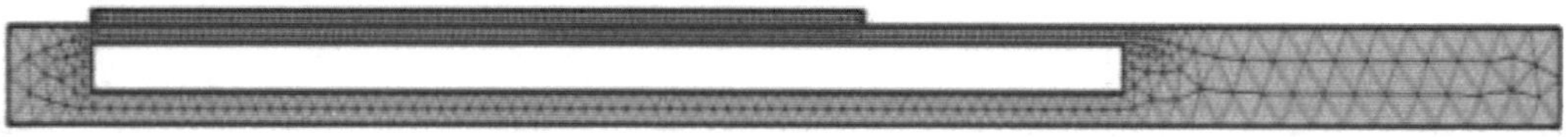

Figure 2. Finite element grid of PEH.

3. Result and analysis

3.1. Effect of cavity thickness

The fundamental frequency, output voltage and power of the harvester without cavity are 254.2 Hz, 18.5 V and 3.4 mW under resonance. In order to investigate the influence of the cavity, different thicknesses of cavity are selected and set $t = \frac{h_b - h_c}{2}$. The cavity middle plane is in coincident with base layer middle plane. Perform modal analysis firstly, the nature frequency is computed, then apply uniform load to base beam and conduct frequency domain analysis, the voltage-frequency curve for different h_c and nature frequency/maximum output voltage-h_c curve are obtained.

It can be seen from figures 3 and 4 that the thickness of the cavity has a significant influence on the characteristic frequency and voltage: the voltage is positively correlated with the thickness of the cavity, while the natural frequency is negatively. Compared with the case without cavity, the introducing of the cavity not only reduces the characteristic frequency of the device, but also increases the output voltage during resonance, especially when h_c is 4 mm, the frequency is reduced by 58%, and the voltage is increased by 35%, which verifies the necessity of cavity introducing.

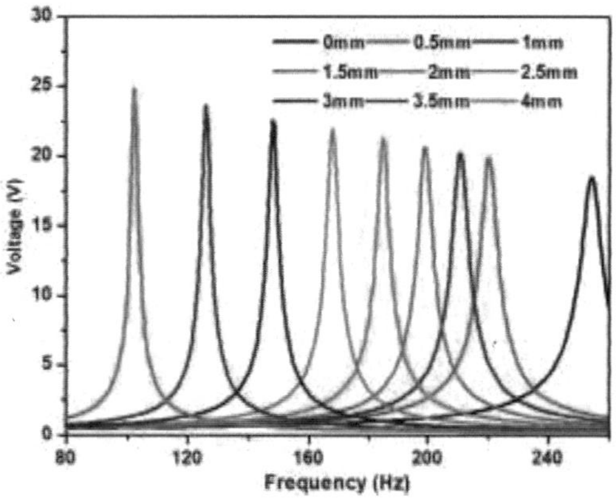

Figure 3. Voltage-frequency response curve at different h_c.

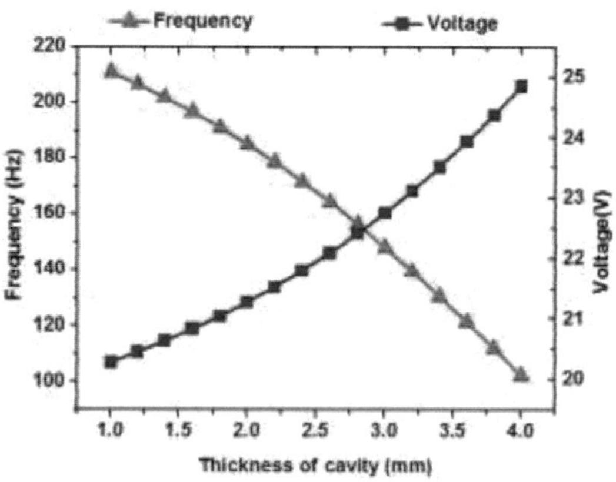

Figure 4. Effect of thickness of cavity h_c.

3.2. Effect of cavity position

The parameter t is the distance between the upper surface of base beam and cavity. Once t has changed, the neutral axis is obviously shifted. The expression of the strain is: $S_{xx} = z \cdot \frac{\partial^2 w}{\partial x}$, where z is the distance of point to neutral axis. Change in the position of neutral axis leads to variation of z, while the strain in piezoelectric layer also changes which has an effect on the output voltage according to the force-electric coupling equation.

The thickness h_c of 1 mm, 2 mm, 3 mm, 4 mm are selected to study the effect of the t comparatively. From figure 5, with the change of the position of the cavity, it can be seen that the natural frequency decreases first and then increases. The frequency is the lowest when the cavity is at the coincident position where the central axis of cavity and beam coincide, and the bigger h_c is, the more obvious variation trend is. The same in figure 6, when the h_c is 1 mm and 2 mm, the voltage is almost constant while h_c is 3 mm and 4 mm, the variation of output voltage reaches 4 V and 6 V. Choosing h_c and t are 4 mm, 0.5 mm respectively, output voltage at resonance reaches 30.3 V, which is 26.2% higher than the case at coincident position and 63.8% higher than that without cavity. From the data it indicates that the position of the cavity has a great influence on the voltage.

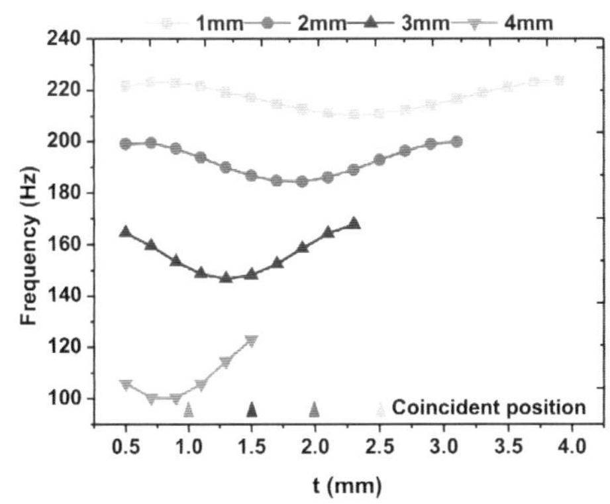

Figure 5. Effect of cavity position on frequency.

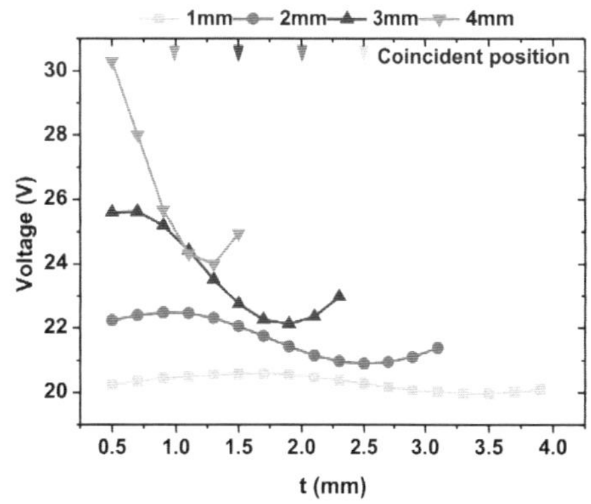

Figure 6. Effect of cavity position on voltage.

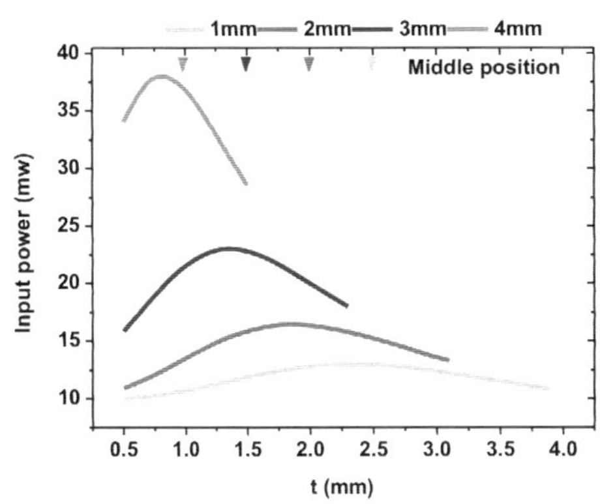

Figure 7. Effect of cavity position on input power.

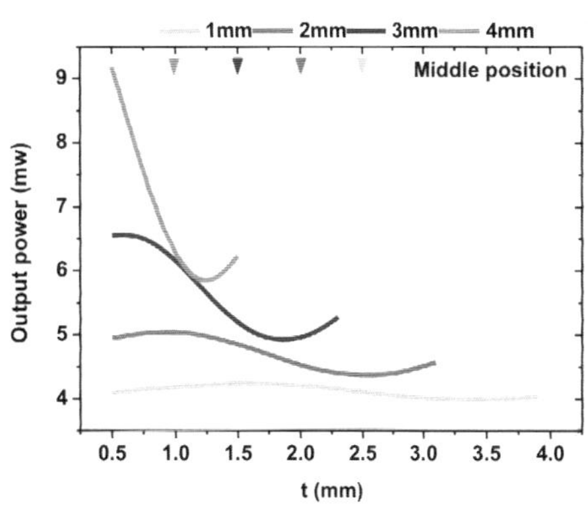

Figure 8. Effect of cavity position on output power.

As shown in figures 7- 9, the conversion rate is changed with t and reaches lowest when cavity is at middle position. From the curve in figure 9, the conversion rate is only improved when it is in rectangle area at the upper left corner compared to the case without cavity. Compared the results in figures 6 and 9, it can be seen that the output voltage is highest at $h_c = 4$ mm, the conversion rate is lowest even much lower than the case without cavity. Therefore, considering the effect of the cavity on the output voltage and the conversion efficiency comprehensively, the thickness is selected to be 3 mm, and the size of t is selected to be 0.5 mm.

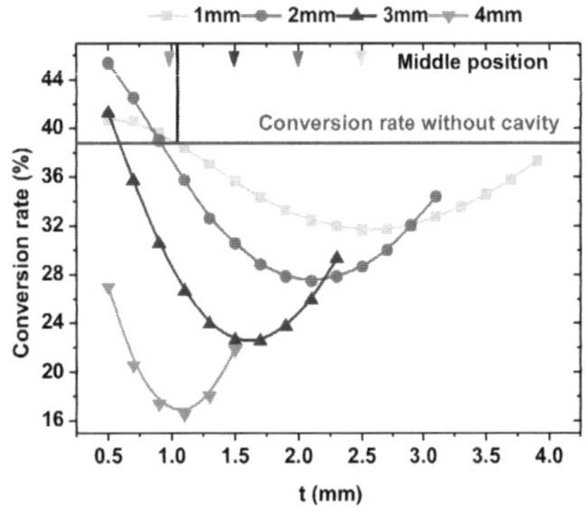

Figure 9. Effect of cavity position on conversion rate.

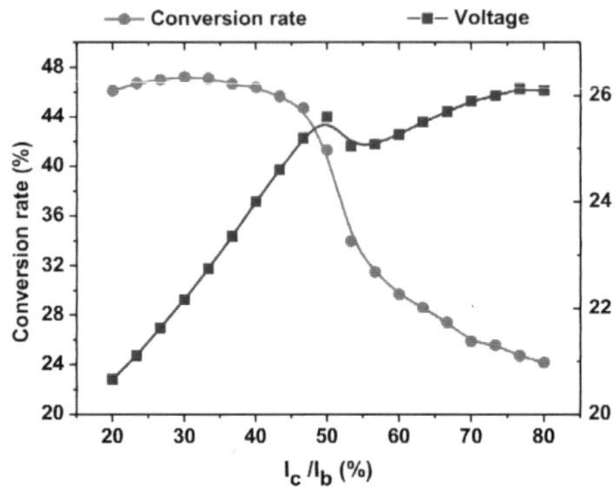

Figure 10. Effect of length on conversion rate and voltage.

3.3. Effect of cavity length

After selecting of h_c and *t*, the influence of the length of the cavity l_c on the energy harvester is further studied. Under the same external excitation, the calculated maximum voltage and conversion rate are shown in figure 10.

The length of the piezoelectric layer is half of the base layer. The output voltage in figure 10 increases almost linearly when the length ratio l_c/l_b less than 50%, while voltage drops first and then slowly increases when the ratio is greater than 50%. For conversion rate, the conversion efficiency increases slowly and then decrease, when the ratio is greater than 50%, the conversion efficiency drops rapidly. It can be seen that the output voltage growth rate is much lower than the conversion rate reduction rate when the ratio is greater than 50%, and the optimal length is determined as 75 mm finally.

4. Conclusion

This paper established a model of PEH with a cavity by COMSOL Multiphysics. Comparative studies show that the introduction of cavity obviously improves the capability of the device, especially when h_c is 4 mm, output voltage is increased by 35% compared with the case without cavity. The thickness, length and position have obvious impact on the output voltage and conversion rate; at last choosing the length, thickness and t are 75 mm, 3 mm, 0.5 mm, corresponding nature frequency is 164 Hz that is reduced by 35.4% and the output voltage, conversion rate reach 25.6 V, 41.3%, which are much higher than the case without cavity.

Acknowledgments

This paper is supported by the national natural science foundation of China, project No. 11472019. We really appreciate the assistance of the committee of the national natural science foundation of China.

References

[1] Roundy S, Wright P K and Rabaey J M 2004 *Energy Scavenging for Wireless Sensor Networks* (US: Springer)
[2] Jeon Y B, Sood R, Jeong J H *et al* 2005 MEMS power generator with transverse mode thin film PZT *Sensor actuat. A. Phys.* **122** 16-22
[3] Priya S and Inman D (Eds.) 2009 *Energy Harvesting Technologies* (New York: Springer)
[4] Marin A, Bressers S and Priya S 2011 Multiple cell configuration electromagnetic vibration energy harvester *J. Phys. D. Appl. Phys.* **44** 295501
[5] Todaro M T and Guido F 2017 Piezoelectric MEMS vibrational energy harvesters: Advances and outlook *Microelectron Eng.* **183-184** 23-36

[6] Roundy S and Wright P K 2004 A piezoelectric vibration based generator for wireless electronics *Smart Mater. Struct.* **13** 1131-42
[7] Roundy S *et al* 2005 Improving power output for vibration-based energy scavengers *IEEE Pervasive Comput.* **4** 28-36
[8] Zhao X Q, Shang Z G, Luo G X and Deng L C 2015 A vibration energy harvester using AlN piezoelectric cantilever array *Microelectron. Eng.* **142** 47-51
[9] Renaud M, Karakaya K, Sterken T, Fiorini P, Van Hoof C and Puers R 2008 Fabrication, modelling and characterization of MEMS piezoelectric vibration harvesters *Sensor actuat. A. Phys.* **145** 380-6
[10] Wang Z and Xu Y 2007 Vibration energy harvesting device based on air-spaced piezoelectric cantilevers *Appl. Phys. Lett.* **90** 263512
[11] Reddy A R and Umapathy M 2014 Improved energy harvesting from vibration by introducing cavity in a cantilever beam *J. Vib. Control* **1** 10
[12] Srinivasulu Raju S, Umapathy M and Uma G 2015 Cantilever piezoelectric energy harvester with multiple cavities *Smart Mater. Struct.* **24** 115023
[13] Ramalingam U and Gandhi U 2018 A new piezoelectric energy harvester using two beams with tapered cavity for high power and wide broadband *Int. J. Mech. Sci.* **142-143** 224-34
[14] Srinivasulu Raju S and Umapathy M 2018 Design and analysis of high output piezoelectric energy harvester using non uniform beam *Mech. Adv. Mater. Struc.* **0** 1-10

Finite element study of tunable cantilever plate structure using position change

L H Chen[1,2], L Q Chang[1,2], J T Xue[1,2] and F H Yang[3,4]

[1]College of Mechanical Engineering, Beijing University of Technology, Beijing 100124, China
[2]Key Laboratory of Nonlinear Vibration and Strength of Mechanical Structures, Beijing 100124, China
[3]School of Statistics and Mathematics, Central University of Finance Economics, Beijing 100081, China

E-mail: yangfh_cufe@163.com

Abstract. In this paper, a tunable cantilever plate structure is presented, in which a box is attached to each corner of its free sides and a ball is placed in each box. Change of the balls position makes natural frequency of the structure variable, which expands working bandwidth of the structure to accommodate the complex ambient vibration. The modal analysis of the tunable cantilever plate is carried out by the finite element software-ABAQUS. Changing the balls position, it is found that there are maximum and minimum values in the natural frequency of the self-tuning structure, which forms a resonance bandwidth. Moreover, there are multiple position combinations of two balls at the same natural frequency in the range of resonance band. The paper studies the influence of mass and geometrical dimensions of boxes and balls on the natural frequency and bandwidth. While ensuring strength and stiffness of the structure, optimal material and size parameters are selected to improve the resonant bandwidth. This provides a theoretical basis for applying the tunable structure to piezoelectric energy harvester.

1. Introduction

Vibration energy widely exists in daily life and practical engineering. Vibration energy harvester is an effective way to convert vibration energy of surrounding environment into electrical energy [1, 2]. The advantages of the piezoelectric energy harvester are high conversion efficiency, simple structure, no additional power supply, environment-friendly, etc., and its application prospect is broad [3-7]. Generally, working bandwidth of piezoelectric energy harvester is relatively narrow, but power of the energy harvester can be maximized when external excitation frequency is consistent with natural frequency of the system. Because of low frequency and wide frequency band of the external vibration, Natural frequency of the piezoelectric energy harvester cannot match the external excitation frequency, which leads to a great reduction of the output power of the harvester [8, 9].

In order to improve efficiency of piezoelectric energy harvester, it is necessary to automatically adjust natural frequency of the device according to the frequency of external environment vibration to achieve resonance. The earliest tunable energy harvester relied on manually varying position of additional mass [10] to adjust natural frequency. After that, some tunable devices using not manual operation but additional energy [11], such as electromagnetic [12] to change the stiffness of the system, so the actual capture efficiency is low. In addition, there was a reference that combined several cantilever beams with different natural frequencies to a complex structure, but it occupies a large space [13]. The devices in the reference [14, 15] have achieved self-tuning, but the frequency is higher and the tunable

IOP Conf. Series: Materials Science and Engineering **629** (2019) 012028 doi:10.1088/1757-899X/629/1/012028

range is narrower. Reference [16] designed a self-tuning device by attaching a box with two cylinders based on cantilever beam structure, and did not further study the relationship between position and frequency/bandwidth. Based on cantilever plate, this paper proposes a tunable cantilever plate structure with additional balls and boxes, no electricity recovery part. By changing position of the balls, tunable function in broad-frequency vibration environment is realized.

2. Finite element modeling

The research is based on a cantilever plate structure with one end fixed and the other end free. A box is attached to each corner of free end of cantilever plate, and a small ball is placed in each box. The model structure is shown in figure 1. Natural frequency of the structure is adjusted by changing position of balls within boxes.

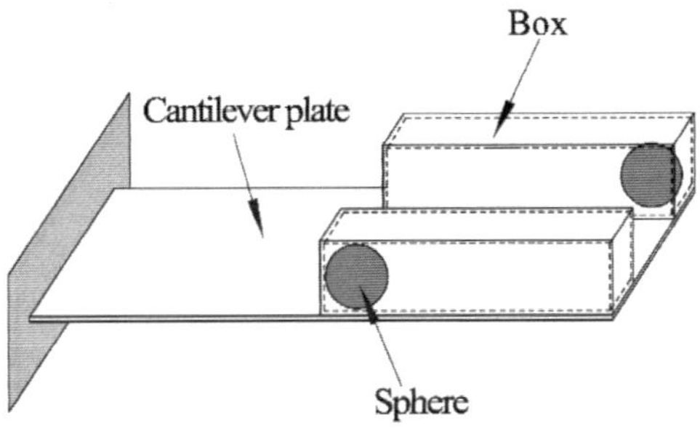

Figure 1. Sketch of the cantilever plate structure.

In order to study the influence of the balls position, size parameters and mass of boxes and the mass of balls on the natural frequency, modal analysis is carried out by using ABAQUS—a common finite element software. ABAQUS has 433 types of units, which provides designers with more choices and can reflect the subtle structural phenomena and differences between phenomena. It is easy to simulate the mechanical behavior of the actual structure. In view of the universality of contact problems in practical engineering, ABAQUS has separately set up an interaction module, which can accurately simulate various contact problems existing in actual engineering. In this paper, cantilever plate and boxes are built with S4R shell elements, and the balls are built with C3D10M solid elements. Boxes and plate are linked by the TIE in ABAQUS. Size of cantilever plate is 120mm×60mm×0.06mm, and diameter of each sphere is 16 mm. Model of finite element simulation is shown in figure 2.

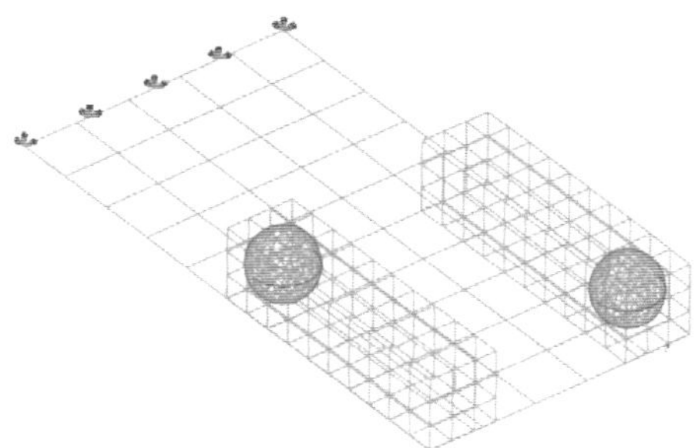

Figure 2. Finite element simulation of the structure.

3. Result and analysis

3.1. Influence of ball location

The purpose is to obtain the influence regulation of the ball position on the natural frequency, first-order natural frequency of the structure is obtained by modal analysis using ABAQUS, and three special change cases of sphere position are studied:

1. Initial positions of both balls are in the left-most side of box near the fixed end of the cantilever plate. Changing the positions of the balls two balls sites equidistantly toward free end of the plate until the balls reach the right-most side of the boxes, modal analysis is carried out at each new situation of two balls.

2. Initial positions of two balls are the same as in case 1. One ball is fixed and position of another ball is changed equidistantly to the right end of box. Modal analysis is performed for each change of the balls' position.

3. Initially, two balls are on the left and right sides of box respectively. The sphere at the free end is fixed, and position of the left ball is changed equidistantly to the right until the ball reaches the free end. Modal analysis is carried out each time the ball position is changed.

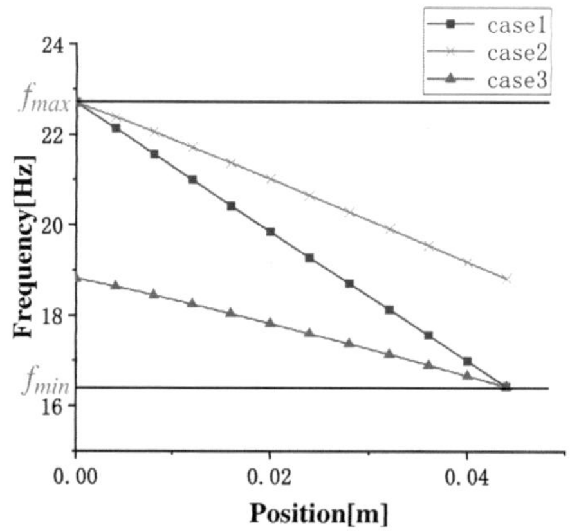

Figure 3. Change curves of natural frequency in three cases.

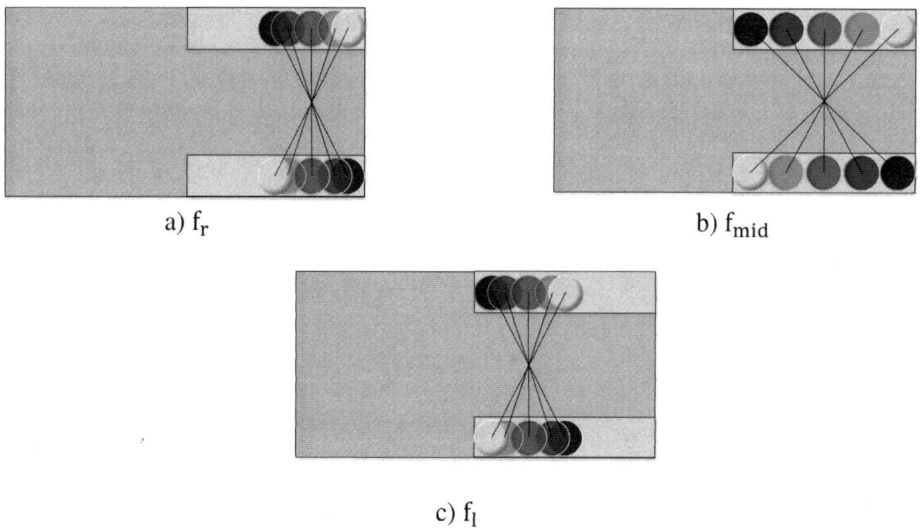

Figure 4. Position distribution of balls at different frequencies.

The curves of the first-order natural frequency following balls position in three cases are shown in figure 3 from ABAQUS.As position of balls changes from left to right (from 0 to 0.04m), the natural frequency gradually decreases. When two balls are at the left-most side of boxes, the value of the natural frequency is the highest, named f_{max}. When the balls are simultaneously at the right-most side near the free end, the value is the lowest and is named f_{min}. In other positions, the natural frequency is in the range of f_{max} and f_{min}, so the difference between f_{max} and f_{min} can be determined as bandwidth, called $f_b = f_{max} - f_{min}$. As the site of two balls changes, the natural frequency varies within the bandwidth, so that tunable function of the structure is achieved.

In the above three cases, the ball position is changed in a special way, that is, either the two balls are in the same situation (case 1), or one ball is fixed at an end, and site of another ball is changed (case 2, 3). In general, during the vibration of the board, two balls can roll freely in the box, and the position is not limited. Taking three frequencies in the bandwidth as examples, they are defined as $f_r = (f_{min} + f_b/4)$, $f_{mid} = (f_{max} + f_{min})/2$ and $f_l = (f_{max} - f_b/4)$respectively , and the positions distribution of two balls at a natural frequency is analyzed. As shown in figure 4a-c.

It is found that within the bandwidth range, a natural frequency corresponds to multiple groups of two balls' positions, and each group is represented by a line between two balls in figure 4. All of the lines corresponding to a certain natural frequency constitute an X-shaped region. When the magnitude of natural frequency approaches the midpoint of bandwidth (f_{mid}), the range in which the balls can roll is maximized, and the area of the region is the largest. Therefore, as the frequency changes from f_{min} to f_{max}, the X-region moves from right to left, and its area increases first and then decreases.

3.2. Effect of ball mass

In order to enlarge the bandwidth, the influence of ball' mass on natural frequency and bandwidth of the system is investigated. Respectively, steel and 4luminium are selected as balls material. Fixed the balls at the right or left end of the boxes, modal analysis is performed to obtain the minimum and maximum natural frequencies (f_{min} and f_{max}) respectively. As shown in table 1, the results of f_{min} are very similar, but the $f_{max}$of steel balls is larger than that of 4luminium balls. Therefore, the bandwidth of the structure with steel balls is superior, and finally steel is selected as the material of the balls.

Table 1. Influence of mass on natural frequency and bandwidth (Hz).

Material	f_{min}	f_{max}	Bandwidth
Steel	16.421	22.707	6.286
Aluminum	16.206	17.634	1.428

3.3. Effect of box

In order to widen the frequency band further, the influence of box mass on the natural frequency is investigated. Steel, aluminum and acrylic commonly used in engineering are selected as box materials respectively to calculate the maximum and minimum natural frequencies. As shown in table 2, the natural frequencies of structures with steel or aluminum boxes are lower. In contrast, although the natural frequency of the structure with acrylic box is higher, its bandwidth is wider. For the purpose of increasing the resonant frequency band, acrylic is chosen as the material of two boxes.

Table 2. Influence of materials on natural frequency and bandwidth (Hz).

Material	f_{min}	f_{max}	Bandwidth
Stainless steel	13.965	16.655	2.690
Aluminum	14.702	18.166	3.464
Acrylic	16.421	22.707	6.286

On the basis of previous studies, the influence of box length on the natural frequencies is studied. The width, height and wall thickness of the box are 20mm, 20mm and 1mm respectively. The maximum and minimum natural frequencies of the structures with box length of 30mm, 60mm and 90mm are calculated respectively, and the results are shown in table 3. Increase in the length of boxes causes the balls' travels to become larger in the box, so the bandwidth gradually increases. Considering that the rigidity of the boxes is much larger than the stiffness of board, the stress concentration near the fixed end of the cantilever plate will be too serious, which may cause the destruction of the cantilever plate, if the box is too long. Therefore, 60 mm box is optimum for the structure.

Table 3. Influence of length on natural frequency and bandwidth (Hz).

Lengths	f_{min}	f_{max}	Bandwidth
30mm	15.971	17.996	2.025
60mm	16.421	22.707	6.286
90mm	20.612	31.341	10.729

4. Conclusions

In this paper, a tunable cantilever plate structure is designed and studied by using the finite element software—ABAQUS. Natural frequency of the structure is changed through the variation of balls' position, so that there is a resonance bandwidth formed by the maximum and minimum natural frequencies. By altering the relative position of the balls, any natural frequency in the bandwidth range corresponds to multiple position combinations of the two spheres, which establish an X-shaped region. Therefore the tunable ability of the structure is excellent. The influence of balls and boxes on the natural frequency is studied, and it is found that in order to widen the bandwidth under the premise of ensuring the structural strength and rigidity, it is necessary to elevate the mass ratio of ball to box, and raise the boxes' lengths. Results of the study contribute to the structural optimization of energy harvester to increase its operating bandwidth.

References

[1] Romero E, Warrington R O and Neuman M R 2009 Energy scavenging sources for biomedical sensors *PHYSIOL MEAS* **30** R35-R62.
[2] Roundy S 2005 On the effectiveness of vibration-based energy harvesting *J INTEL MAT SYST STR* **16** 809-23
[3] Roundy S, Wright P K and Rabaey J 2003 A study of low level vibrations as a power source for wireless sensor nodes *COMPUT COMMUN* **26** 1131-44
[4] Mateu L and Moll F 2005 Review of energy harvesting techniques and applications for microelectronics (Keynote Address) [C]//VLSI Circuits and Systems II *International Society for Optics and Photonics* **5837** 359-74
[5] Chen S N, Wang G J and Chien M C 2006 Analytical modeling of piezoelectric vibration-induced micro power generator *MECHATRONICS* **16** 379-87
[6] Sodano H A, Inman D J and Park G 2004 A review of power harvesting from vibration using piezoelectric materials *Shock and Vibration Digest* **36** 197-206
[7] Ramlan R, Brennan M J, Mace B R *et al* 2010 Potential benefits of a non-linear stiffness in an energy harvesting device *NONLINEAR DYNAM* **59** 545-58
[8] Marzencki M, Basrour S, Charlot B *et al* 2005 Design and fabrication of piezoelectric micro power generators for autonomous microsystems *Symposium on Design, Test, Integration and Packaging of MEMS/MOEMS (DTIP'05)* 299-302
[9] Cottone F, Vocca H and Gammaitoni L 2009 Nonlinear energy harvesting *PHYS REV LETT* **102** 080601
[10] Wu X, Lin J, Kato S *et al* 2008 A frequency adjustable vibration energy harvester *Proceedings of PowerMEMS* 245-8
[11] Williams C B and Yates R B 1996 Analysis of a micro-electric generator for microsystems *SENSOR ACTUAT A-PHYS* **52** 8-11
[12] Boudaoud A, Couder Y and Amar M B 1999 A self-adaptative oscillator *The European Physical Journal B-Condensed Matter and Complex Systems* **9** 159-65
[13] Shahruz S M 2006 Limits of performance of mechanical band-pass filters used in energy scavenging *J SOUND VIB* **293** 449-61

[14] Kozinsky I 2009 Study of passive self-tuning resonator for broadband power harvesting *Proc. Power MEMS (Washington, DC, Dec.)* 388-9

[15] Jackson N, Stam F, Olszewski O Z *et al* 2015 Broadening the bandwidth of piezoelectric energy harvesters using liquid filled mass *Procedia engineering* **120** 328-32

[16] Somkuwar R, Chandwani J and Deshmukh R 2018 Wideband auto-tunable vibration energy harvester using change in centre of gravity *MICROSYST TECHNOL* **24** 3033-44

IOP Conf. Series: Materials Science and Engineering **629** (2019) 012029 doi:10.1088/1757-899X/629/1/012029

New adhesive strength evaluation method based on the singular stress field considering three-dimensional geometry

N-A Noda[1,2]

[1]Mechanical Engineering Department, Kyushu Institute of Technology, 1-1 Sensui-cho Tobata-ku, Kitakyushu-shi, 804-8550, Japan

E-mail: noda.naoaki844@mail.kyutech.jp

Abstract. Adhesive joints are widely used although different materials properties cause the singular stress field whose intensity is controlled by the adhesive joint geometry. Our previous studies showed that debonding strength can be expressed as a constant value of the critical intensity of singular stress field (ISSF) by applying two-dimensional modelling. By considering the real adhesive geometry, in this study, the ISSFs along the interface edge of three-dimensional prismatic butt joints are considered by varying the adhesive thicknesses. It is found that the critical ISSF in 3D modelling is almost constant independent of the adhesive thickness. The magnitude and position of the maximum ISSF are discussed by varying the corner fillet radius in comparison with two-dimensional modelling.

1. Introduction

Adhesive joints have several advantages such as light weight and low cost compared with the other traditional joints. Therefore, adhesive joints are widely used in various industrial fields [1-3]. However, different material properties cause singular stresses at the interface end, which may lead to debonding failure in structures [4-12].

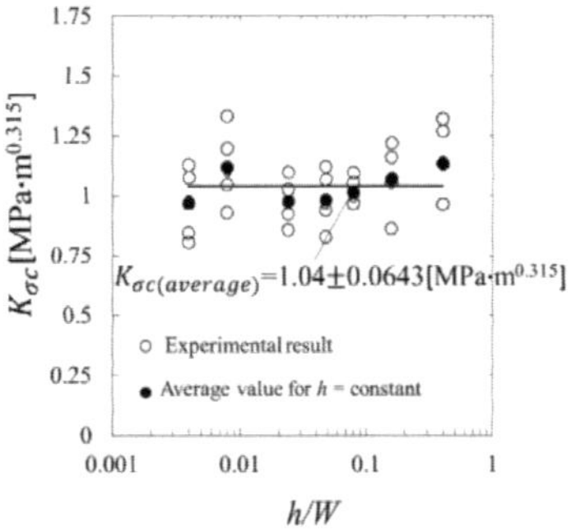

Figure 1. Adhesive strength for S35C / Epoxy resin expressed as a constant critical ISSF $K_{\sigma c}$ by using 2D model.

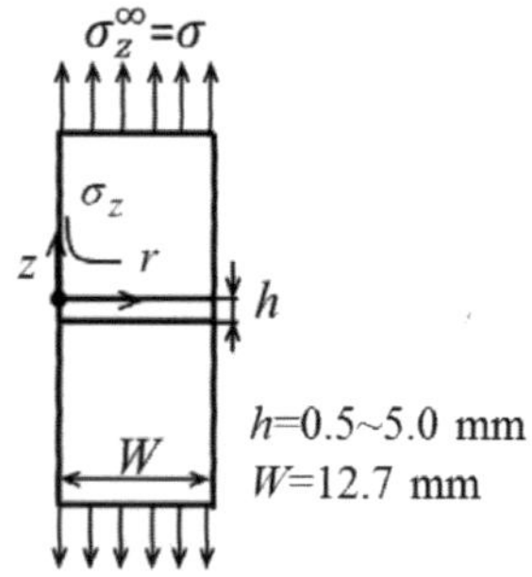

Figure 2. 2D butt joint modelling.

The bonded strength is closely related to the intensity of the singular stress field (ISSF). Recent previous studies showed that the debonding strength can be expressed as a constant value of ISSF as

IOP Conf. Series: Materials Science and Engineering **629** (2019) 012029 doi:10.1088/1757-899X/629/1/012029

shown in figure 1 [12, 13]. In those studies, the ISSF was analyzed by using two-dimensional modelling assuming plane strain as shown in figure 2. Suzuki [14] studied the adhesive bonded strength when medium carbon steel plates are bonded by epoxy resin. His experimental results are used widely in research since the experiment was conducted carefully with high accuracy. In this study, therefore, Suzuki's specimen in figure 3 [14] will be analysed by considering the three-dimensional prismatic butt joint geometry. Then, the ISSF distribution will be discussed along the adhesive interface edge by extending the mesh independent technique proposed previously [7-13].

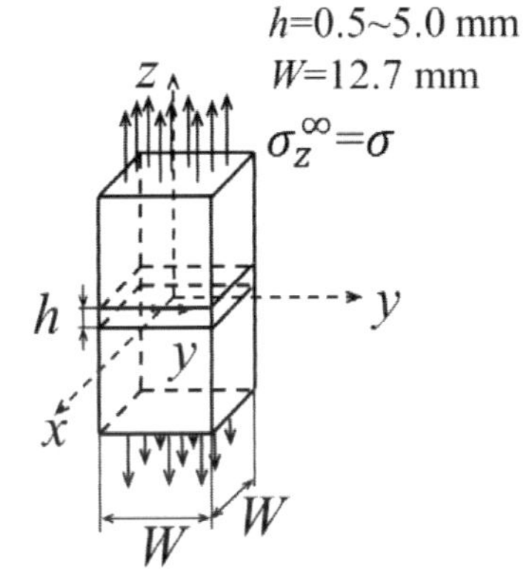

Figure 3. 3D prismatic butt joint.

2. FEM stress distribution along the adhesive interface

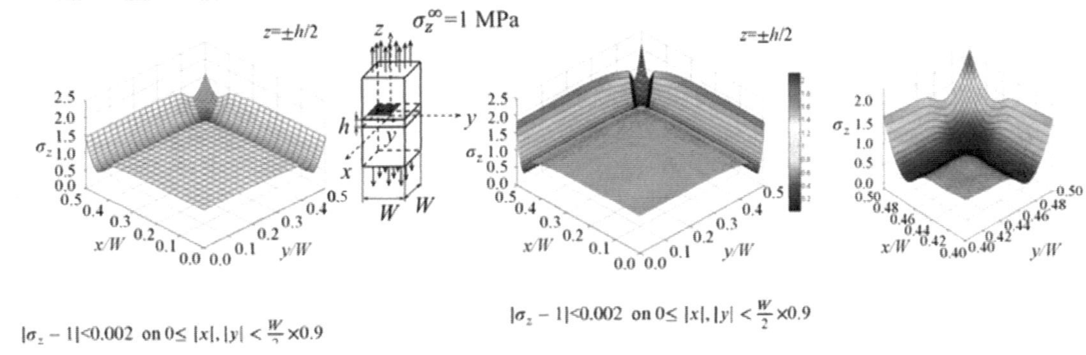

(a) Large mesh (b) Small mesh (c) Detail of x/W, y/W = 0.4~0.5

Figure 4. FEM stress distribution σ_z of 3D prismatic butt joint on $z = \pm h/2$.

In this study, the finite element method (FEM) is applied to adhesive strength analysis. The FEM has been used in wide engineering fields such as composite materials [7-13], impact strength [15], and complicated structures including traditional joints [16]. figure 4 shows an example of FEM stress distributions along the adhesive interface in prismatic butt joint when h/W=0.1. This result is corresponding to one of Suzuki's specimens [14]. The mild steel S35C has Young's modulus E=210GPa and Poisson's ratio ν=0.3 and the epoxy resin has E=3.14GPa and ν=0.37. As shown in figure 4, in the interior region of the interface $0 \leq x$, y<0.45, FEM stress is accurate since they are independent of the minimum mesh size e_{min} and satisfy $|\sigma_z - 1|$ <0.002 under the remote tensile stress σ_z^∞=1. However, FEM stress values are not accurate near the interface side $|x|$ =0.5 and $|y|$ =0.5 since they varies depending on the mesh size e_{min}. It should be noted that the real interface stress should go to infinity along the interface side $|x|$ =0.5 and $|y|$ =0.5. In this study, the intensity of this singular stress field (ISSF) is obtained by applying mesh independent technique. Since the ISSF varies depending on the location and also the adhesive thickness, the ISSF distributions will be discussed in the next section. In this study, the mesh independent technique coupled with FEM is extended to the three dimensional prismatic butt joints. The details may be found in [7-13, 17]. This new method can be applied to real adhesive joints to evaluate the adhesive strength.

3. ISSF distribution and critical ISSF distribution when debonding occurs

figure 5 shows the ISSF distributions along the interface side K_σ^{Side} under remote tensile stress $\sigma_z^\infty = 1$. The ISSF K_σ^{Side} is defined by eq.(1).

$$K_\sigma^{Side} = \lim_{r\to 0}\left[r^{1-\lambda} \times \sigma_z^{Side,Real}(r,y)\right] \tag{1}$$

The ISSF K_σ^{Side} decreases with decreasing the adhesive thickness. Each ISSF distribution is almost constant at the most portion except near the vertex. The detail of ISSF around the vertex is shown in figure 5(b), (c). In figure 5 (c), as shown by the dotted lines in the range of $0.4995 \le y \le 0.5$, the ISSFs go to infinity because different singular stress field exists at the vertex [18, 19].

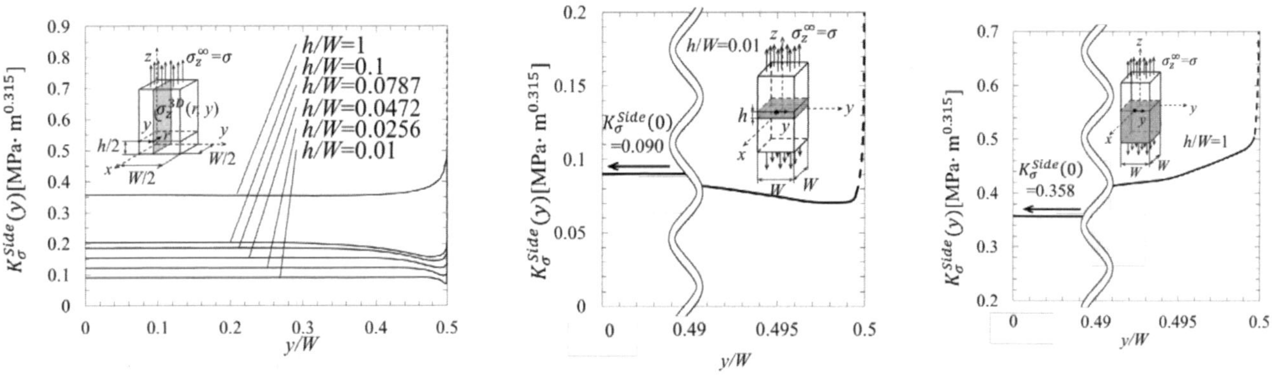

(a) ISSF distributions of 3D butt joint (b) Detail of around the vertex when $h/W = 0.01$ (c) Detail of around the vertex when $h/W = 1$

Figure 5. ISSF distribution $K_\sigma^{Side}(y)$ of 3D butt joint under $\sigma_z^\infty = 1$ MPa.

figure 6(a) shows the critical ISSF distributions of $K_{\sigma c}^{Side}(y)$ under remote tensile stress $\sigma_z^\infty = \sigma_c$ when the debonding occurs. figure 6(b) shows the detail of $K_{\sigma c}^{Side}(y)$ at $0.49 \le y \le 0.5$. As shown in figure 6(a), the critical ISSF distributions are quite similar and they are in a narrow band region. It should be noted that the critical value of the ISSF in the middle side region in figure 6 (a) is almost the same of the value obtained by 2D modelling in figure 1.

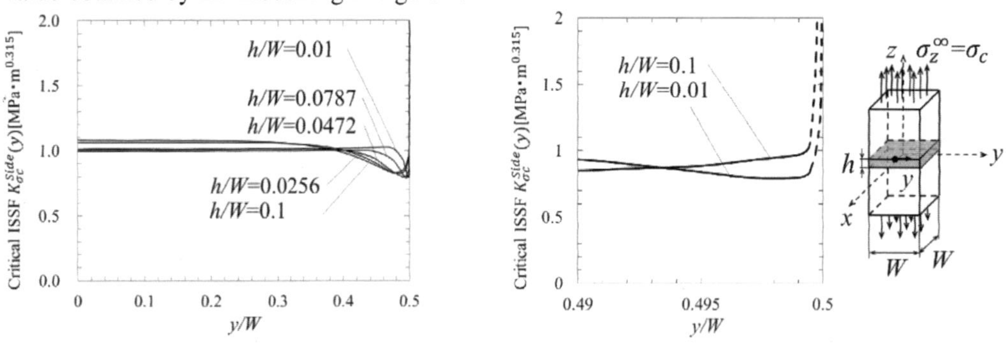

(a) Critical ISSF distribution (b) Detail of around the vertex when h/W=0.1 and 0.01

Figure 6. Critical ISSF distributions $K_{\sigma c}^{Side}(y)$ when debonding occurs under $\sigma_z^\infty = \sigma_c$.

IOP Conf. Series: Materials Science and Engineering **629** (2019) 012029 doi:10.1088/1757-899X/629/1/012029

4. ISSF distribution of fillet corner

In reality, there is no sharp corner and usually there is a small fillet radius at the corner as shown in figure 7. The local polar coordinate along the interface as shown in figure 7 is used to describe the position along the fillet by changing the fillet radius ρ under the same adhesive thickness h/W= 0.01.

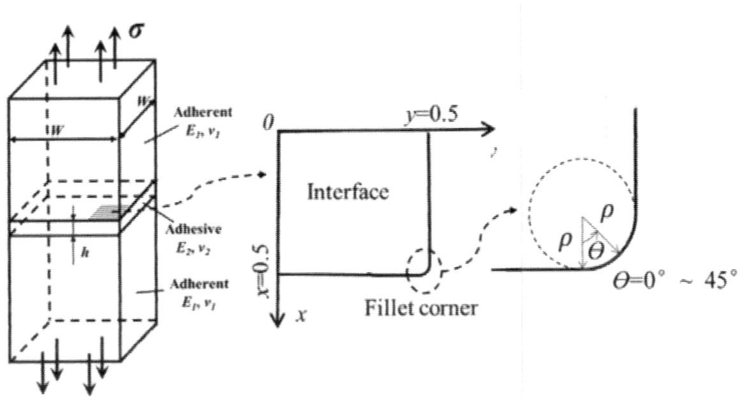

Figure 7. Prismatic butt joint model with fillet considered in this study.

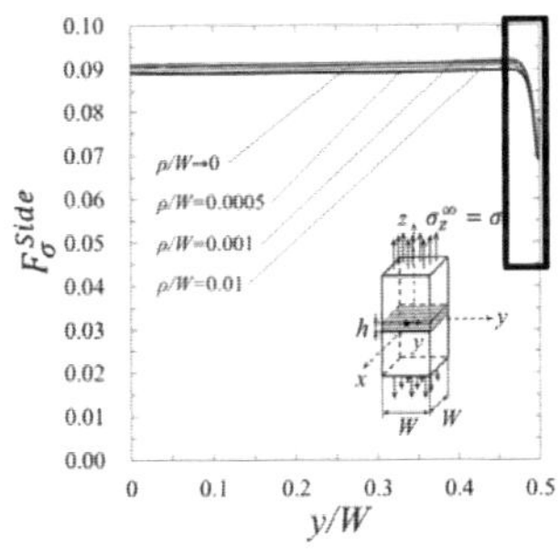

(a) When ρ/W=0~0.01

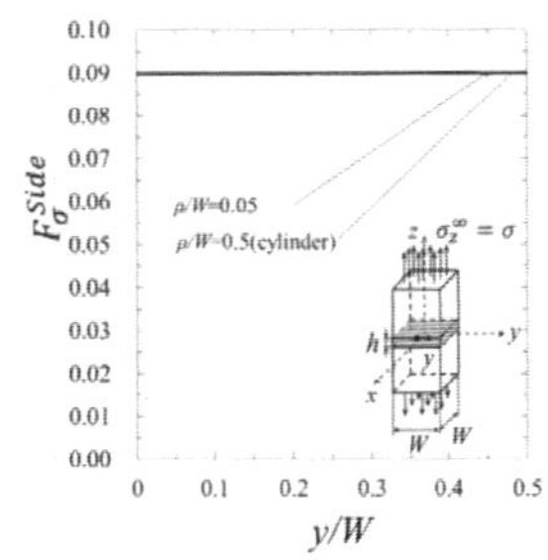

(b) When ρ/W=0.05~0.5

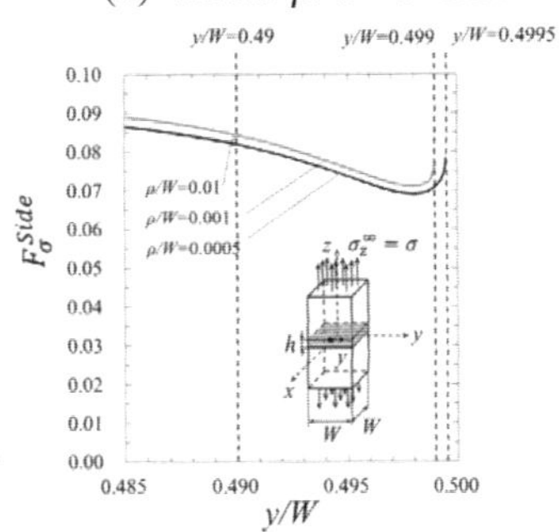

(c) Detail in figure 8(a)

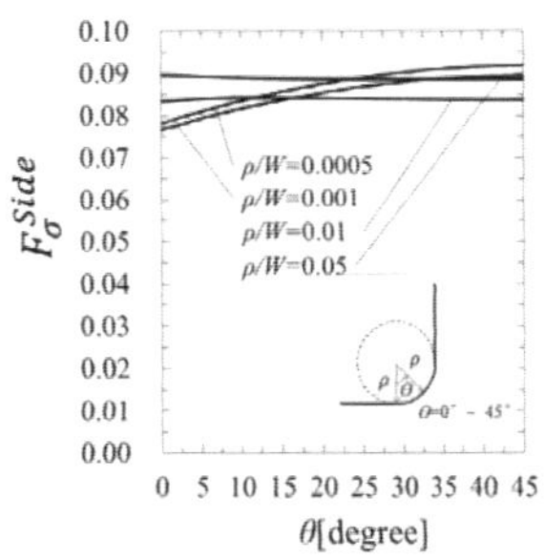

(d) Detail at fillet in figure 8(a)

Figure 8. ISSF distributions in figure 7 along the edge of fillet.

figure 8 show the ISSF distribution along the interface corner when the relative roundness of the fillet ρ/W is changed as ρ/W = 0, 0.0005, 0.001, 0.01, 0.05. It is found that the maximum value of the intensity ISSF occurs around $y/W\approx0.46$ in most cases. The corner of the specimen is always chamfered, and ρ can be regarded as the minimum chamfer size in practical use, therefore $\rho \geq 0.2$ mm should be considered. Therefore, for Suzuki's specimen width W = 12.7 mm in figure 3, $\rho/W \geq 0.016$. figure 9

IOP Conf. Series: Materials Science and Engineering **629** (2019) 012029 doi:10.1088/1757-899X/629/1/012029

shows an example of the fractured surface in [14]. The fractured origin looks close to the position of the maximum ISSF. figure 8 shows the minimum ISSF occurs around $y/W \approx 0.498$ in most cases. The variation of the ISSF is less than 1% when $\rho/W \geq 0.05$ and less than 10% even when $\rho/W \geq 0.0005$. It may be concluded that the effect of the fillet radius ρ of the ISSF is relatively small.

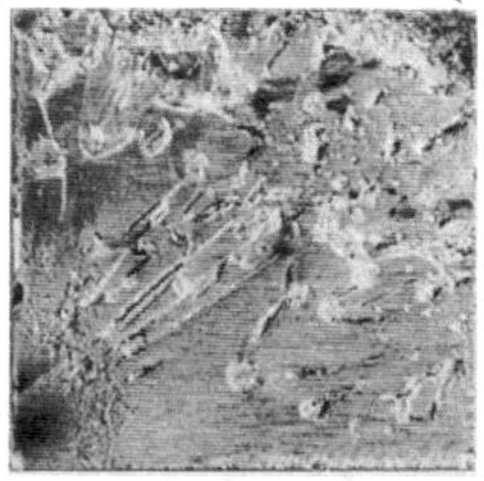

Figure 9. Fracture surface when h/W=0.00787 (h = 0.1 mm, W = 12.7 mm) in figure 3.

5. Conclusion

In this study, the ISSF distributions along the interface edge of three-dimensional joints were discussed by considering the real specimen geometry for various adhesive thicknesses. The magnitude and position of the maximum ISSF were investigated by varying the corner fillet radius in comparison with two-dimensional modelling. The following conclusions can be drawn from that discuss.

(1) The ISSF distribution along the interface edge is almost constant except near the vertex. The ISSF value decreases with decreasing the adhesive thickness.
(2) The critical ISSF distributions when the debonding occurs are almost the same independent of the adhesive thickness.
(3) It is found that the maximum value of the intensity ISSF occurs around $y/W \approx 0.46$ in most cases.

References

[1] Barnes T A and Pashby I R 2000 Joining techniques for aluminum spaceframes used in automobiles: part II - adhesive bonding and mechanical fasteners *J Mater Process Technol* **99** 72-9
[2] Higgins A 2000 Adhesive bonding of aircraft structures *Int J Adhes* **20** 367-76
[3] Petrie E M 2008 Adhesives for the assembly of aircraft structures and components: decades of performance improvement, with the new applications of the horizon *Metal Finish* **106** 26-31
[4] Qian Z and Akisanya A R 1998 An experimental investigation of failure initiation in bonded joints *Acta Mater* **46** 4895-904
[5] Akisanya A R and Meng C S 2003 Initiation of fracture at the interface corner of bi-material joints *J Mech Phys Solids* **51** 27-46
[6] Mintzas A and Nowell D 2012 Validation of an Hcr-based fracture initiation criterion for adhesively bonded joints *Eng Fract Mech* **80** 13-27
[7] Zhang Y, Noda N A, Takaishi K T and Lan X 2011 Effect of adhesive thickness on the interface of singular stress at the adhesive dissimilar joint *Transactions of the Japan Society of Mechanical Engineers Series A* **77** 360-72
[8] Zhang Y, Noda N A, Wu P Z and Duan M L 2015 A mesh-independent technique to evaluate stress singularities in adhesive joints *Int J Adhes Adhes* **57** 105-17
[9] Zhang Y, Noda N A, Wu P and Duan M 2015 Corrigendum to "A mesh-independent technique to evaluate stress singularities in adhesive joints" *Int J Adhes Adhes* **60** 130
[10] Miyazaki T, Noda N A, Li R, Uchikoba T and Sano, Y 2013 Debonding criterion for single lap joints from the intensity of singular stress field *Transactions of The Japan Institute of Electronics Packaging* **16** 143-51 (in Japanese)

IOP Conf. Series: Materials Science and Engineering **629** (2019) 012029 doi:10.1088/1757-899X/629/1/012029

[11] Miyazaki T, Noda N A, Uchikoba T, Li R and Sano Y, 2014 Proposal of a convenient and accurate method for evaluation of debonding strength *Trans Soc Autom Eng Jpn* **45** 895-901 (in Japanese)
[12] Noda N A, Miyazaki T, Li R, Uchikoba T, Sano Y and Takase Y 2015 Debonding strength evaluation in terms of the intensity of singular stress at the corner with and without fictitious crack *Int J Adhes Adhes* **61** 46-64
[13] Noda N A, Miyazaki T, Uchikoba T, Li R, Sano Y and Takase Y 2014 Convenient debonding strength evaluation based on the intensity of singular stress for adhesive joints *Transactions of The Japan Institute of Electronics Packaging* **17** 132-42 (in Japanese)
[14] Suzuki Y 1987 Adhesive tensile strengths of scarf and butt joints of steel plates (relation between adhesive layer thicknesses and adhesive strengths of joints) *JSME. Int. J.* **30** 1042-51
[15] Noda N A, Shen Y, Takaki R, Akagi D, Ikeda T, Sano Y and Takase Y 2017 Relationship between strain rate concentration factor and stress concentration factor *Theor Appl Fract Mec* **90** 218-27
[16] Noda N A, Chen X, Sano Y, Wahab M A, Maruyama H, Fujisawa R and Takase Y 2016 Effect of pitch difference between the bolt-nut connections upon the anti-loosening performance and fatigue life *Mater Design* **96** 476-89
[17] Marc Mentat team 2012 Theory and User Information A *MSC Software* Tokyo 713
[18] Koguchi H and Nakazima M 2010 Influence of adhesive layer thickness on intensity of stress singularity at a vertex in three-dimensional joints with three layers: In case of external loading *Transactions of the Japan Society of Mechanical Engineers Series A* **76** 1110-8 (in Japanese)
[19] Miyazaki T, Inoue T, Noda N A and Sano Y 2018 A precise and efficient method for evaluating singular stress field at the vertex in three dimensional bonded body *Transactions of the JSME* **84** 1-17 (in Japanese)

Analysis of oil droplet deformation model in oil-water dynamic hydrocyclone

G D Guo[1,2], C Q Cao[1] and H Y Shao[1]

[1]Naval Logistics Academy of China, Tianjin, China

E-mail: ggdong2007@126.com

Abstract. The flow field in the rotating drum of dynamic oil-water hydrocyclone separator is a high turbulent shear flow field. Oil droplets are easily deformed, broken and even emulsified when the oil droplets enter the drum. Because oil droplets are affected by high turbulent shear flow field. If oil-water emulsification occurs, the separation performance of dynamic oil-water hydrocyclone will be dramatically reduced. Therefore, it is an important subject to study dynamic oil-water hydrocyclone separator and improve its separation performance to carry out oil drop deformation model research, which can reduce oil drop deformation and fragmentation and formulate measures to avoid oil-water emulsification.

1. Introduction

Dynamic oil-water hydrocyclone separator is a new type of oil-water separator developed on the basis of static oil-water hydrocyclone separator. Dynamic oil-water hydrocyclone separator has a wide range of advantages, such as wide processing range and strong separation capacity. It has been widely used in mining, petrochemical industry, environmental protection and other fields. The dynamic Liquid-liquid Hydrocyclone is based on the principle that the centrifugal force is different when the density of two or more phases is different in rotating motion. The centrifugal force field needed to realize separation is established by external force to realize the separation of two or more phases.

2. Dynamic oil-water hydrocyclone separation process

From the sketch of total dynamic liquid-liquid hydrohydrocyclone as shown in figure 1, high-speed rotating motion of moving parts (drum, grille, etc.) is driven by external power (motor). The oil-water mixture is transported to the inlet of the dynamic hydrocyclone under the action of dynamic pressure (pump). Under certain pressure, the liquid flow at the inlet enters the central hole of the rotating grate along the axial inlet and flows into its runner, then flows through the guide cone at the end of the rotating grate. The liquid flow passing through the rotating grate is actually a process of diversion and pre-rotating acceleration of the liquid flow. The liquid flow rotates at high speed and then enters the swirl chamber in the rotary drum. The larger and stronger swirl velocity field is formed by the friction between the liquid flow and the inner wall of the rotary drum. Because of the density difference between the two phases, the light dispersed phase is compressed to the center of the cylinder under centrifugal force; the heavy continuous phase moves to the inner wall of the cylinder, and the liquid flow moves to the other end of the swirl chamber under the action of axial force along the same direction of the cylinder. At this time, the oil in the central part is collected by the overflow nozzle and discharged through the outlet, while the water around the cylinder is discharged through the outlet. Finally, the two-phase media can be separated stably.

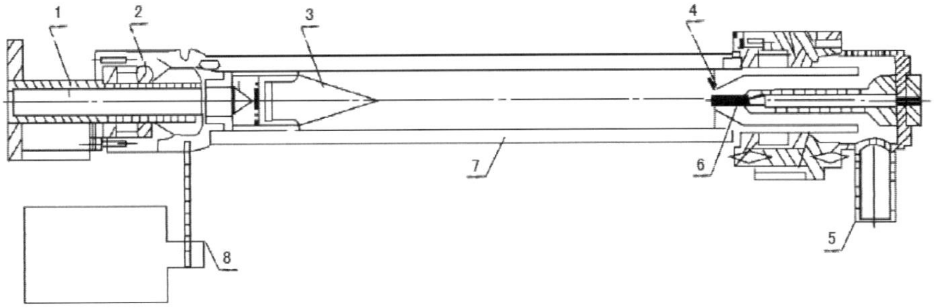

1-entrance; 2-mechanical seal; 3-rotating grate; 4-oil collecting sleeve; 5-underflow; 6-overflow outlet; 7-drum; 8-motor

Figure 1. The sketch of Total dynamic liquid-liquid hydrohydrocyclone.

3. Analysis of oil drop deformation model

During the operation of the dynamic oil-water hydrocyclone separator, the rotary drum rotates at a high speed driven by the motor, and the oil-water two-phase turbulent shear flow field is formed in the rotary drum. Serious oil-water emulsification may occur in the cylinder, which results in separation failure. This is mainly due to the deformation and fragmentation of oil droplets under the force of the cylinder. In order to reduce the decrease of separation efficiency caused by oil-water emulsification, the stress and deformation of a single oil drop are analyzed, and the critical conditions of oil drop deformation and fragmentation are discussed. Then it is applied to the turbulent shear flow field of a dynamic oil-water hydrocyclone separator with low or medium concentration. For the deformation of single dispersed phase oil droplets in shear flow field, affine deformation [1], shear deformation [2] and three-dimensional deformation [3] have been deeply studied. When the oil droplets undergo affine deformation, the oil droplets deform into ellipsoids with the same two short axes, while when the oil droplets undergo shear deformation, the oil droplets deform into ellipsoids with different two short axes, and the three-dimensional deformation model shows that the three main axes deform.

3.1. Affine deformation

In the turbulent shear flow field of a dynamic oil-water hydrocyclone separator, assuming that the dispersed phase oil droplets undergo affine deformation under viscous shear force, the oil droplets will deform along the OX axis under viscous shear force f, the deformation is δx .

According to the axisymmetry of Taylor's theoretical deformation, there will be deformation in the other two directions OY and OZ, the deformation is $-k\delta x$, k is the shrinkage coefficient of oil drop deformation, $k \in (0,1)$, According to the incompressibility of fluid and the volume invariance of oil droplets. The following relations can be obtained:

$$\frac{4}{3}pR^3 = \frac{4}{3}p(R + dx)(R - k\,dx)^2 \tag{1}$$

According to the solution of quadratic equation:

$$k = \frac{R - R\sqrt{1 - \dfrac{dx}{R + dx}}}{dx} = \frac{1}{e}(1 - \sqrt{\frac{1}{1+e}}) = \frac{1}{1 + e + \sqrt{1+e}} \tag{2}$$

Where, ε is the elongation of oil droplet deformation, $\varepsilon = \delta x / R$

The expression of affine deformation of a single oil drop is as follows:

$$\frac{x^2}{(R+\delta x)^2}+\frac{y^2}{(R-k\,\delta x)^2}+\frac{z^2}{(R-k\,\delta x)^2}=1 \tag{3}$$

Where, k is the coefficient of deformation and shrinkage of oil droplets.

From equation (2), it can be seen that k decreases with the increase of ε, the relationship curve was shown in figure 2. Through experiments Taylor found that small deformation of oil droplets occurs when $\varepsilon < 0.2$. If $\delta x << R$, then $\varepsilon \to 0$, $k = 0.5$, If, $\varepsilon = 10$ then $k = 0.07$, At this time, the oil droplets will undergo large deformation, showing a slender filament.

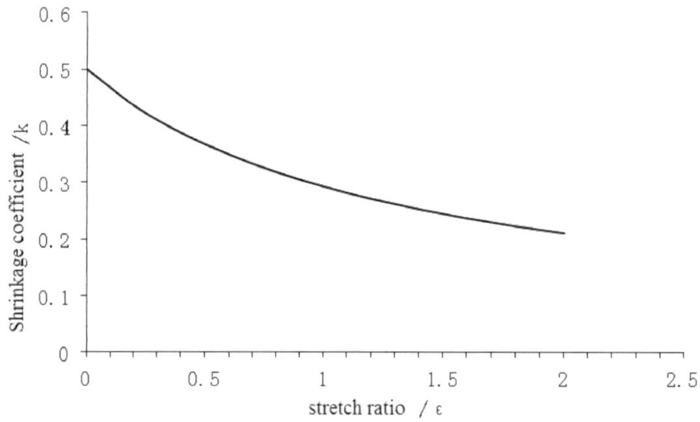

Figure 2. Shrinkage coefficient and stretch ratio of oil droplets deformation.

3.2. Shear deformation

The rotary drum of the dynamic oil-water hydrocyclone separator rotates at high speed driven by an electric motor. Assuming that the force and deformation of oil droplets in the turbulent shear flow field inside the rotary drum are shear deformation, under the action of viscous shear force *f*, the oil droplets have a deformation along the OX axis direction and the deformation is δx.

According to the characteristics of shear deformation, the deformation along the direction of *OY* is $-k\delta x$, where *k* is the coefficient of deformation and contraction of oil droplets, $k \in (0,1)$, and there is no deformation along the direction of *OZ*, According to the incompressibility of fluid and the volume invariance of oil droplets, the following relations can be obtained:

$$\frac{4}{3}pR^3 = \frac{4}{3}p(R + dx)(R - k\,dx)R \tag{4}$$

Thus the following formula can be obtained.

$$k = \frac{1}{1+e} \tag{5}$$

Where, ε is the elongation of oil drop deformation, $\varepsilon = \delta x / R$.

The expression of oil droplets after shear deformation is as follows:

$$\frac{x^2}{(R + dx)^2} + \frac{y^2}{(R - k\,dx)^2} + \frac{z^2}{R^2} = 1 \tag{6}$$

3.3. 3D deformation

Assuming that the interfacial tension σ is constant and the initial state of oil droplets is spherical, ellipsoidal deformation occurs in the turbulent shear flow field in the rotating drum of a dynamic oil-water hydrocyclone separator in a Newtonian oil-water mixture system, as shown in figure 3.

The relationship between point *A*, point *B* and point *C* after oil droplet deformation is as follows:

$$\text{point } A\text{: } p-\tau=\sigma(a/b^2+a/c^2);$$

$$\text{point } B\text{: } p=\sigma(b/a^2+b/c^2);$$

$$\text{point } C\text{: } p+\tau=\sigma(c/a^2+c/b^2).$$

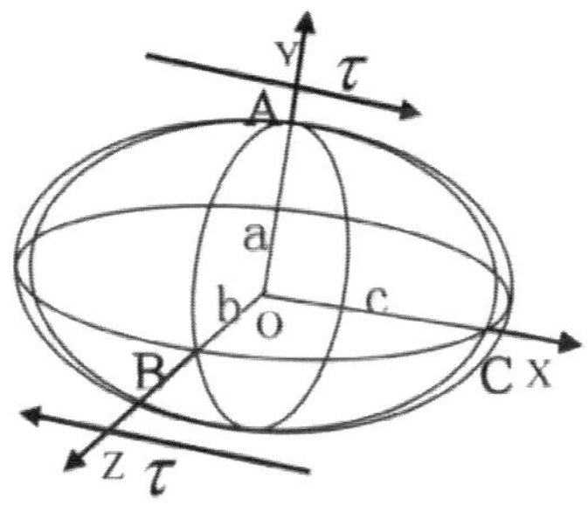

Figure 3. Three-dimensional deformation of single droplet.

In order to further study the deformation of oil droplets under shear conditions, it is assumed that the deformation of oil droplets in *OX, OY* and *OZ* directions is $R+\delta x$, $R-k_1\delta x$ and $R-k_2\delta x$, respectively. Because of the incompressibility of fluid, and considering that the pressure at point B does not change, it can be obtained that:

$$R^3=(R+\delta x)(R-k_1\delta x)(R-k_2\delta x) \tag{7}$$

$$2/R=(R-k_2\delta x)/(R-k_1\delta x)^2+(R-k_2\delta x)/(R+\delta x)^2 \tag{8}$$

Simultaneous equations (7) and (8), from which we can get:

$$\begin{aligned} &k_1^3[-2\delta x^2(R+\delta x)^3]+k_1^2(6R\delta x^4+18R^2\delta x^3+18R^3\delta x^2+5R^4\delta x) \\ &+k_1(-6R^2\delta x^3-18R^3\delta x^2-18R^4\delta x-5R^5)+R^3(2\delta x^2+5R\delta x+4R^2)=0 \end{aligned} \tag{9}$$

If $R>>\delta x$, the deformation of oil droplets is small and can be approximately obtained:

$$k_1=(9\varepsilon-2\sqrt{1+4\varepsilon+14\varepsilon^2}+2)/(5\varepsilon) \tag{10}$$

where, ε is the elongation of oil drop deformation, $\varepsilon=\delta x/R$, k_1 and k_2 are shrinkage coefficients of oil drop deformation, or shrinkage coefficients.

If the deformation of oil droplets is large, it can be calculated accurately:

$$\begin{aligned} k_1=\frac{1}{6\varepsilon(\varepsilon+1)}\Big[&\sqrt[3]{-54(\varepsilon+1)^8-1+3\sqrt{6}(\varepsilon+1)^4\sqrt{54(\varepsilon+1)^8+2}}+ \\ &\sqrt[3]{-54(\varepsilon+1)^8-1-3\sqrt{6}(\varepsilon+1)^4\sqrt{54(\varepsilon+1)^8+2}}+6(\varepsilon+1)^3-1\Big] \end{aligned} \tag{11}$$

From equation (7), it can be concluded that:

$$k_2 = \left[k_1(\varepsilon+1)-1\right] / \left[(\varepsilon+1)(k_1\varepsilon-1)\right] \tag{12}$$

Assuming that the initial state of oil droplets is spherical and elliptical deformation occurs, the relationship between the tensile rate of oil droplets and the shear stress can be calculated.

$$\tau = (\sigma / R)[(\varepsilon+1)/(1-k_1\varepsilon)^2 + (\varepsilon+1)/(1-k_2\varepsilon)^2 - 2] \tag{13}$$

After affine deformation of oil droplets, the expression of their shape is as follows:

$$\frac{x^2}{(R+\delta x)^2} + \frac{y^2}{(R-k_1\delta x)^2} + \frac{z^2}{(R-k_2\delta x)^2} = 1 \tag{14}$$

4. Model selection

In the oil drop deformation model established by Taylor [4,5] and Cox [6], it is assumed that the oil drop deforms into an ellipsoid with circular cross section, i.e. affine deformation, and the prediction formulas of deformation degree and orientation angle are given as follows.

$$D = \frac{L-B}{L+B} = \frac{Ca}{2} \cdot \frac{19\lambda+16}{16\lambda+16} \cdot \frac{1}{\sqrt{(19\lambda Ca/40)^2+1}}$$

$$\alpha = \frac{\pi}{4} + \frac{1}{2}\tan^{-1} 19\lambda Ca/40 \tag{15}$$

Choi S J [7] predicts the maximum cross-sectional radius of a single slender oil drop, with longitudinal and semi-axial ratios of longitudinal to transverse, respectively:

$$b_2 = 0578\frac{\tau}{\sigma\eta_c}Ca^{3/4},\ a_2 = 3.45\frac{\tau}{\sigma\eta_c}Ca^{2/3},\ p = \frac{a_2}{b_2} \cong 6Ca^{3/4} \tag{16}$$

It can be obtained that $b_2{}^2 a_2 = 1.15r^3$. Therefore, the predicted oil droplets are affine deformation, which coincides with the theoretical geometric deformation.

5. Conclusion

In this study, the deformation, fracture and coalescence of oil droplets in the rotating drum of a dynamic oil-water hydrocyclone separator are analyzed. The deformation of oil droplets in the shear flow field is assumed to be affine. For a certain concentration of oil-water mixture, the dispersed phase of oil droplets may exhibit deformation, fracture and aggregation in the flow field, but the deformation model should be consistent with that of a single oil droplet.

Reference

[1] Zhang X 1999 Dynamics of drop formation in viscous flows *Chem Eng Sci* **54** 1759-74
[2] Krüger T, Kaoui B and Harting J 2014 Interplay of inertia and deformability on rheological properties of a suspension of capsules *J Fluid Mech* **56** 211-33
[3] Li G, McKinley G H and Ardekani A M 2015 Dynamics of particle migration in channel flow of viscoelastic fluids *J Fluid Mech* **785** 486-505
[4] Taylor G I 1932 The viscosity of a fluid containing small drops of another fluid *Proc. Royal Society of London Series A* **138** 41-8
[5] Taylor G I 1934 The formation of emulsions in definable fields of flow *Proc. Royal Society of London Series A* **146** 501-23

[6] Cox R G 1969 The deformation of a drop in a general time-dependent fluid flow *J Fluid Mech* **37** 601-23
[7] Kunze A, Che J, Karimi A and Carlo D 2015 Research highlights: Cell separation at the bench and beyond *Lab Chip* **15** 605-9

Experimental study on inductive method for online material loss detection with high debris concentration

T Y Li[1,2], S P Wang[1], E Zio[2,3], J Shi[1,5] and W Hong[4]

[1]School of Automation Science and Electrical Engineering, Beihang University, Beijing 100191, China
[2]Energy Department, Politecnico di Milano, Via La Masa 34, 20156 Milano, Italy
[3]MINES ParisTech, PSL Research University, CRC, Sophia Antipolis, France
[4]MOE Key Laboratory of Fundamental Physical Quantities Measurement & Hubei Key Laboratory of Gravitation and Quantum Physics, PGMF and School of Physics, Huazhong University of Science and Technology, Wuhan 430074, China

E-mail: soredsun@163.com

Abstract. Online Oil debris detection is a complex and systemic issue that multiple requirements should be met for a practical application. When severe wear or tear happens, more accurate detection is needed for higher debris concentration. Aiming at promoting detection accuracy under the practical high debris concentration conditions, in this work, an experimental study is carried out to test the performance of an inductive oil debris detection method using a multi-channel sensor based on dual excitation sources with serial layout. Ferrous powders with different sizes and amounts are used for the experiment. Data processing methods including band-pass filter and anti-aliasing techniques are used for optimizing the estimation. The tested results indicate that the counting accuracy of coarse iron powder with 2.5 g/L concentration hydraulic oil at a flow rate of 40 L/min is higher than 85.8% and by data processing, the accuracy is promoted by 21.4% averagely.

1. Introduction

Wear and tear unavoidably brings about material loss [1]. By ferrography [2], people started to establish the relationship between debris features and characteristics of material loss [3]. Nowadays online debris detection methods were introduced and debris analysis has become a powerful approach for interpreting material loss and providing prognostic information. Existing methods include optics, capacitance, resistance, ultrasonic, x-ray and inductive methods [4]. Compared with the other methods, the inductive sensors are usually easily-installed on oil pipes and able to differentiate ferrous debris from non-ferrous debris [5].

Research studies [4,6] indicate that for engineering applications the sizes of debris usually range from 1μm to 150μm under normal work conditions. When severe wear happens, the sizes of debris will follow a different distribution. Several inductive sensors have been designed to promote the detection precision. Hong *et al* [7] proposed a radial inductive sensor which can detect 20μm thick ferromagnetic debris in 20 mm diameter pipes. Li *et al* [8] introduced a high throughput inductive pulse sensor which is able to detect 50μm debris in lubrication oil with a high throughput. Zhu *et al* [9] utilized a sensor array which is able to detect debris at a flow rate of 460 mL/min. High precision means identifying more debris from the same contaminative oil. High throughput results in short time-interval, which will

IOP Conf. Series: Materials Science and Engineering **629** (2019) 012031 doi:10.1088/1757-899X/629/1/012031

influence the detecting performance. Large diameter pipes will reduce the intensity and uniformity of magnetic field. To cope with these problems, data processing methods are employed. Band-pass filter and correlation algorithm may be used to improve signal-to-noise ratio [10]. Degenerate unmixing estimation technique can be used for aliasing signals separation [11]. The accurate detection for high concentration of debris should, then, be regarded as a systematic issue instead of focusing on structure designing or other individual aspects.

In this work, we designed an experiment setup to test the systematic performance of a former proposed inductive method [5] with a practical high throughput. The test results are compared with results processed using the aforementioned data processing techniques, aiming to identify potentially problems for practical online applications. The experimental procedure is presented in Section 2 and the results are elaborated in Section 3. Conclusion is drawn in Section 4.

2. Experimental procedure

The prototype of the inductive debris detection sensor based on dual excitation sources [5] is shown in **Figure 1**(a). Each sensor is designed with two channels whose outputs are theoretically equal but practically different because of noise. As is shown in **Figure 1**(b), hydraulic oil flows in the system driven by an axial piston pump. Motor connected to the pump is used to control the flow rate. Two sensors are installed serially on a 10 mm diameter pipe. Debris are injected into an injection device with the switch off. When the test begins, the switch is turned on and the debris flow with hydraulic oil into the tested part. Before and after the tested part, two oil filters are set to exclude the interaction among test groups. Signals are collected by a data acquisition device connecting to the computer.

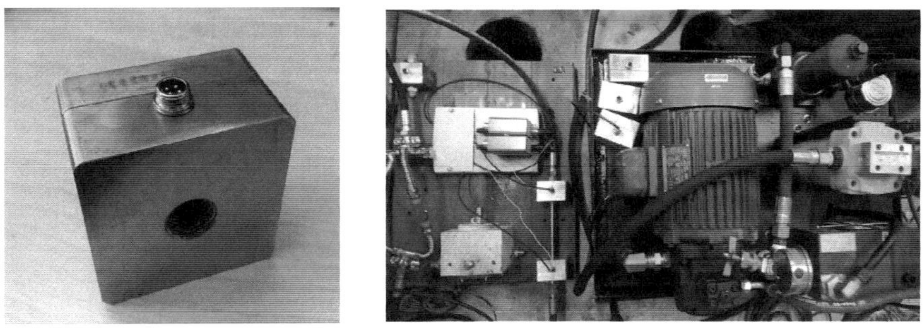

(a) (b)

Figure 1. Sensor and experiment setup.

The output pressure of the pump is 0.7 MPa. The flow rate is approximately 40 L/min which is close to the practical work condition and much larger than the laboratorial test conditions. The sample frequency of the data acquisition device is 10 kHz. Two types of iron powders are used for the test. The size of medium powder ranges from $50\mu m$ to $100\mu m$ and that of coarse power ranges from $150\mu m$ to $500\mu m$. The injection switch is turned on after the system runs 10 seconds for each test. An instantaneous fluid with high debris concentration will pass through the tested pipe. After each test group, the system runs for a while, so that the remaining powders flow away from the injection device and the powders are well-filtered by the oil filters.

3. Experimental results

Iron powders were employed to replace real abrasive debris. For the first test group, the pump was not started, signals from the two sensors were collected as shown in **Figure 2**. **Figure 2**(a) displays the signal from sensor 1. By converting the time domain data into frequency domain data, the frequency spectrum can be obtained as shown in **Figure 2**(b). It can be seen that the time domain data are noisy, but there are two strong peaks at 50 Hz and 150 Hz in the frequency domain. Obviously, the peak at 50 Hz is caused by the electromagnetic interference from commercial electrical power. Since no other device runs during the test, 150 Hz is caused by the harmonic interference from three-phase electric. Weak peaks can be found at each harmonic frequency like 250 Hz. **Figure 2**(c) displays the signal from sensor 2 and the corresponding converting result by discrete Fourier transformation can be found in **Figure 2**(d).

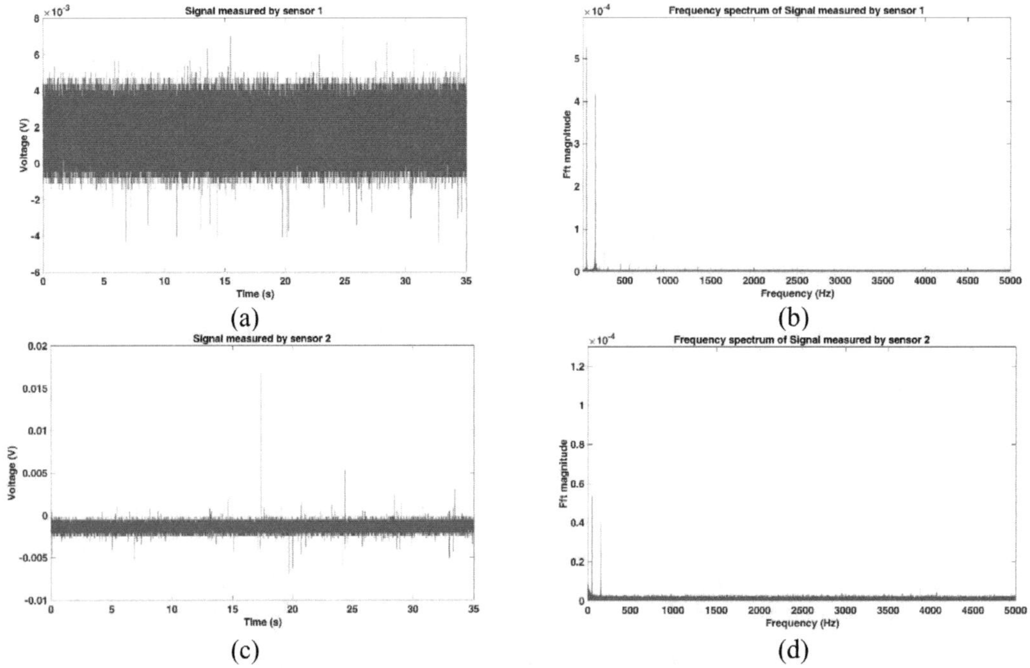

Figure 2. Experiment results when the pump stops.

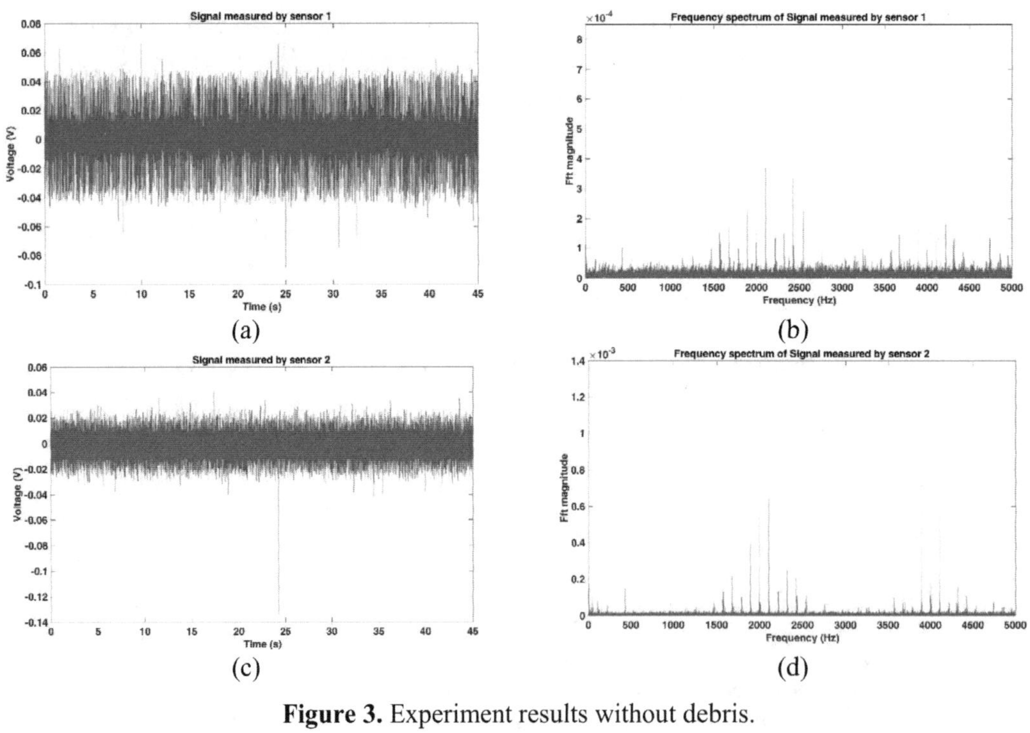

Figure 3. Experiment results without debris.

For the second test group, the pump started running but no powders were injected to the system. Signals measured by sensor 1 can be seen in **Figure 3**(a) and the corresponding frequency spectrum is shown in **Figure 3**(b). Compared with the first group, strong peaks occur at higher frequency near 2000 Hz, which is caused by the periodic rotation of the piston pump. Its harmonic interference occurs near 4000 Hz. **Figure 3**(c) displays the signal from sensor 2 and the corresponding convert result by discrete Fourier transformation can be found in **Figure 3**(d). Obviously, the method is sensitive to the vibration which is commonly seen in practical conditions.

Table 1. Experiment results.

Group	Debris	Concentration	Reference	Sensor 1	Sensor 2	Optimized result	Error
3	1g Medium	250mg/L	1.6×10^4	5893	5416	7504	0.531
4	2g Medium	500mg /L	3.2×10^4	6798	6545	12960	0.595
5	4g Medium	1g/L	6.4×10^4	15890	16341	27008	0.578
6	10g Medium	2.5g/L	1.6×10^5	25621	26848	49120	0.693
7	1g Coarse	250mg/L	600	436	401	552	0.08
8	2g Coarse	500mg /L	1200	726	792	1055	0.121
9	4g Coarse	1g/L	2400	1489	1450	2100	0.125
10	10g Coarse	2.5g/L	6000	4256	4534	5148	0.142

For group 3 to group 10, iron powders with different amounts were injected into the system. The parameters for each group are listed in **Table 1**. In group 3 to group 6, medium powders were tested and in group 7 to group 10, coarse powders were tested. The theoretical amount of powders is estimated by assuming that the density of the iron powders is 3 g/cm^3. The results of group 7 are shown in **Figure 4**. Unlike the results collected in laboratory conditions, all that can be seen is a noisy waveform that conveys little information. From the frequency domain, no evident difference can be found compared with the result shown in **Figure 3**.

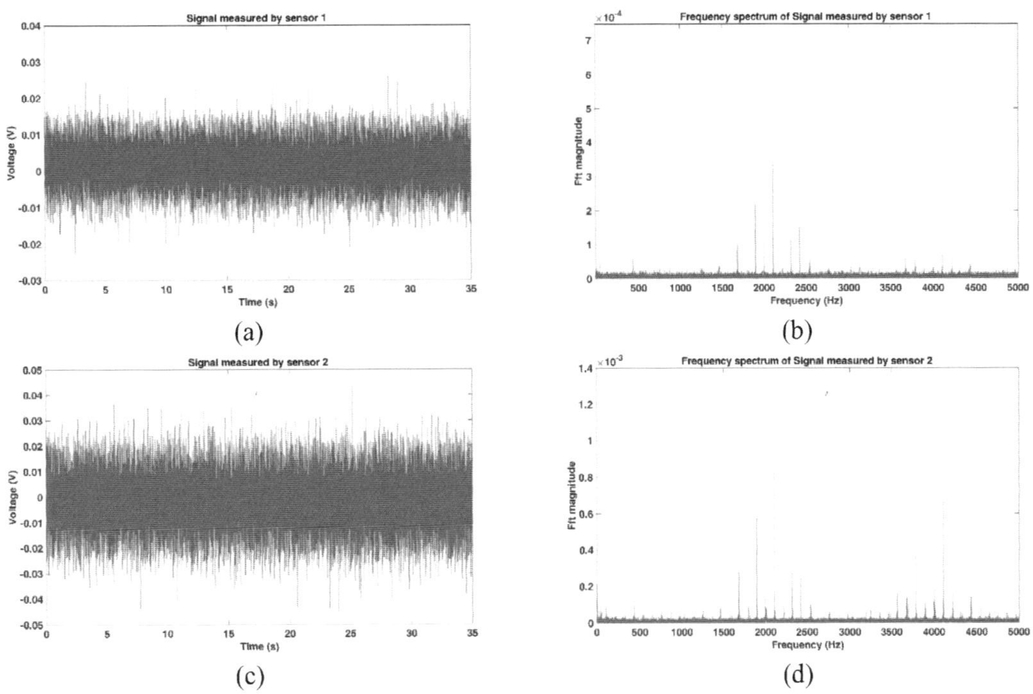

Figure 4. Experiment results with debris.

The induced voltages by debris are aperiodic sine-like waveforms whose frequency is around 100 Hz. According to the aforementioned methods, a 200-order digital band-pass filter using hamming window is designed as is shown in **Figure 5**. The range of the pass band is from 70 Hz to 130 Hz. The filter is implemented by MATLAB Filter Designer. The filtered results of the signals shown in **Figure 4**(a) and **Figure 4**(c) are shown in **Figure 6**(a) and 6(b), respectively. The filtered results are much distinct than the unfiltered results. It can be seen that at 10s, the powders were injected into the oil and about 2 seconds later, both the two sensors detected the debris. Obviously, some induced voltages are superimposed and noise by the harmonic interference still exists. By counting the peaks, the amount of debris can be obtained directly. The corresponding results are listed in **Table 1**.

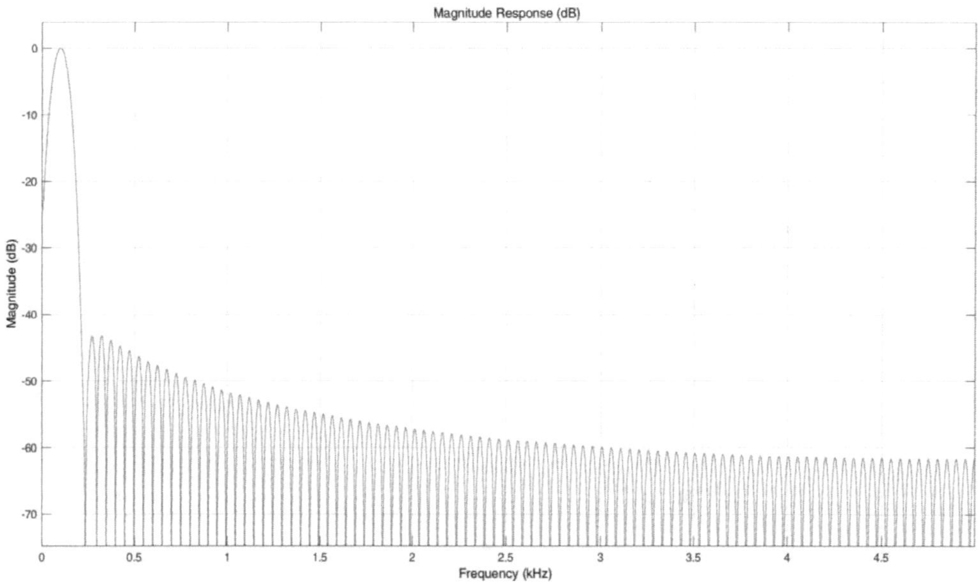

Figure 5. The designed band-pass filter.

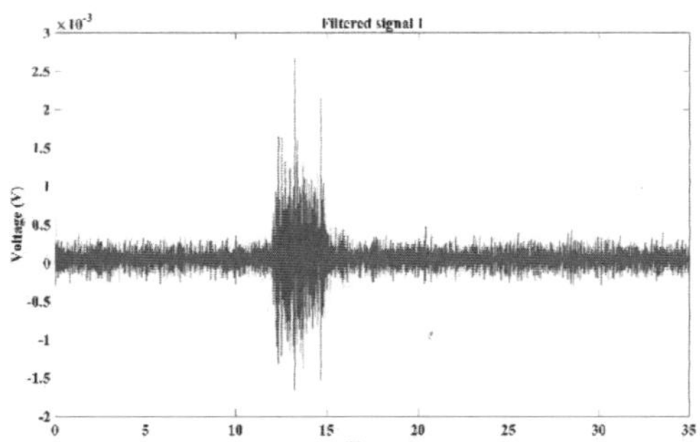

(a)

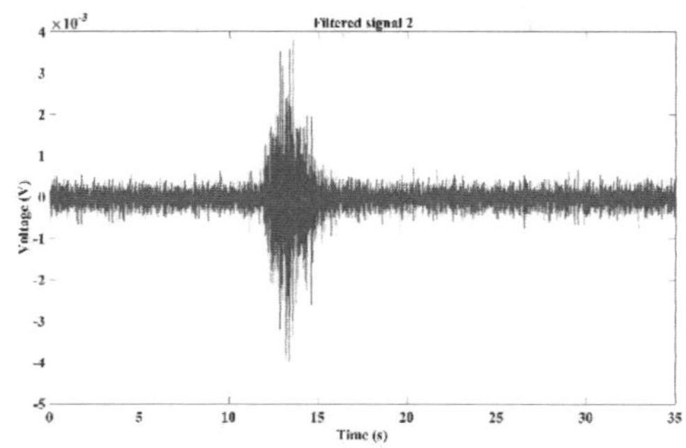

(b)

Figure 6. Filtered signals.

The estimated amount of debris work is as the references. The concentration is calculated by assuming that all the debris pass through the two sensors in 3 seconds. By counting the peaks directly, results from the two sensors are similar. Generally, the accuracy for coarse powder detection is better than the accuracy for medium powder detection. The performance for less debris is better than the performance for larger amount of debris, which is in accordance with the former experimental result. But the amounts estimated by sensor 1 are not always larger than sensor 2, which means that the differences between the two sets of results are not caused by the inherent discrepancy of the two sensors. More probably, the relative phase differences among debris change with the flowing viscous fluid. Methods structured with sensor arrays may suffer from the same problem, while the multi-channel methods [9] display good performance.

Well-designed sensors may integrate multi-channel and band-pass techniques, which will resist extraneous interference largely. However, the intrinsic aliasing problem caused by the high throughput cannot be precautionary. Post-processing is an indispensable part of this systematic issue. Here, we employed an anti-aliasing method based on degenerate unmixing estimation technique [11]. The superimposed signals are firstly separated and then, the amounts of debris are estimated from the optimized results. The error is calculated by

$$\text{Error}=1-\frac{\text{Optimized result}}{\text{Reference}}. \tag{1}$$

By conducting the anti-aliasing method, the detecting accuracy is improved compared with counting the peaks directly. The error increases with the increasing of amount of debris. The average accuracy promotion for the coarse powder detection is about 21.4%, and that for the medium powder detection is about 15.6%. The anti-aliasing optimization method is likely to be more effective for larger debris with lower concentration. This also indicates that the aliasing problem is more severe in high-throughput and high-precision conditions. After optimization, the method provides us acceptable results for large debris particles. However, the existing methods nearly fails in detecting medium powder with high-aliasing accurately.

4. Conclusion

Online inductive oil debris detection method is able to measure material loss in oil flow, which provides high-efficiency interpretation of material properties. By promoting the detection precision, small debris can be detected, in the meanwhile, the stronger effect of noise and the higher possibility of debris superimposition can no more be ignored. Therefore, simply pursuing the higher precision will provide less value for practical applications. The intrinsic aliasing of debris should also be considered as well as identifying useful signals from noise. Meanwhile, in practical application, the oil flow with high debris concentrations are commonly seen and external vibration and harmonic interference will be mixed in the ideally laboratorial signals. The practical signals may need more engineering techniques to recover the laboratorial level.

In this study, the inductive method is adapted on a real engineering application. Ferrous powders are used to present material loss. Combining the band-pass filter and anti-aliasing technique, the method is able to count coarse powder with minimum accuracy of about 85.8%. The method does not display good performance in counting medium powders that the maximum accuracy is about 46.9%.

Future works may concern promoting small debris detection accuracy with a relative high debris concentration by more effective noise control means.

Acknowledgments

This study was co-supported by the National Natural Science Foundation of China (51575019, 51620105010), Natural Science Foundation of Beijing Municipality (L171003) and Program 111 of China.

References

[1] Li T, Shi J, Wang S, Zio E and Ma Z 2019 Mesoscale numerical modeling for predicting wear debris generation *Tribol Lett* **67** 38
[2] Roylance B J 2005 Ferrography-then and now *Tribol Int* **38** 857-62
[3] Anderson D P 1982 Wear particle atlas (revised) *Foxboro Analytical Burlington Ma*

IOP Conf. Series: Materials Science and Engineering **629** (2019) 012031 doi:10.1088/1757-899X/629/1/012031

[4] Hong W, Cai W, Wang S and Tomovic M M 2018 Mechanical wear debris feature, detection, and diagnosis: A review *Chinese J Aeronaut* **31** 5-20
[5] Hong W, Wang S, Tomovic M M, Liu H and Wang X 2015 A new debris sensor based on dual excitation sources for online debris monitoring *Meas Sci Technol* **26**
[6] Hong W, Wang S, Tomovic M M, Liu H, Shi J and Wang X 2017 A novel indicator for mechanical failure and life prediction based on debris monitoring *IEEE T Reliab* **66** 161-9
[7] Hong W, Wang S, Tomovic M, Han L and Shi J 2013 Radial inductive debris detection sensor and performance analysis *Meas Sci Technol* **24** 5103
[8] Du L and Zhe J 2011 A high throughput inductive pulse sensor for online oil debris monitoring *Tribol Int* **44** 175-9
[9] Zhu X, Li D and Jiang Z 2017 A 3×3 wear debris sensor array for real time lubricant oil conditioning monitoring using synchronized sampling *Mech Syst Signal Pr* **83** 296-304
[10] Hong W, Wang S, Liu H, Tomovic M M and Chao Z 2016 A hybrid method based on band pass filter and correlation algorithm to improve debris sensor capacity *Mech Syst Signal Pr* **82** 1-12
[11] Li T, Wang S, Zio E, Shi J and Hong W 2018 Aliasing signal separation of superimposed abrasive debris based on degenerate unmixing estimation technique *Sensors* **18** 866

IOP Conf. Series: Materials Science and Engineering **629** (2019) 012032 doi:10.1088/1757-899X/629/1/012032

Effects of using metal granules on strength and stiffness

N U Kockal[1,2], T Z Kocaer[1]

[1]Department of Civil Engineering, Akdeniz University, Antalya, 07070, Turkey

E-mail: ukockal@yahoo.com

Abstract. Using metal granules, which are waste products of the metal industry, as a replacement material in mortars, is economically productive and environmentally beneficial because it contributes to reduction of the amount of the metal waste. It was attempted to develop and produce metal granule reinforced cementitious composite instead of using conventional fibers. Metal wastes bonded to the matrix with the help of adhesion effect is aimed to improve the mechanical and deformation properties of the mortar. As the amount of metal waste increased in the mortar, the compressive strength of the mortars except those with 10% granules enhanced. The rise in the maximum stress values of the samples and the growth in the modulus of elasticity were similar. 30% metal waste mortar demonstrated the highest value of compressive strength and modulus of elasticity in this study.

1. Introduction

As the human population continue to increase, there is a boost of productions. This leads to a rise of production waste as well. For this reason, many substances are found in soil, air and water that threat the health of living things. As time passes by, difficulty appears in eliminating and storing such materials. This pollution in nature also causes economic hardship. Consequently, the reuse of these substances, which industrial wastes (such as slag and metal granules) are the part of, has become the subject of scientific studies. Recently, various studies have been carried out especially on the use of metal granules instead of metal fibers. This may be attributed to the low price of metal granules, that is a by-product of the lathes, compared to metal fibers [1].

Depending on their characteristic, the recycled aggregate produced from construction and demolition wastes can be used in different engineering works that can help circuitously the economic and environmental sustainability of the various countries. With further research and development to overcome technical and market barriers, a meaningful increase in improvement rates can be achieved with current technologies in developed economies [2].

In pig iron production, the impurities in the iron sprout, coke ash and the residues from burning of limestone form slag. The slag obtained from the furnace takes the grain structure according to the cooling method. When it is cooled slowly in the air, it has a crystal structure. It is almost completely glassy when it is cooled in water. The total amount of CaO, MgO and SiO_2 in the slags is expected to be at least 67% [3].

For the safety, durability and utility of structural materials, compressive strength and the elastic properties are the most important parameters [4]. They are used in some cases to fulfill eventual requirements and serviceability limit position such as in the design of pre-stressed concrete structures. Particularly, these parameters come into prominence in risky earthquake zones. Obtaining the compressive strength of a building material is not only important input for the strength of the materials.

For example, approximations might be made about the volume of pores in a known material if the compressive strength is attained. The modulus of elasticity allows us to get information about stiffness of a material. The stiffness enhances due to the tangent of the slope in linear area of the strain-stress curve rises.

The slag obtained by different methods are used in cement based composites as binder or aggregate considering the grain size. The granulated blast-furnace slag had been used in concrete as binder instead of cement, resulted in high performance in terms of long-term strength in concrete [5]. Electric arc furnace (EAF) steel slag produced which currently account for more than 40% of global steel production is still recognized as a solid waste by-product that needs to dispose of. Several studies have been accomplished to characterize EAF steel slag with respect to its application in the construction industry such as its attributes as an additive material and having a great potential for development in the existence of cement mineral compounds [6].

On a study, metal granules in a cementitious composite were determined the roles to ensure what points they can replace traditional fibers. As a result of the tests, it was achieved that compressive strength of the cementitious composite reinforced metal granules was more 18% than the traditional concrete [7].

Mortar, which is a building material obtained at the desired consistency, granulated aggregate (which its maximum size is 4,75 mm), cement as a binder, enough water and additive supplemented as required are homogeneously mixed. The properties of the mortar depend on the aggregate, matrix, metal reinforcement as well as the zone between the aggregate and matrix, and the metal and matrix. These zones state many of the important properties of mortar, such as strength and cracking behavior.

The aim of this study was to explore new fields by using waste materials, to obtain values by analyzing the effects of metal granules on strength and stiffness of the mortar, and to be able to lead to further studies if efficient results were achieved. It might be also economically beneficial to reuse wastes instead of stocking or disposing. Experimental scope was that the results were collected about the strength and hardness of the samples by compressive strength test.

2. Materials and method

2.1. Materials

The materials involving in the mixture were first put in the oven at 105±5 °C until the moisture was removed before mixing. In this study, slag was provided by Ferrokrom Factory. The slag with a maximum grain size of 4 mm was used as aggregate in the mortar samples. The metal granules were substituted at certain ratios. The metal granules with the same chemical compositions and properties were used at ratios of 10% and 30% by volume in 10% intervals. CEM I 42,5 R was used as a binder in the blends. Table 1 presents the mix proportions of mortars.

The blend was mixed in a benchtop laboratory mixer. Due to the different physicochemical properties of the components, super plasticizer was used as needed. As a result of the test mixes for the production, suitable component ratios, strength and workability were taken into consideration and in order to be able to compare results of the mixtures, while the flow diameter was kept at a certain range in the consistency table test. Thus, the cement dosage and water / cement ratio were determined as 400 kg / m^3 and 0.7, respectively. Beyond 30% of metal granules addition caused segregation. Thus, the maximum limit of the amount of granule replacement was determined as 30%.

Each mixture was placed in the steel molds after it became homogeneous in the mixer. The surface of the mold was covered with plastic foil and the samples were kept in for 24 hours to let them set and harden. In this step, plastic foil was used to prevent the loss of water required for hydration. The dimensions of the solid samples were 40x40x160 mm.

As soon as demoulding, solid samples were allowed to cure in the tank filled with water, which its temperature was 20±2 °C. After curing for 28 days, the samples were taken from the curing tank. Then, they were stored in laboratory conditions at 22±2 °C for 4 hours before the test.

Table 1. Mix proportions of mortars.

	Cement (kg/m³)	Water (kg/m³)	Metal granules (kg/m³)	Slag (kg/m³)	Super plasticizer (kg/m³)
M0	400,0	417,4	-	1716,4	-
M10	400,0	376,8	419,0	1209,0	-
M20	400,0	390,6	837,9	1381,7	4,0
M30	400,0	376,8	1256,9	1554,4	6,8

2.2. Method

The prism samples were used to determine the compressive strength and modulus of elasticity. The compressive strength tests were performed and the modulus of elasticity calculated at 28 days for each mortar which were divided into two pieces by flexural strength test. The compressive strength was examined by displacement controlled hydraulic press. The pace rate was 2 mm per a minute. At least, 3 specimens were determined each test.

The modulus of elasticity was concluded by stress – strain curve. The modulus can be determined by some other calculation methods as well [8]. Equation (1) is eligible to find the modulus of elasticity in this study [9].

$$E = \frac{\sigma_2 - \sigma_1}{\varepsilon_2 - \varepsilon_1} \quad (1)$$

Where:
σ_1: stress at a longitudinal strain
σ_2: stress at 40% of ultimate load
ε_1: of $5\text{x}10^{-05}$
ε_2: longitudinal strain produced by stress σ_2

The bulk density and volume of permeable pore space rates were obtained by testing prism samples according to ASTM C 642 [10]. ASTM C 109 was used to plot the stress-strain curves [11].

3. Results and discussion

3.1. Compressive strength results

The compressive strength test results are presented in Table 2. The compressive strength values of M0, M10, M20 and M30 were specified 14.15, 13.76, 17.91 and 18.61 MPa, respectively. In comparison with M0, while compressive strength of M30 and M20 incremented by 31.52% and 26.57% respectively, M10 resulted in 2.76% reduction.

Using metal waste powders didn't perform well in a study [12]. Replacement aggregate with certain percentages made the mortars compressive strength regularly down. %8 metal existence caused the most reduction. Also in another study [1], it was declared that %3 metal presence lowered the compressive strength by %1.51.

Compressive strength is related to the quantity of voids in a sample. It can be assumed that a rise in compressive strength can be generally achieved by lessening of porosity in the mortars [13]. The existence of metal granules in the mortar resulted in development at the strengths of other mixtures, except M10. Although the volume of permeable porosity gives information about the spaces inside the mortars, it does not reflect actual situation about the total porosity. Because the rate of closed porosity is not exactly known by the method. The strength reduction in the M10 might be caused by a large number of closed porosity compared to the reference samples'. Besides that, these results were only data obtained by curing for 28 days. Although it is not indicated here, M10 was slightly higher than M0 in the test results of 90-day samples. Standard deviation of M0 and M10 are 1.00, 0.82, respectively. Since the values are close to each other, it is possible that the results remain unchanged.

Table 2. Test results of mortar samples.

	Oven dry bulk density	Volume of permeable pore space (%)	Compressive strength (MPa)	Modulus of elasticity (GPa)
M0	2.18	22.95	14.15	7.23
SD[a]	0.04	0.34	1.00	-
M10	2.21	28.92	13.76	10.27
SD	0.02	1.01	0.82	-
M20	2.48	27.09	17.91	12.76
SD	0.03	0.11	0.43	-
M30	2.61	25.96	18.61	14.16
SD	0.02	0.35	0.30	-

[a] Standard deviation

3.2. The modulus of elasticity

Stress – strain curves, which were presented in figure 1, were obtained as a result of the compressive strength test. The modulus of elasticity was calculated by equation (1) for each mixture. The values of the modulus were exhibited on table 2.

Considering the modulus of elasticity at the samples, M30 had the highest rate as 14.16 GPa. The ascent rates of M20, M10 in modulus of elasticity were 76.65% and 42.15%, respectively, while comparing the other mortars with the reference mortar (M0). This rate reached 95.96% for M30.

While comparing slag aggregated concrete with traditional concrete, slag affected slightly decrease on the modulus of elasticity [14]. It was explained that slag porosity was higher than the natural aggregate.

The modulus of elasticity of M10 reached higher ratios while the strength was low than M0, because the stiffness of the metal granules was higher than the slag.

Compressive strength is not the only influence on achieving for the modulus of elasticity. Such as volumetric ratio, considerably affects cementitious composites strain due to the anisotropic structure of materials [15]. The interface between aggregate and cement paste is an important parameter in the elastic behavior of the composites as well.

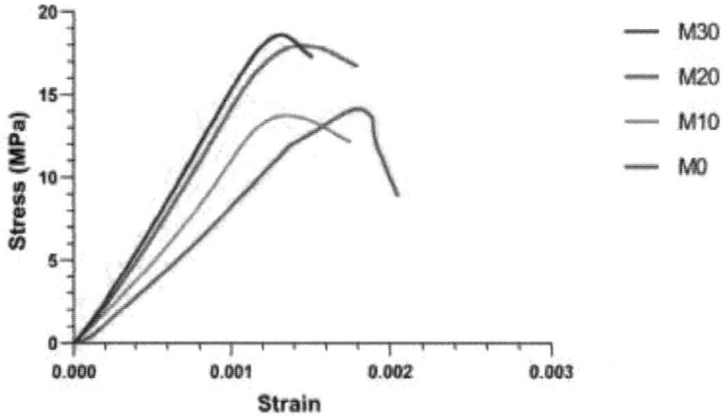

Figure 1. Stress - strain curves of the mortars.

4. Conclusions

In the study, using metal granules in mortars remarkably enhanced their modulus of elasticity values. Especially, M30 had the highest values under the load. Although the metal existence in the mortar enhanced the volume of permeable pore spaces, the modulus of elasticity did not lower their performance. As the strain parameter in the graph was taken into consideration, metal granules allowed to more rigid property gain for mortars. If the 28 days M10 slight strength drop was ignored, the presence metal granules caused an effective rise in terms of strength and stiffness. While the curves of M20 and M30 on figure 1 were observed, the strength and the stiffness approached the same behaviours.

The optimum solution for this study is related to workability and compressive strength. Maximum strength and stiffness were achieved at 30% volume fraction within an acceptable range.

Acknowledgements

This research was supported by TUBITAK of the project number of 216M078.

References

[1] Hong S, Sato Y, Kaneko Y and Kimura S 2015 *Journal of Advanced Concrete Technology* **13** 393-402
[2] Tam V W, Soomro M and Evangelista A C J 2018 *Construction and Building Materials* **172** 272-92
[3] Tokyay M 2013 *Beton 2013 Hazır Beton Kongresi* 201-38
[4] Kozul R and Darwin D 1997 Effects of aggregate type, size, and content on concrete strength and fracture energy (*Kansas: University of Kansas Center for Research Report*)
[5] Li G 2004 *Cement and Concrete research* **34** 1043-9
[6] Jahangiri N, Koseoglu K and Cengizler H 2017 *Dicle University Institute of Natural and Applied Science Journal* 6, 29-38
[7] Djebali S Bouafia Y Larbi S and Bilek A 2014 *In Key Engineering Materials* **592** 672-5
[8] Erdogan T Y 2007 *Beton* (Ankara: METU Press Publishing Company) **2** 566-8
[9] Kockal N U 2015 *IOP Conference Series: Materials Science and Engineering* **103** 1-6
[10] ASTM C 618 1998 *Annual Book of ASTM Standards*
[11] ASTM C 109 2001 *Annual Book of ASTM Standards*
[12] Chowdhury S, Roy S, Maniar A T and Suganya O 2014 *APCBEE procedia* **9** 317-22
[13] Chen X, Wu S and Zhou J 2013 *Construction and Building Materials* **40** 869-874
[14] Saravanan J and Suganya N 2015 *International Journal of Engineering Inventions* **4** 7-16
[15] Topcu I B and Ugurlu A 2007 *Teknik Dergi* **18** 4055-67

IOP Conf. Series: Materials Science and Engineering **629** (2019) 012033 doi:10.1088/1757-899X/629/1/012033

Characterization of Fe-Cu-Ni composites sintered by hot-press

J Borowiecka-Jamrozek[1,3] and J Lachowski[2]

[1]Faculty of Mechatronics and Mechanical Engineering, Department of Metal Science and Materials, Kielce University of Technology, Kielce, Al. Tysiąclecia PP 7, Poland
[2]Faculty of Management and Computer Modelling, Department of Applied Computer Science and Applied Mathematics, Kielce University of Technology, Kielce, Al. Tysiąclecia PP 7, Poland

E-mail: jamrozek@tu.kielce.pl

Abstract. The main objective of the work was to determine the effects of powder composition on a microstructure and properties of iron-base materials used as matrices in diamond impregnated tools. Mixed powders and 30 hours Fe-Cu-Ni milled powders were used for the experiments. The influence of manufacturing process parameters on microstructure and mechanical properties of sinters was investigated. Sintering was performed in a graphite mould using a hot-pressing technique. The powders were consolidated to a virtually pore-free condition during 3 min holding under 35 MPa and at 900°C. The investigations included: density, hardness, static tensile test and X-ray diffraction (XRD) analysis. Observations were also made of microstructure and fracture surfaces of broken samples using a scanning electron microscope (SEM) were also conducted. The SEM observations revealed an evident dependence of grain size and microstructural homogeneity on milling time. The obtained test results indicate that the sinters are virtually densified, have good plasticity with relatively high hardness and yield strength, but sinters of non-ground powders are characterized by a coarse-grained microstructure.

1. Introduction

Cutting tools, such as circular and wire saws, drills, milling cutters or grinding wheels are traditionally fabricated by powder metallurgy (PM) consolidation of diamond-cobalt mixtures [1,2]. The progress in the tools manufacturing routes, using PM techniques, resulted in developing diamond grit impregnated saw blades, which were employed around 1940. Metal bonded diamond composites were applied after World War II [3-5].

A tool selection depends both on properties of a workpiece (its hardness and abrasion resistance) and cutting conditions (linear speed, cooling efficiency and cutting rate, etc.). Such factors affect both the tool geometry as well as the composition and structure of diamond segments, which allows engineers to design a tool with a desired shape and diamond segments with an appropriate form and composition. The lifespan of diamond-impregnated segments primarily depends on diamond crystals retention of and tribological properties of the metal matrix [6,7].

Recent advanced development in cutting tools manufacturing are due the progress in PM technology, but an increasing industrial production of synthetic diamond has also been of prime importance. China is the largest manufacturer of synthetic diamond in the world. In 2010, China's production exceeded of 7 billion carats. The annual average increase rate has reached nearly 20% for the last 10 years. It accounts for about 90% of the world's production. China's diamond is being exported to 58 countries and regions [8].

The best and the most common metal bond material is cobalt. It has been widely used in diamond tools for several decades. Cobalt combines good chemical compatibility with diamond at the processing temperatures, an excellent diamond retention, associated with a satisfactory wear resistance after some cutting operations. One of the major shortcoming of cobalt, however, is its high and unstable price, which has significantly affected tool production costs since the beginning of the new millennium, when diamond became a commodity product [9]. Nevertheless, cobalt is a strategic metal – produced only by few countries – and therefore cobalt is no longer the best choice for some applications. In addition, cobalt is toxic substance.

Significant changes in the cost of raw materials as well as a relative decrease in the other production costs have forced toolmakers to find cheaper alternatives [10].

Considering all the negative aspects, researchers have recently proposed new iron-based alloys which could be used as a matrix in diamond tools [11].

The main objective of the study was to determine the suitability of ball-milled Fe-Cu-Ni powder mixtures for fabrication of diamond-impregnated metal-matrix composites. The combined effects of chemical composition, powder milling conditions and sintering parameters on the as-consolidated microstructure and mechanical properties were studied. The obtained results were compared with properties of a hot pressed SMS (sub-micron size) grade cobalt powder.

2. Experimental procedure

The experimental powder mixture was made from:

- Höganäs NC100.24 grade, carbon-reduced iron powder (20-180 μm)
- ECKA CH-L10 grade, electrolytic copper powder (<45 μm)
- Vale T255 grade, carbonyl nickel powder (Fisher Sub-Sieve Size = 2.4 μm).

Morphologies of the starting powders are shown in figure 1.

Figure 1. Experimental powders: (a) NC100.24; (b) CH-L10 and (c) T255.

Prior to consolidation, the powder mixture containing 60% Fe, 28% Cu and 12% Ni was prepared by blending the constituent powders in a Turbula-type mixer for 30 minutes. Then, the powders were ball-milled for 30 hours in ethyl alcohol with a small amount of glycerol using the EnviSense RJM-102 laboratory mill. The milling vial was filled to half of its volume with 12 mm diameter 100Cr6 steel balls. The ball-to-powder weight ratio was 10:1. The milling vial was set to run at 150 rpm, which is about 70% of the critical rotational speed. The particle shape of as-mixed and the ball-milled powder is shown in figure 2.

(a) (b)

Figure 2. SEM micrograph of the particle shapes of a) mixed powders and b) the 30-hour ball milled powders.

The mixed powders and ball-milled powders were tested for their particle size distribution, by means of the HELOS (H2769) & RODOS laser particle sizer with WINDOX 5 software enabling the measurement of particle sizes in the range of 0.1 μm-2000 μm. The results are shown in figure 3.

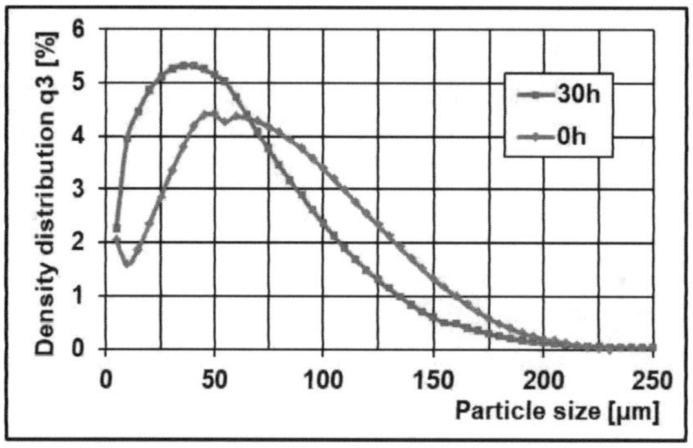

Figure 3. Particle size distribution of the starting mixed powders and ball milled for 30 h.

The primary mixture and the ball- milled powders were subjected to consolidation by hot pressing in a graphite mould. The hot pressing process was performed in the *ARGA CAR1001* hot press furnace in nitrogen. The pressing temperature was properly selected to obtain the as-sintered porosity lower than 3%. Hence, the powder was held at 900°C and 35 MPa for 3 minutes.

The sinters made from the mixed and the ball-milled powders were subjected to an X-ray phase analysis using a PANalytical Empyrean X-ray diffractometer using a copper anode lamp (Cu Kα1= 1.5406 Å). The identified phases of the samples are shown in figure 4.

The sintered specimens were first tested for density and hardness. The density measurements involved weighing the specimens in air and water using the WPA120 hydrostatic weighing system,

IOP Conf. Series: Materials Science and Engineering **629** (2019) 012033 doi:10.1088/1757-899X/629/1/012033

according to the PN EN ISO 2738:2001 standard. The results were also used to assess the as-sintered porosity. The hardness of the sinters was measured using the Vickers method for the load of 10 kG. The results are summarized in table 1.

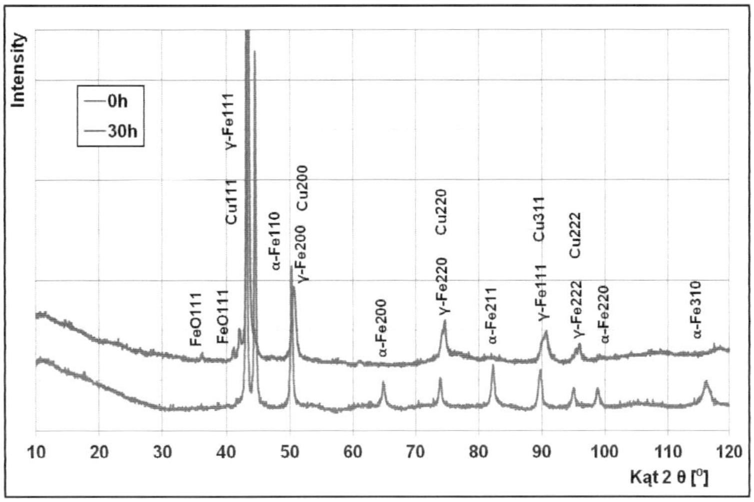

Figure 4. X-ray spectrum of the sinters obtained from (a) the mixed powders, (b) the ball milled powders.

Table 1. Density, porosity and hardness [a].

Fe-Cu-Ni material	Density [g/cm^3]	Theoretical density [g/cm^3]	Porosity [%]	HV10
Mixed powders	7.82 ±0.03	8.25	5.18 ± 0.20	157.6 ± 12.1
Milled for 30 h	8.10± 0.02	8,25	1.79 ± 0.25	300.9 ± 16.3

[a]scatter intervals estimated at 90% confidence level

The specimens were then machined, by turning, to produce non-standard specimens for static tensile tests. The tensile strength tests were carried out using the INSTRON 4502 universal testing machine The diameter of the gage section of specimens was 3.5 mm. The cross head speed was set to 0.5 mm/min. The specimen elongation was registered by means of an extensometer with a gauge length of 10 mm. The results were then used to calculate the offset yield strength $R_{0.2}$, ultimate tensile strength R_m and elongation ε. The results of the static tensile strength test and a typical stress-strain curve are given in table 2 and plotted in figure 5, respectively.

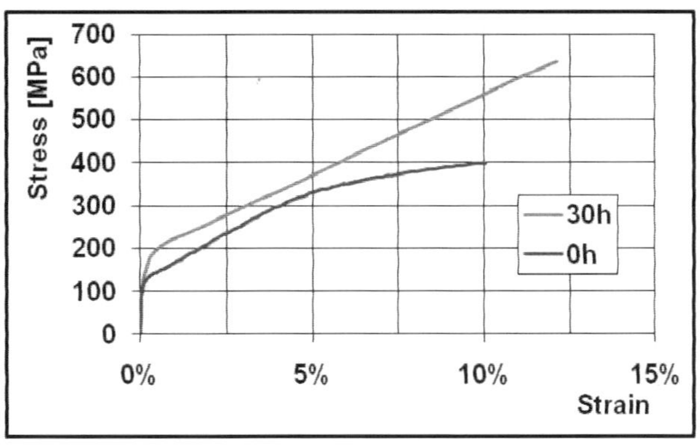

Figure 5. Stress-strain curve.

Table 2. Static tensile test results[a].

Fe-Cu-Ni material	Offset yield strength $R_{0.2}$ [MPa]	Ultimate tensile strength R_m [MPa]	Elongation ε [%]
Mixed powders	137 ±10	396.9 ±7.5	10.0 ±0.20
Milling for 30 h	229 ±15	634.9 ±13.3	12.1 ±0.45

[a] scatter intervals estimated at 90% confidence level

The fractured specimens were examined fractographically using the JSM-7100F scanning electron microscope fitted with an OXFORD INSTRUMENTS X-Max-AZtec EDS system. The grip sections of tensile specimens were also used to produce metallographic specimens for microstructural observations. A typical fracture surface and microstructure are shown in figure 6(a) and figure 6(b), respectively.

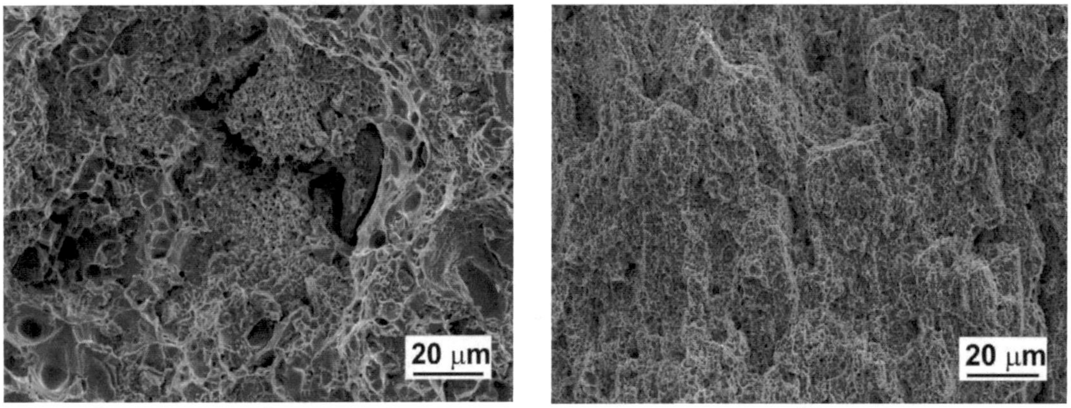

(a) (b)

Figure 6. Typical fracture surface after a tensile strength test obtained for (a) the mixed powders and (b) the milled powders.

The microstructures of the experimental alloys were examined and analyzed by means of a SEM fitted with an energy dispersive spectroscopy (EDS) system (figures 7(a) and 7(b)). Table 3 presents chemical compositions obtained from the EDS analysis in micro-areas indicated in figure 7.

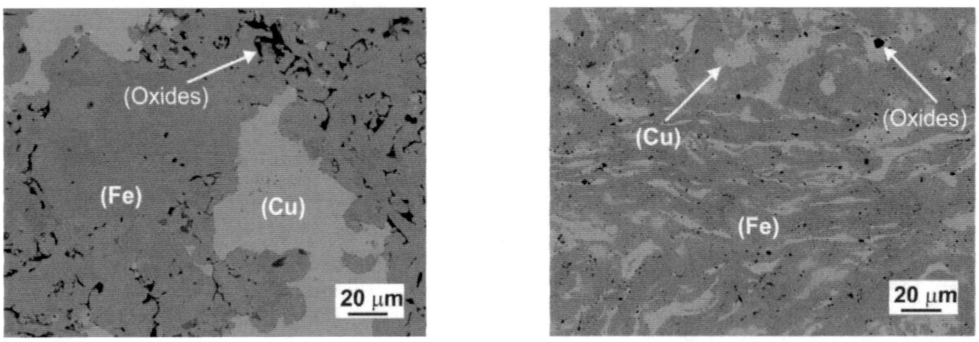

(a) (b)

Figure 7. Microstructure of the sintered Fe-Cu-Ni material, (a) mixed powders and (b) ball-milled powders.

Table 3. Chemical compositions in selected areas shown in figure 7 (weight %).

	Mixed powders				30-hour milled powders			
	O	Fe	Cu	Ni	O	Fe	Cu	Ni
(Fe)	---	73.9	21.2	4.9	---	85.97	3.99	10.04
(Cu)	---	2.31	97.5	0.19	---	5.75	89.38	4.87
(Oxides)	17.84	75.08	1.86	5.22	19.04	72.73	4.25	3.98

3. Discussion

The objective of the study was to fabricate sintered materials using inexpensive iron-based powders and to assess their potential application as a sintered matrix in diamond impregnated tools. The hot-press parameters were carefully selected to obtain the as-sintered density approaching the nearly full density (table 1). It was found that the mixed and ball-milled powders could be consolidated to a virtually pore-free state by the hot press route at the temperature of 900°C during 3 m under the pressure of 35 MPa. During milling, the powder particles are repeatedly flattened, work-hardened and fractured, and welded together. After 30-hour ball milling, they became flaky in shape (figure 2).

After consolidation, the alloys possess high hardness, which considerably increases during milling. The ball-milled material shows a good combination of high strength and ductility. The obtained test results indicate that the produced sinters have roughly theoretical density, good plasticity with relatively high hardness and yield strength, but sinters premixed are characterized by a coarse-grained microstructure.

Nevertheless, the microscopic examination of the metallographic sections revealed marked differences in microstructural inhomogeneity of the tested specimens. The ones manufactured from the mixed powders showed a highly inhomogeneous distribution of structural components. After milling, a strong grain refinement was discovered. The sintered microstructure of the material made of the milled powder for 30 h was generally fine sized (figure 7(b)). Both the ground and unground material revealed the presence of following phases: solid solution-type Fe, Cu, and oxides.

The XRD analysis showed that the α-Fe peaks of the sinters obtained from the milled powders disappeared. There are lines indicating the occurrence of γ-Fe iron which are connected with copper lines. The sinters obtained from milled powders revealed weak lines of FeO (II) oxide. However, there is no line of Ni, which indicates the diffusion of Ni atoms to Fe and Cu phases (figure 4).

The fractographic examination of the sinters produced from the milled powders showed ductile dimpled fracture surfaces. Whereas, for the specimens obtained from the mixed powders the fracture surfaces varied, with ductile fracture being predominant (figure 6).

The addition of nickel and copper to iron resulted in high hardness and tensile strength of the alloy and acceptable ductility.

4. Conclusion

The experimental data imply that the material obtained from the powders ball-milled for 30 h and hot-pressed at 900°C is characterized by a fine-grained microstructure and inhomogeneous microstructure. It should be emphasized that although the analysed material shows slightly lower strength and ductility as compared with cobalt [9-11], its hardness and offset yield strength are sufficiently high to meet the criteria for less demanding applications, such as professional general-purpose tools.

Undoubtedly, the tested Fe-Cu-Ni material deserves further attention because of its attractive price and the ease of consolidation by hot-pressing. Presumably its strength and plastic properties can be improved by fine tuning its chemical and powder composition as well as the powder processing and consolidation conditions.

References

[1] Hughes F H 1980 The early history of diamond tools *IDR* **11** 405-7
[2] Xu P, Hong Y B and Chen S 2004 Performance of diamond segments in different machining processes *Mater Sci Forum* **471-472** 77-81

[3] Tönshoff H K *et al* 2002 Diamond tools in stone and civil engineering industry: Cutting principles, wear and applications *Diam Relat Mater* **11** 736-41
[4] Oliveira L J *et al* 2007 Processing and characterization of impregnated diamond cutting tools using a ferrous metal matrix *Int J Refract Met H* **25** 328-35
[5] Oliveira F A C, Anjinho C A, Coelho, Amaral P M and Coelho M 2017 PM materials selection: The key for improved performance of diamond tools *Metal Powder Report* **72** 339-44
[6] Mechnik A 2014 Production of diamond-(Fe-Cu-Ni-Sn) composites with high wear resistance *Powder Metall Met C+* **52** 577-87
[7] Barbosa A P *et al* 2010 Structure, microstructure and mechanical properties of PM Fe–Cu–Co alloys *Materials and Design* **31** 522-6
[8] Li Z H, Zhao B, Li L and Wong J 2011 China's diamond in the world market Key Engineering *Materials* **487** 526-32
[9] Borowiecka-Jamrozek J and Lachowski J 2016 Properties of sinters produced from commercially available powder mixtures *Archives of Foundry Engineering* **16** 37-40
[10] Borowiecka-Jamrozek J and Konstanty J 2014 Microstructure and properties of a new iron-base material used for the fabrication of sintered diamond tools *Adv Mat Res S-Switz* **1052** 520-3
[11] Borowiecka-Jamrozek J 2013 Engineering structure and properties of materials used as a matrix in diamond impregnated tools *Arch Metall Mater* **58** 5-8

IOP Conf. Series: Materials Science and Engineering **629** (2019) 012034 doi:10.1088/1757-899X/629/1/012034

Influence of hot isostatic pressing on mechanical response of as-built SLM titanium alloy

M Roudnicka[1,3], K Mertova[2] and D Vojtech[1]

[1]Department of Metals and Corrosion Engineering, University of Chemistry and technology Prague, Technicka 5, 166 28 Prague 6, Czech Republic
[2]COMTES FHT a.s., Prumyslova 995, 334 41 Dobrany, Czech Republic

E-mail: michaela.roudnicka@vscht.cz

Abstract. Hot Isostatic Pressing (HIP) can be used to increase the fatigue limit of additively manufactured metals. However, with no further surface treatment, its contribution might be restricted by a rough surface. We thus tested Ti6Al4V alloy prepared by selective laser melting in the as-built surface condition to evaluate the utility of HIP. The influence of HIP on microstructure, static mechanical properties and fatigue was evaluated in comparison with untreated and conventionally heat treated state.

1. Introduction

Selective laser melting (SLM) belongs into the group of powder-bed additive manufacture. A powder bed is selectively exposed to a laser beam. The laser beam is moved by a scanning system according to the 2D section of the produced part corresponding to the layer being processed. The powder is locally melted and original particles fuse into a solid part. Further layers of powder are gradually deposited and processed until the product is complete [1].

In general, the properties of components made by means of SLM are mainly determined by the parameters of the SLM process [2]. Material density is one of the first characteristics that are investigated in the search for an optimal SLM processing windows. Varying scanning speed and laser power, for example, changes the dimensions of melt pools, leading to the generation of different kinds of defects such as spheroidal keyhole pores or crack-like voids due to lack of fusion. Such defects can significantly influence mechanical response of SLM parts, especially under fatigue loading [3]. But even with optimized parameters, completely dense parts cannot be obtained [4].

To enhance the density and improve the properties of SLM parts, hot isostatic pressing (HIP) is increasingly applied for post-processing after SLM. HIP is a technique used commonly for sintering of powders, diffusion bonding or densification of castings. It involves the simultaneous application of a high pressure of a gas and high temperature, which leads to diffusion and plastic deformation [5]. However, it has been shown that the effectiveness of HIP depends on the type of pores. As the SLM chamber is filled with an inert gas, pores formed by gas encapsulation can be formed in final SLM parts. The complete closure of such pores by HIP is hard to achieve. Also, if the porosity is not closed, pressure equalization prevents pores from closing [6].

In numerous studies, HIP was used to improve the fatigue of SLM parts [7-9]. Haan *et al* [7] showed extraordinary high values of fatigue limit for a Co-based alloy F75 prepared by combination of SLM+HIP. Such improvement was attributed not only to the closure of porosity but also to complete recrystallization. Leuders *et al* [8] showed a significant improvement in fatigue life of Ti6Al4V samples.

Yu *et al* [10] studied the effect of post-processing in a broader scale, looking at the effect of surface finishing and heat treatment including HIP. All these post-processings were contributory, however, a vital role in fatigue performance was played by surface condition. With samples polished to Ra = 0.04 μm and HIPed, the highest fatigue limit of 450-500 MPa was reached. Similarly, Vayssette *et al* [11] showed an increase from 222.5 to 512.5 MPa after machining HIPed samples of Ti6Al4V alloy.

Although there were several papers dealing specifically with HIP of SLM Ti6Al4V alloy [3,8,10,12], not many among them discussed microstructural changes associated with elevated temperature of HIP and their share in changing mechanical properties. Moreover, fatigue tests were mostly performed on machined samples. Nevertheless, the main potential of SLM lies in the preparation of complex shaped products which machining might not be possible. Therefore, our paper is focused on the characterization of SLM Ti6Al4V alloy in the as-built condition with no surface treatment. The aim is to assess whether the application of HIP is reasonable in such case.

2. Materials and methods

This paper is focused on titanium alloy Ti6Al4V, a material widely used in medicine and aerospace industry [13]. For such applications, selective laser melting (SLM), as an additive manufacture technology, is useful as it provides geometrical freedom of manufactured products. We applied SLM to fabricate testing samples. As SLM does not yield completely defectless products, their mechanical behavior is influenced by present defects. Hot isostatic pressing (HIP) was thus applied to achieve ideal defect-free structure. The influence of HIP on the structure and mechanical response of testing samples was then assessed.

2.1. SLM

Selective laser melting of Ti6Al4V alloy was carried out on an M2 cusing machine by ConceptLaser. SLM setting was adjusted according to the manufacturer's recommendation, table 1. Island scanning strategy, with island size of 5x5 mm^2, was applied. The whole manufacturing was carried out under a protective argon atmosphere (4.6 purity) to prevent oxidation. Ti ELI Grade 23 powder supplied by ConceptLaser was used. Its composition meets the ASTM F136-02a standard.

Round dog-bone samples with a gauge diameter of 5 mm, suitable for both tensile and fatigue tests, were manufactured. The longitudinal axis of the samples was oriented in the building direction Z. All samples were kept in the as-built state with no surface treatment.

Table 1. SLM setting.

Laser power	Scanning speed	Hatching distance	Layer thickness
200 W	800 mm/s	112 μm	30 μm

2.2. HIP

Hot isostatic pressing was done commercially at an argon pressure of 102 MPa at a temperature of 920°C for 2 hours.

In order to provide a comparison, a recommended standard heat treatment for Ti6Al4V alloy SLM products was done for one batch of the samples at 820°C for 1.5 hours. Because we registered a diffusion of oxygen during heat treatment under argon atmosphere (technical argon of 4.6 purity) [14], we carried out the heat treatment in vacuum (a vertical vacuum furnace Xerion XRETORT, 10^{-6} MPa).

For simplification, three studied states will be denoted further on as 'no HT', 'sHT' (standard HT) and 'HIP'.

2.3. Metallography

Microstructure and porosity were studied on metallographic sections prepared by a standard metallographic way parallel to the building direction of SLM. The whole sections in un-etched condition were captured by optical light microscope (Olympus PME3) and processed by image analysis in ImageJ software to determine porosity. After etching in Kroll's reagent, microstructure was observed by a scanning electron microscope Tescan Vega-3 LMU. The thickness of α-lamellae in the α+β microstructure was determined by means of image analysis.

2.4. Mechanical properties
Mechanical properties were studied under uniaxial tensile loading, under compression and during symmetric cyclic loading tension-compression with asymmetry ratio R = -1. Static loading was carried out on a universal testing machine LabTest 5.250SP1-VM at room temperature at a constant deformation rate of 0.001 s^{-1}. High-cycle fatigue tests were measured at a frequency of 50 Hz according to ASTM E466-7 standard (universal servo-hydraulic testing machine Instron, force capacity 50 kN). The fatigue limit was determined after 10^7 cycles. To evaluate only the effect of microstructure on mechanical performance of the Ti6Al4V alloy (neglecting influence of porosity), Vicker's hardness was measured on a semi-automatic hardness tester Future-Tech FM-700 with a load of 1 kg.

3. Results and discussion
The results section presents microstructural changes between the as-built state, after standard heat treatment and after the HIP of SLM Ti6Al4V alloy. The effect of HIP on porosity is shown. Static mechanical properties are compared for the initial state and both post-process treatments. In fatigue tests, only the effect of HIP is evaluated by testing sHT samples and HIPed samples.

3.1. Microstructure
Microstructures of the studied states of SLM Ti6Al4V alloy are presented in figure 1. In the as-built state, Ti6Al4V shows fully martensitic microstructure with acicular morphology of α'-martensite (figure 1(a)). The diffusionless transformation results from high cooling rates during SLM, which can reach up to 10^{6}°C/s [1]. When heat treatment is applied, with annealing temperature higher than the temperature of martensitic decomposition (650-800°C [15]), martensitic phase transforms into a more stable α+β microstructure (figures 1(b) and 1(c)). As the HIP treatment was characterized by a temperature of 100°C higher than the standard HT and 30 minutes longer dwell at such temperature, the thickness of α-lamellae has grown to 2.1±0.8 μm compared to 0.8±0.3 μm in the sHT state.

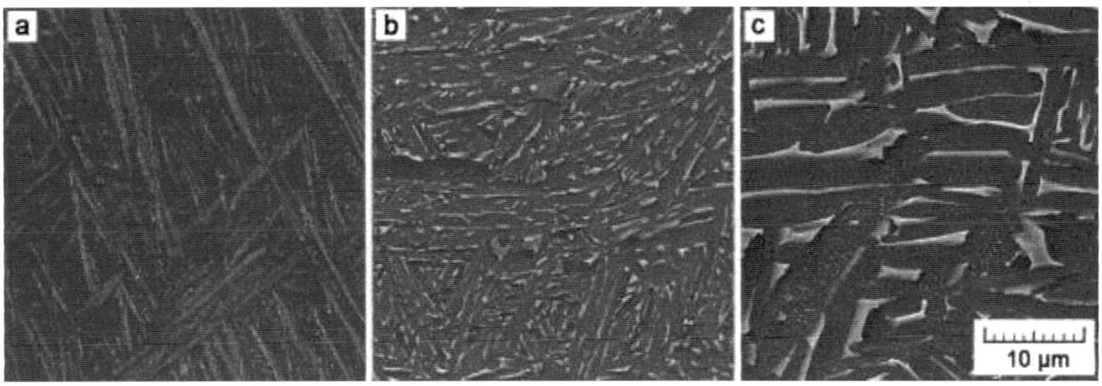

Figure 1. Microstructure of SLM Ti6Al4V alloy: (a) noHT, (b) sHT, (c) HIP.

3.1.1. Porosity. The primary purpose of HIP is to eliminate porosity. The samples prepared by SLM showed porosity of 0.49±0.17%. The pores were of irregular shape, elongated in the scanning direction, sometimes containing particles of unmelted powder (figure 2). The origin of such pores is attributed to the lack of fusion due to an insufficient energy. For proper fusion between individual scan lines in one layer, it is necessary that melt pools formed by adjacent scan lines overlap. Similarly, proper fusion between two layers is ensured if the depth of melt pools is larger than the thickness of a deposited layer. Good melt pool overlapping is primarily determined by hatching distance which approximately tells what is the overlapping between adjacent scan lines. However, the actual melt pool width is dictated by actual energy density (energy given to a volumetric unit of the powder). Energy density is a function of laser power, scanning speed, layer thickness, hatching distance, but the actual melt pool size is also dependent on powder properties, scanning pattern, preheating etc. [16]. HIP closed most of the pores, resulting in minimal porosity of 0.01±0.01%.

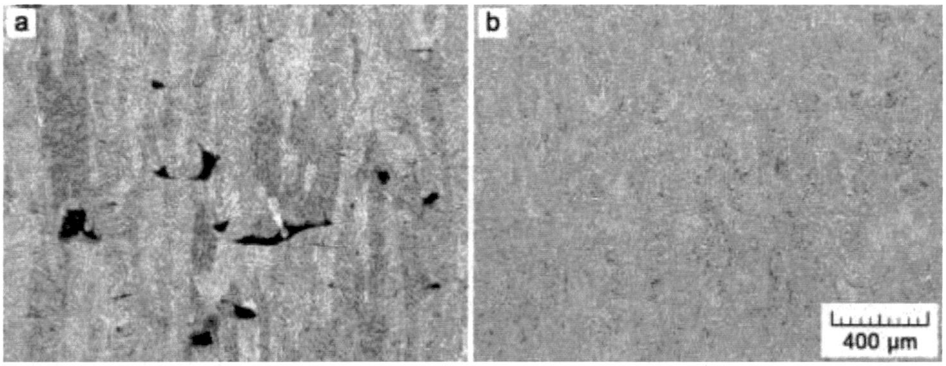

Figure 2. Longitudinal metallographic sections showing porosity in: (a) non-HIPed samples and (b) HIPed samples.

3.2. Mechanical properties

HIP affected mechanical properties of the SLM Ti6Al4V alloy by change in the microstructure as well as in porosity. Porosity lowers the load-bearing area and acts as stress concentrator, so negatively affects mechanical performance in both static and dynamic loading. Microstructural fineness of SLM as-built Ti6Al4V alloy and high internal stresses give it high strength, often higher than in conventional wrought condition [17,18].

3.2.1. Static properties. Mechanical properties determined by static loading are summarized in table 2 (TYS – tensile yield strength, UTS – ultimate tensile strength, A – elongation, CYS – compressive yield strength, UCS – ultimate compressive strength, HV1 – Vicker's hardness with 1 kg load). The highest yield strength, both in tension and compression, was reached in the as-built state with no HT. Very fine martensitic needles and high internal stresses associated with martensitic structure contribute to material strengthening. On the other hand, such material condition suffers from low elongation. Generally, heat treatment leads to the coarsening of the microstructure (transformation of martensite into lamellar α-phase and occurrence of interlamellar β-phase) and release of internal stresses, which reduces the material strength but promotes plasticity. In addition, the application of pressure during HIP closes the present porosity and promotes plasticity further. After sHT the elongation doubled, while it tripled after HIP. A minimal elongation required by ASTM F1472 standard for wrought Ti6Al4V alloy of 10% was surpassed only after HIP although standard HT brought the plasticity close to this limit.

Table 2. Mechanical properties of the studied states of SLM Ti6Al4V alloy.

	TYS (MPa)	**UTS (MPa)**	**A (%)**	**CYS (MPa)**	**UCS (MPa)**	**HV1 (-)**
no HT	1000±21	1193±35	4.3±1.5	1235±12	1763±30	389±9
sHT	973±4	1023±4	9.0±2.0	1079±6	1942±33	355±5
HIP	946±12	1043±2	13.9±1.8	1041±5	1766±30	330±6

* TYS – tensile yield strength, UTS – ultimate tensile strength, A – elongation, CYS – compressive yield strength, UCS – ultimate compressive strength, HV1 – Vicker's hardness with 1 kg load

Despite a slight decrease in TYS, UTS, CYS and UCS due to the coarsening of microstructure, measured properties still satisfy the requirements by ASTM F1472 standard (min. TYS of 860 MPa and min UTS of 930 MPa). The difference of about 30 MPa between TYSs and CYSs of sHT and HIP state are caused by larger thickness of α-lamellae after HIP (figure 1), although this difference is partially offset by the removal of porosity.

To assess only the influence of microstructure without the influence of porosity, we measured Vicker's hardness with only 1 kg load. Increase in α-lamellae thickness from 0.8 to 2.1 μm after HIP led to a decrease in hardness by 25 HV. The transformation from martensitic needles (no HT) into α+β lamellar microstructure (sHT) brought a higher decrease by 34 HV as internal stresses were relieved. Generally, Vicker's hardness can be used to roughly estimate material strength. The value of TYS should be approximately 3 HV [19]. If we look at the HIP samples and sHT samples, ΔHV of 25 should yield in a difference of ~ 75 MPa but the real difference is only 27 MPa because 0.49% porosity, which has the contradictory effect on strength, was present in non-HIPed samples.

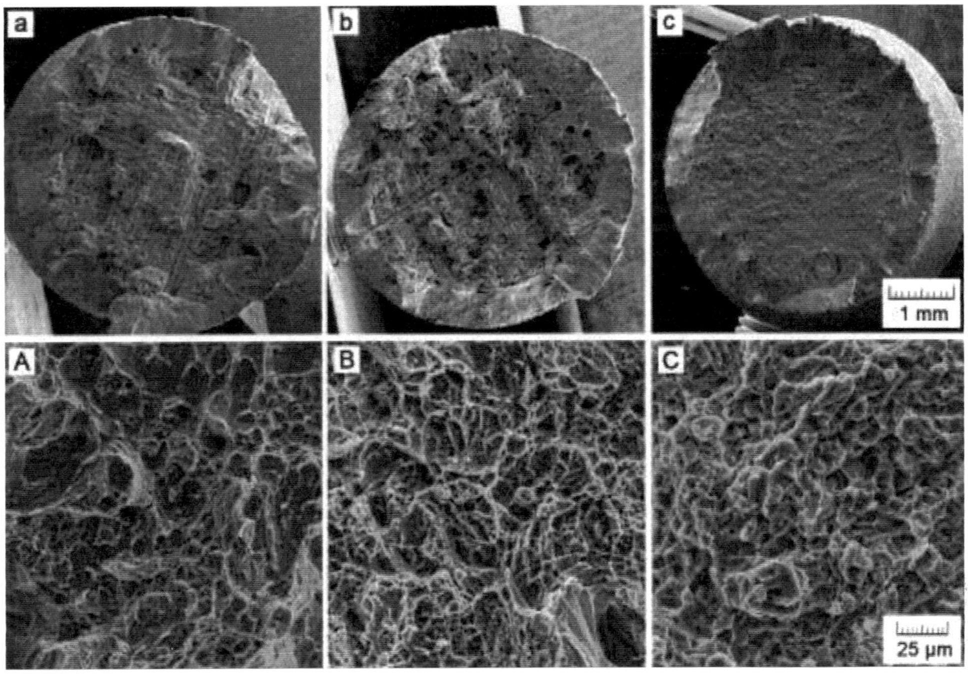

Figure 3. Fracture surfaces: (a,A) no HT, (b,B) sHT, (c,C) HIP.

Images of fracture surfaces (figure 3) clearly depict the above-described differences. With no porosity, the fracture surface of HIP samples (figure 3(c)) resembles a standard fracture surface of wrought material. The fracture surface is plane while in non-HIPed samples (figures 3(a) and 3(b)), it is rough because the crack propagated through present lack-of-fusion defects. It is nicely visible that these lack-of-fusion defects are arranged in a grid of parallel rows due to the incomplete overlap of melt pools in adjacent scan lines. In detail images (figure 3A-3C), a dimple-like morphology of fracture can be seen, with dimple depth increasing with increase in plasticity.

3.2.2. Fatigue. Although HIP is recommended to increase the fatigue limit of SLM Ti6Al4V alloy, our measurements on samples in as-built condition showed that the contribution of HIP is negligible. Non-HIPed samples in sHT state showed a fatigue limit of 212±10 MPa for the required lifetime of 10^7 cycles. After HIP, the fatigue limit was 222±19 MPa. The explanation is clear from figure 4 showing fracture surfaces after the rupture. Samples loaded at a stress level of 300 MPa are compared. In both sample groups, initiation of fatigue crack occurred at the sample surface (as indicated by the arrows in figures 4(a) and 4(b) and detail images of initiation site in figures 4A and 4B). Machining the as-built surface is thus needed to achieve the desired effect of HIP.

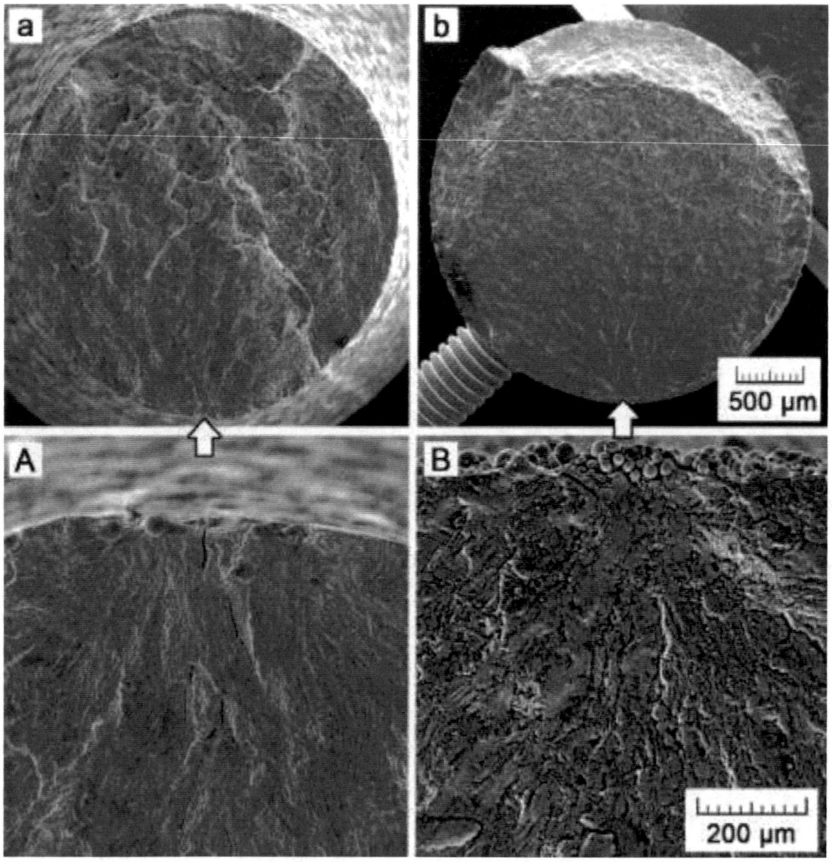

Figure 4. Fracture surfaces after fatigue tests: (a,A) sHT, (b,B) HIP.

4. Conclusion

The key asset of SLM is the preparation of complex structures which cannot be achieved by conventional fabrication techniques. Such structures include porous structures (fully porous parts, surface porosity, internal porous structures) or hollow parts (enclosed, intricately shaped) that cannot be machined at all or with great difficulty. In such case, additional treatment by HIP is questionable. We have shown that without surface treatment the contribution of HIP to the fatigue limit of SLM Ti6Al4V alloy is negligible. The main contribution was an increase in elongation (13.9±1.8%) above the desired limit of 10% thanks to the closure of lack-of-fusion defects. Nevertheless, the coarsening of the $\alpha+\beta$ lamellar microstructure due to the higher temperature of HIP brought about a slight decrease in material strength and hardness.

Acknowledgment

The authors wish to thank the specific university research (MSMT no. 21-SVV/2019) for the financial support of this research.

References

[1] Gokuldoss P K, Kolla S and Eckert J 2017 Additive manufacturing processes: Selective laser melting, electron beam melting and binder jetting-selection guidelines *Materials (Basel, Switzerland)* **10** 672
[2] Hanzl P, Zetek M, Bakša T and Kroupa T 2015 The influence of processing parameters on the mechanical properties of SLM parts *Proc. Eng.* **100** 1405-13
[3] Mishurova T, Artzt K, Haubrich J, Requena G and Bruno G 2019 New aspects about the search

for the most relevant parameters optimizing SLM materials *Addit. Manuf.* **25** 325-34
[4] Talib Mohammed M 2018 Mechanical properties of SLM-titanium materials for biomedical applications: A review *Mater. Today Proc.* **5** 17906-13
[5] Atkinson H V and Davies S 2000 Fundamental aspects of hot isostatic pressing: An overview *Metall. Mater. Trans. A* **31** 2981-3000
[6] Tillmann W, Schaak C, Nellesen J, Schaper M, Aydinöz M E and Hoyer K P 2017 Hot isostatic pressing of IN718 components manufactured by selective laser melting *Addit. Manuf.* **13** 93-102
[7] Asseln M, Zivcec M, Eschweiler J, Radermacher R and Broeckmann C 2015 Effect of subsequent hot isostatic pressing on mechanical properties of ASTM F75 alloy produced by selective laser melting AU - Haan *Powder Metall.* **58** 161-5
[8] Leuders S, Thöne M, Riemer A, Niendorf T, Tröster T, Richard H A and Maier H J 2013 On the mechanical behaviour of titanium alloy TiAl6V4 manufactured by selective laser melting: Fatigue resistance and crack growth performance *Int. J. Fatigue* **48** 300-7
[9] Benedetti M, Torresani E, Leoni M, Fontanari V, Bandini M, Pederzolli C and Potrich C 2017 The effect of post-sintering treatments on the fatigue and biological behavior of Ti-6Al-4V ELI parts made by selective laser melting *J. Mech. Behav. Biomed. Mater.* **71** 295-306
[10] Yu H, Li F, Wang Z and Zeng X 2019 Fatigue performances of selective laser melted Ti-6Al-4V alloy: Influence of surface finishing, hot isostatic pressing and heat treatments *Int. J. Fatigue* **120** 175-83
[11] Vayssette B, Saintier N, Brugger C, Elmay M and Pessard E 2018 Surface roughness of Ti-6Al-4V parts obtained by SLM and EBM: Effect on the high cycle fatigue life *Proc. Eng.* **213** 89-97
[12] Zhou B, Zhou J, Li H and Lin F 2018 A study of the microstructures and mechanical properties of Ti6Al4V fabricated by SLM under vacuum *Mater. Sci. Eng. A* **724** 1-10
[13] Wycisk E, Emmelmann C, Siddique S and Walther F 2013 High cycle fatigue (HCF) performance of Ti-6Al-4V alloy processed by selective laser melting *Adv. Mater. Res.* **816-817** 134-9
[14] Fousova M and Vojtech D 2018 Thermal treatment of 3D-printed Titanium alloy *Manuf. Technol.* **18** 227-32
[15] Xu W, Lui E W, Pateras A, Qian M and Brandt M 2017 In situ tailoring microstructure in additively manufactured Ti-6Al-4V for superior mechanical performance *Acta Mater.* **125** 390-400
[16] Dilip J J S, Zhang S, Teng C, Zeng K, Robinson C, Pal D and Stucker B 2017 Influence of processing parameters on the evolution of melt pool, porosity, and microstructures in Ti-6Al-4V alloy parts fabricated by selective laser melting *Prog. Addit. Manuf.* **2** 157-67
[17] Fousova M, Vojtech D and Kubasek J 2016 Titanium alloy Ti-6Al-4V prepared by selective laser melting (SLM) *Manuf. Technol.* **16** 691-7
[18] Liu S and Shin Y C 2019 Additive manufacturing of Ti6Al4V alloy: A review *Mater. Des.* **164** 107552
[19] Busby J T, Hash M C and Was G S 2005 The relationship between hardness and yield stress in irradiated austenitic and ferritic steels *J. Nucl. Mater.* **336** 267-78

IOP Conf. Series: Materials Science and Engineering **629** (2019) 012035 doi:10.1088/1757-899X/629/1/012035

Crack tip shape effect on stress-strain fields in plastically compressible materials

M I Alam[1], D Khan[1,2], Y Mittal[1] and S Kumar[1]

[1]Department of Mechanical Engineering, Indian Institute of Technology (BHU) Varanasi, Varanasi - 221005, India

E-mail: dkhan.mec@itbhu.ac.in

Abstract. In the present study, the effects of crack tip shape on near tip deformation and fields are numerically investigated for a mode I crack under plane strain and small scale yielding conditions. The material is characterized by finite strain elastic-viscoplastic constitutive relation with bi-linear hardening and hardening-softening-hardening hardness functions. Both plastically incompressible and compressible solids have been considered for analyses. It has been observed that the combination of crack tip profile and plastic compressibility has significant effect on the near tip deformation and plastic fields.

1. Introduction

In several numerical simulations based on initially sharp crack tip it has been observed that severe blunting takes place in the very early stages of crack tip plasticity and consequently it leads to a highly distorted mesh. Therefore, in numerical simulation, people mostly use originally blunted crack having a circular arc tip [1]. However, in practical situation it is observed that the crack tip is neither exactly circular nor completely sharp, but of arbitrary shape with a finite amount of curvature. In recent times, in the study of fracture in solids, the effect of crack-tip shape has attracted notable research attention.The notch root radius effect on fracture toughness was examined for a variety of brittle and ceramic materials in [2-4] and it was noticed that the root radius can have impact on the fracture toughness significantly. Further, experimentation of notch tip radius on fatigue crack growth for two different alloys was performed in [5] and it was concluded that for blunter notch the fatigue crack growth rate is more. Also, it is reported in literature that with increase in the crack-tip curvature radius, near crack-tip plastic strain as well as stresses increase considerably in glassy polymer [6].For elastomers, some studies on crack growth and the associated changes in crack-tip shapes have been reported in [7].

Some of the results obtained from the experimentation analysis on the behavior of materials like metallic foams, transformation toughened ceramics and plastics suggest that the constitutive equations of these materials comprise pressure-sensitive yielding. Due to some impressive mechanical properties, these materials have attracted numerous researchers over the past several years. For understanding the deformation and stress-strain fields near a crack tip in pressure sensitive and/or plastically compressible solids, a large amount of research works has been done, in recent times, for both monotonic and fatigue loadings. It should be mentioned here that most of these studies are based on small deformation and only a few studies are founded on large deformation formulation [8-13]. Considering potential applications of the above mentioned materials, it appears that it is required to look into the influence of notch or crack-tip shape on near tip deformation and fields for the above mentioned materials. Plastic normality; plane strain and small scale yielding conditions have been assumed in the present computations.

2. Constitutive equation

The material model used here is same as that given in [8-13].The elastic part of the deformation is expressed as

$$\mathbf{d}^e = \mathbf{L}^{-1}:\hat{\boldsymbol{\tau}} = \frac{1+\nu}{E}\hat{\boldsymbol{\tau}} - \frac{\nu}{E}\mathrm{tr}(\hat{\boldsymbol{\tau}})\mathbf{I} \tag{1}$$

Where, $\mathbf{L}$ is the elastic moduli tensor,$\hat{\boldsymbol{\tau}}$ is the Jaumann rate of Kirchhoff stress, E and ν are the Young's modulus and Poisson's ratio, respectively, tr(.)indicates the trace and $\mathbf{I}$ the identity tensor, $\mathbf{A}$: $\mathbf{B}$=$A_{ij}B_{ij}$.

The plastic part of the deformation is expressed as

$$\mathbf{d}^p = \frac{3}{2}\frac{\dot{\varepsilon}_p}{\sigma_e}\mathbf{p} \tag{2}$$

Where, $\dot{\epsilon}_\mathrm{p}$ is plastic strain rate and σ_e is effective stress corresponding to the instantaneous $\dot{\epsilon}_\mathrm{p}$ and $\mathbf{p}$ is deviatoric part of Kirchhoff stress tensor. The $\mathbf{p}$ is given by

$$\mathbf{p} = \boldsymbol{\tau} - \alpha\mathrm{tr}(\boldsymbol{\tau})\mathbf{I} \tag{3}$$

When $\alpha = 1/3,$ the equation (3) is reduced to constitutive relation for isotropic hardening viscoplastic Mises solid. In this work,we use the following viscoplastic strain rate relation

$$\dot{\epsilon}_\mathrm{p} = \dot{\epsilon}_0\left(\frac{\sigma_\mathrm{e}}{\mathrm{g}}\right)^{1/m} \tag{4}$$

Where, $\dot{\varepsilon}_0$ and m represent the reference strain rate and the hardening exponent, respectively; the hardening function g is written as

$$\mathrm{g}(\epsilon_\mathrm{p}) = \sigma_0\begin{cases} 1 + h_1\epsilon_\mathrm{p}, & \epsilon_\mathrm{p} < \epsilon_1 \\ 1 + h_1\epsilon_1 + h_2(\epsilon_\mathrm{p} - \epsilon_1), & \epsilon_1 < \epsilon_\mathrm{p} < \epsilon_2 \\ 1 + h_1\epsilon_1 + h_2(\epsilon_2 - \epsilon_1) + h_3(\epsilon_\mathrm{p} - \epsilon_2), & \epsilon_\mathrm{p} > \epsilon_2 \end{cases} \tag{5}$$

Where, σ_0 is a reference stress. The effective stress, σ_e is defined by

$$\sigma_\mathrm{e}^2 = \frac{3}{2}\boldsymbol{\tau}:\mathbf{p} = \frac{3}{2}\left[\boldsymbol{\tau}:\boldsymbol{\tau} - \alpha\left(\mathrm{tr}(\boldsymbol{\tau})\right)^2\right] \tag{6}$$

3. Problem formulation

The field equations, used in this work, are functions of convected coordinates, $y^\mathrm{i}$and time (t) [8-13]. The principle of virtual work in absence of body force and for quasi-static deformation can be written as

$$\int_\mathrm{V} \tau^\mathrm{ij}\,\delta E_\mathrm{ij} dV = \int_\mathrm{S} T^\mathrm{i}\,\delta u_\mathrm{i} dS \tag{7}$$

Where, V and S represent the volume and surface of the body in the reference configuration, τ^ij are the Kirchhoff stress in the deformed convected coordinate, E_ij the Lagrangian strain, T^i the nominal traction and u_i are the displacement components.

From the above equation (7), the incremental equilibrium is obtained as

$$\Delta t\int_\mathrm{V}\left(\dot{\tau}^\mathrm{ij}\delta E_\mathrm{ij} + \tau^\mathrm{ij}\dot{u}^\mathrm{k}_{,\mathrm{i}}\delta u_\mathrm{k,j}\right) dV = \Delta t\int_\mathrm{S} \dot{T}^\mathrm{i}\,\delta u_\mathrm{i} dS - \left[\int_\mathrm{V}\tau^\mathrm{ij}\,\delta E_\mathrm{ij} dV - \int_\mathrm{S} T^\mathrm{i}\,\delta u_\mathrm{i} dS\right] \tag{8}$$

In equation (8), the term $\left[\int_\mathrm{V}\tau^\mathrm{ij}\,\delta E_\mathrm{ij} dV - \int_\mathrm{S} T^\mathrm{i}\,\delta u_\mathrm{i} dS\right]$ stands for an equilibrium correction term.

4. Finite element model and some numerical results

IOP Conf. Series: Materials Science and Engineering **629** (2019) 012035 doi:10.1088/1757-899X/629/1/012035

In the present finite element simulation, the numerical results are generated for the geometry as shown in figure 1, where the outer radius (R_∞) is 2.0 in any arbitrary units. Motivated by the works of Li *et al.* [6], here, we use a similar type of elliptical arc to describe the crack tip profile. The semi-major and semi-minor axes of the ellipse are represented by a and b in the same arbitrary units, respectively and the ratio a/b specifies the crack shape as shown in figure 1. The constant loading rate is $\dot{K_I}/\sigma_0\dot{\varepsilon_0}\sqrt{b} = 31.62$, where $\dot{K}_I = 1\ MPa\sqrt{m}s^{-1}$. In order to calculate the deformation history, linear increments are used with time step size of 0.0002.

Quadrilateral elements are used for grid generation and each quadrilateral is made of four crossed triangular elements with constant strain. To reproduce the localized deformation pattern at finite strains, such elements are widely used [10-13]. In the present analysis, after carrying out the mesh convergence study with meshes consisting of 28 x 64, 28 x 74 and 28 x 84 quadrilateral elements, the mesh density finally has been fixed at28 x 74 quadrilateral elements. The near tip plastic strain contours were compared and it was noticed that the results from 28 x 74 and 28 x 84 were in good agreement.

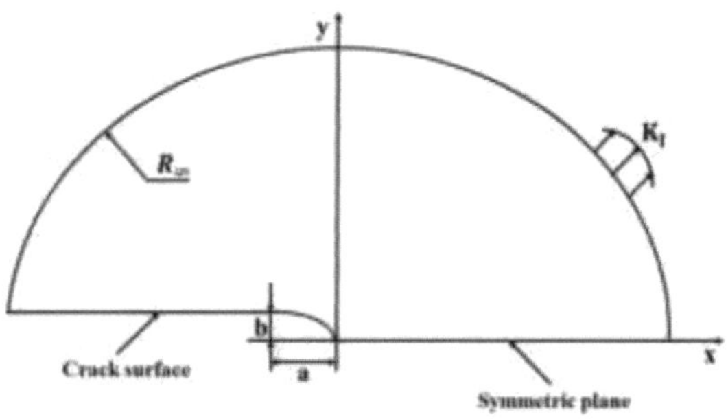

Figure 1. Schematic view of the elliptical crack tip used for simulation in finite element analyses.

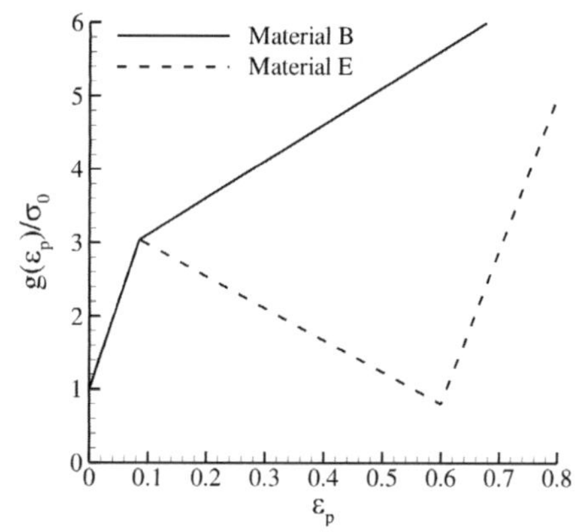

Figure 2. Graph of hardness function $g(\varepsilon_p)$for materials B and E, [10].

In this study, materials B and E from [10] are considered and the hardness functions of those materials are shown in figure 2. Material B is a bilinear hardening material whereas material E is a trilinear hardening-softening-hardening solid. At $\varepsilon_1$the material B starts re-hardening whereas at the same plastic strain softening takes place in material E. The value of ε_2 defines the plastic strain at which material E starts again hardening after the softening. The values of h_1 and ε_1 are same for the two materials and those are 24 and 0.085, respectively; also h_2 and h_3 are same for material B and it is 5.0 whereas for material E, h_2 is -3.90 and h_3 is 15.0. The other constant material parameters are E/σ_0=100, $\dot{\varepsilon}_0$=1 and m = 0.02.

At the remote outer boundary of the semi-circle, K_I field displacements are imposed. Here, traction free crack surfaces are assumed and on line $y^2 = 0$, symmetry boundary conditions are applied; in the present case $y^2 = y$. For the constitutive update, rate tangent modulus method by Peirce *et al* [14] is considered here.

4.1. Near-tip normal stress (σ_{yy}) distributions

Figures 3 – 4 illustrate the normal stress (σ_{yy}) variations in front of the crack tip along the x-axis for materials B and E where the three crack shapes specified by a/b = 1 ,4 and 8 are considered. Both plastically incompressible and compressible solids have been considered for analyses. It is reflected from all the figures that near tip stresses vary with the crack tip distance and also the three crack tip shapes influence these stress fields differently. The stress quantities become almost constant as we move far from the crack tip. There is considerable amount of increase in the values of σ_{yy} as we increase the curvature radius at crack tip and also the far field stress from the tip becomes independent of the crack tip shape. For both the materials, when these are plastically compressible, the maximum stress is reduced than that of plastically incompressible solid. Also for plastically compressible solids, there are kinks, which are slightly away from the tip, and it may be due to the formation of shear band as would be shown in the subsequent figures. The location of these kinks depends on the crack tip shapes. For material E, because of the softening, the maximum stress is further reduced as compared to that of material B.

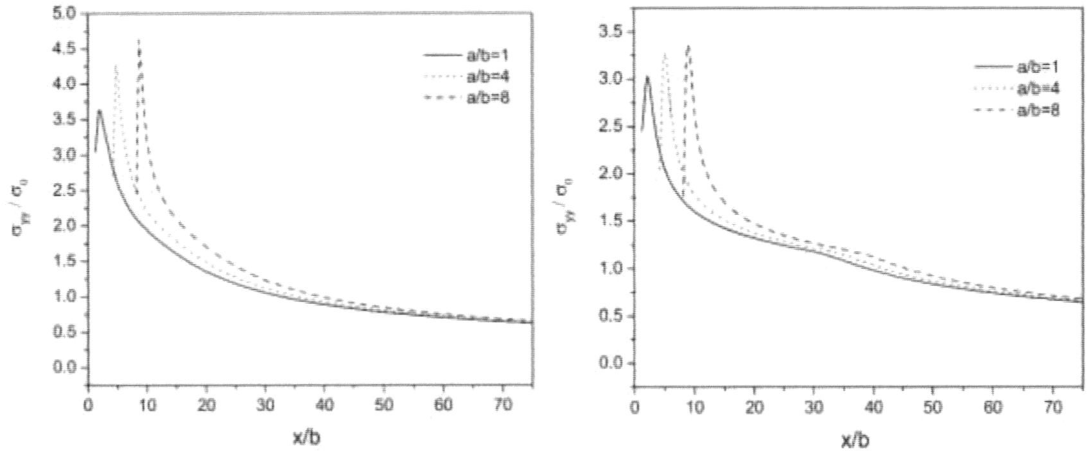

(a) (b)

Figure 3. Variation of normal stress (σ_{yy}) just ahead of the tip, material B, under $J_{app}/(\sigma_0 b)$ =1.6; (a) plastically incompressible solid and (b) plastically compressible solid ($\alpha = 0.28$).

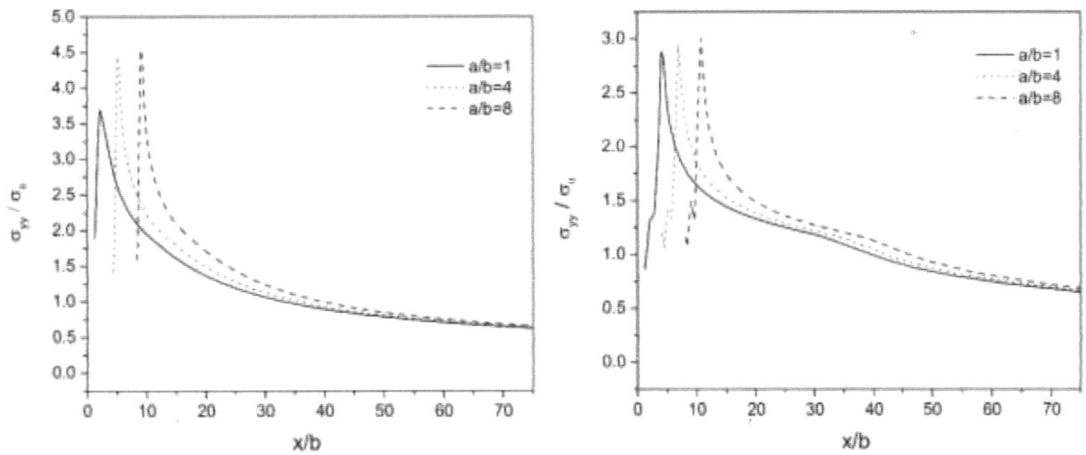

(a) (b)

Figure 4. Variation of normal stress (σ_{yy}) just ahead of the tip, material E, under $J_{app}/(\sigma_0 b)$ =1.6; (a) plastically incompressible solid and (b) plastically compressible solid ($\alpha = 0.28$).

4.2. Accumulated plastic strain distributions

Figures 5 – 8 illustrate the distributions of plastic strain in materials B and E for the two crack shapes, i.e. with a/b= 1 and 8 and at same load level, i. e. $J_{app}/(\sigma_0 b) = 1.6$. It is apparent that crack tip shape, plastic compressibility and material softening have significant effect on the near tip plastic deformation. In figure 5, the intense plastic deformation is limited to a very small region around the tip due to the

IOP Conf. Series: Materials Science and Engineering **629** (2019) 012035 doi:10.1088/1757-899X/629/1/012035

material strain hardening. As, the ratio of a/b increases, the zone of intense plastic deformation decreases along the vertical axis, and the stress concentration is increased on the crack tip. It is also to be noted here that with increase of the curvature radius at tip, the outer kidney shaped contour is tilted in a clockwise sense for the incompressible solid. For plastically compressible solid with $\alpha = 0.28$, it is observed from figure 6 that the extent of severe stain region increases at the crack tip when the curvature radius is more.

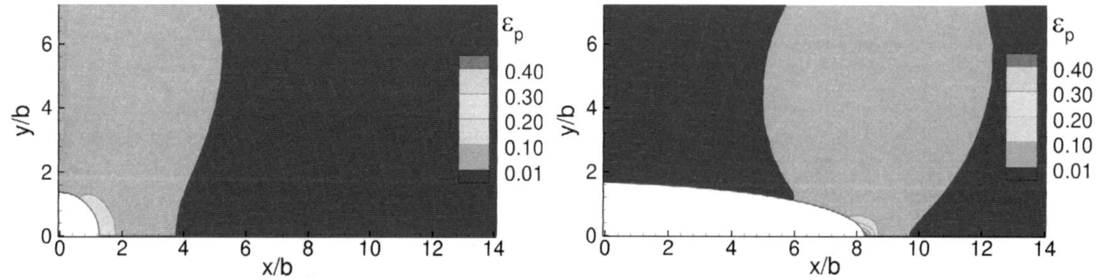

(a) (b)

Figure 5. Near tip plastic strain field, material B at $J_{app}/(\sigma_0 b)) = 1.6$ for plastically incompressible solid; (a) a/b= 1and (b) a/b= 8.

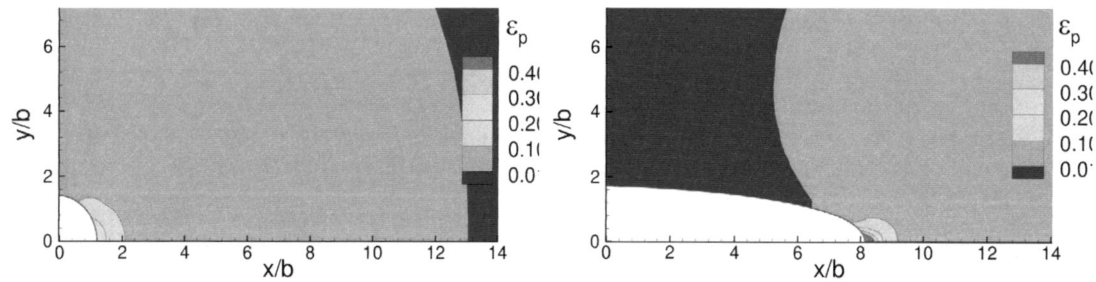

(a) (b)

Figure 6. Near tip plastic strain field, material B at $J_{app}/(\sigma_0 b)) = 1.6$ for plastically compressible solid ($\alpha = 0.28$); (a) a/b= 1 and (b) a/b= 8.

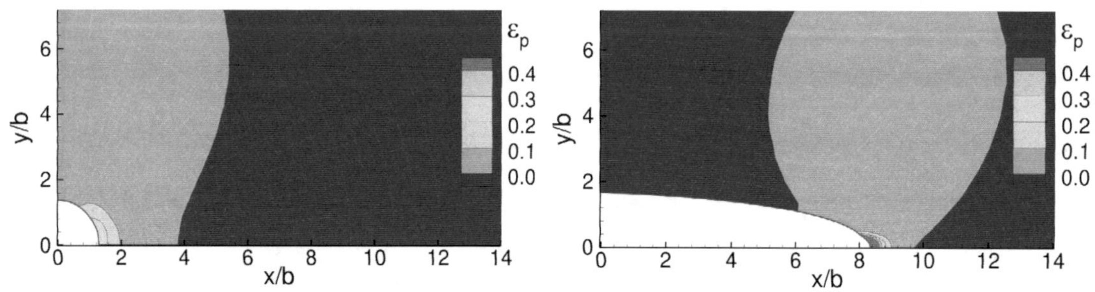

(a) (b)

Figure 7. Near tip plastic strain field, material E at $J_{app}/(\sigma_0 b)) = 1.6$ for plastically incompressible solid; (a) a/b= 1 and (b) a/b= 8.

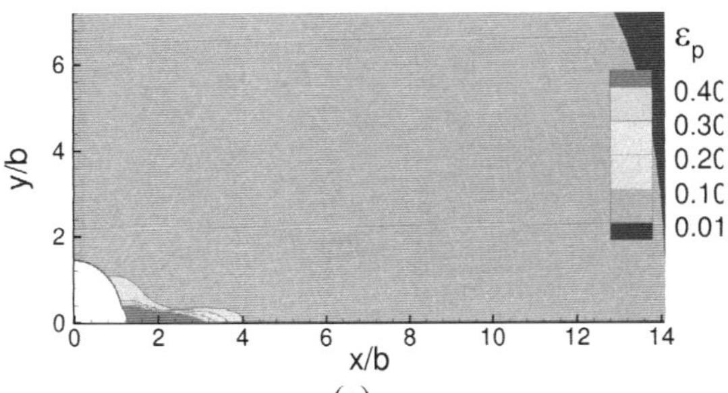

(a)

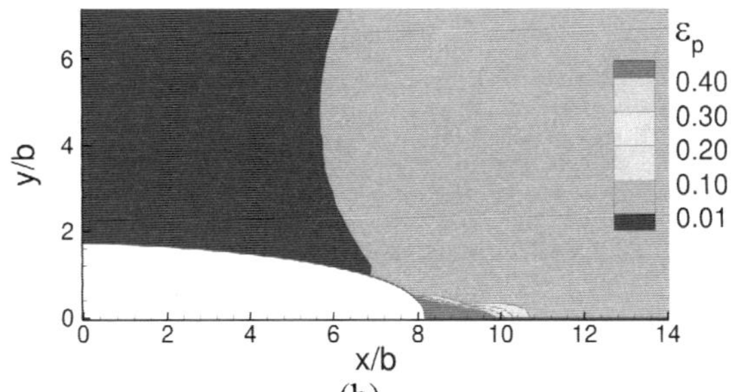

(b)

Figure 8. Near tip plastic strain field, material E at $J_{app}/(\sigma_0 b)) = 1.6$ for plastically compressible solid ($\alpha = 0.28$); (a) a/b= 1 and (b) a/b= 8.

While in figures 5, 6 and 7, the plastic strain is maximum on the blunting surface of the crack, however, the plastic compressibility, material softening and the shape of crack tip have a major consequence on the plastic strain distribution in material E. The zone of intense strain region in material E originates from the crack surface near the symmetry plane. Even though, in material E due to the material softening, localization of the shear band is common to both the two crack shapes, figures 7 and 8, however, it is reflected that for the sharper crack tip, the plastic strain concentration near the tip is slightly more at the present load level.

5. Conclusions

In the present study, the influence of the shape of a crack tip on the near tip deformation and stress-strain fields has been examined. For the investigation purpose, finite element finite deformation study has been conducted where the material is characterized by an elastic viscoplastic constitutive relation with a hardening and hardening-softening-hardening hardness functions. In the analysis, plane strain deformation, mode I loading and small scale yielding conditions have been assumed. The combination of crack tip shape, plastic compressibility and material softening shows the way to main variations in the stress and deformation fields near the tip. The finite element results show that there is considerable amount of increase in the values of σ_{yy} as we increase the curvature radius at crack tip. Also because of plastic compressibility as well as material softening, the maximum normal stress is reduced significantly. It is also revealed that as the curvature radius increases, the intense plastic deformation approaches to crack tip. Furthermore, the presence of plastic compressibility and material softening influences the localization near the crack tip differently depending on the crack tip shape.

References

[1] McMeeking RM 1977 Finite deformation analysis of crack tip opening in elastic-plastic materials and implications for fracture nanotubes *J. Mech. Phys. Solids* **25** 357-81
[2] Nishida T and Hanaki Y 1994 Effect of notch-root radius on the fracture toughness of a fine-grained alumina *J. American Cer. Soc.* **77** 606-8
[3] Damani R, Gstrein R and Danzer R 1996 Critical notch-root radius effect in SENB-S fracture toughness testing *J. Euro. Cer. Soc.* **16** 695-702
[4] Picard D, Leguillon C and Putot C 2006 A method to estimate the influence of tge notch-root radius on the fracture toughness measurement of ceramics *J. Euro. Cer. Soc.* **26** 1421-7
[5] Rozumek D, Macha E, Lazzarin P and Meneghetti G 2006 Influence of notch (tip) radius on fatigue crack growth rate *J. Theo. Appl. Mech.* **44** 127-37
[6] Li H, Wang G and Wang T J 2008 Effect of crack-tip-shape on the near-tip field in glassy polymer *Int. J. Solids Struct.* **45** 1087-100
[7] Morishita Y, Tsunoda K and Urayama K 2017 Crack-tip shape in the crack growth rate transition of filled elastomers *Polymer***108** 230-41
[8] Hutchens S B, Needleman A and Greer J R 2011 Analysis of uniaxial compression of vertically

aligned carbon nanotubes *J. Mech. Phys. Solids* **59** 2227-37

[9] Needleman A, Hutchens S B, Mohan N and Greer J R 2012 Deformation of plastically compressible hardening–softening–hardening solids *Acta Mech. Sin.* **28** 1115-24

[10] Mohan N, Cheng J, Greer J R and Needleman A 2013Uniaxial tension of a class of compressible solids with plastic non-normality *J. Appl. Mech.* **80** 040912-1-8

[11] Singh S and Khan D 2018 On fatigue crack growth in plastically compressible hardening and hardening-softening-hardening solids using crack tip blunting *Int. J. Fract.* **213** 139-55

[12] Khan D, Singh S and Needleman A 2017 Finite deformation analysis of crack tip fields in plastically compressible hardening – softening – hardening solids *Acta Mech. Sin.* **33** 148-58

[13] Singh S and Khan D 2018 Quasi-statically growing crack tip fields in plastically compressible hardening-softening-hardening solid *Int. J. Struct. Integ.* **9** 532-47

[14] Peirce D, Shih C F and Needleman A 1984 A tangent modulus method for rate dependent solids *Comp. Struct.* **18** 875-87

IOP Conf. Series: Materials Science and Engineering **629** (2019) 012036 doi:10.1088/1757-899X/629/1/012036

Dynamic response of gradient cellular materials under high velocity impact

X K Lan[1], Q Huang[1], T Zhou[1,2] and S S Feng[1]

[1]State Key Laboratory of Explosion Science and Technology, Beijing Institute of Technology, Beijing 100081, China

E-mail: zhoutong@bit.edu.cn

Abstract. The dynamic behavior of gradient cellular materials subject to high velocity impact is investigated theoretically and numerically. A parameter refined rigid-perfectly-plastic-locking model is developed and employed in the one-dimensional shock-wave theory to predict the response of gradient cellular materials under high velocity impact. The finite element analysis is carried out based on the periodic Voronoi structure and shows good agreement with the analytical prediction. The dynamic response and energy absorption are examined for different gradient cellular rods with the same mass and length but different density gradient. It is found that, at the first part of the high velocity impact crushing process, the cellular rods with negative density gradient show higher energy absorption ability and lower pressure at the distal end compared to the positive and uniform ones. This advantage can be taken when the partial crushing happens and vanishes when the cellular rods are fully crushed.

1. Introduction

Cellular materials have been used as energy absorption materials [1-17]. To analysis the dynamic response of cellular material, the finite element method has been developed to reveal the deformation of microstructure of cellular materials and the shockwave propagating in it [17-22]. The well-known voronoi structure has been widely adopted to simulate the foam materials, which shows a good agreement with the experiment date [23-26].

To meet the demand of optimal impact performance, the dynamic response of cellular material with density gradient were studied [27,28]. With the development of manufacturing technology, foams with continuous density variation can be produced. He *et al* prepared a density-graded aluminum foam, examined the mechanical properties through quasi-static compression tests and impact tests [29].

However, the impact velocities in the researches mentioned above (few over than 200 m/s) are relative low compared to the impact and incident shock emerged during the using in aerospace and defense industries. Few researches focused on high-speed impact. Lopatnikov investigate the high velocity (140 m/s-1000 m/s) plate impact of homogenous metal foams [30]. Zhou emphasized the threat of fragments induced by explosive and theoretically studied the influence of density variation on energy absorption capability by using the R-PP-L model [31]. The dynamic response of density graded cellular materials under high velocity impact needs further investigations.

2. Problem formulation

As shown in figure 1 the cellular rod with density gradient is fixed at the distal end. And a rigid mass impacting on the proximal end, four types of linear density distribution (k=0.6; k=0; k=-0.6; k=-1.2;

shown in figure 2) are constructed by changing the density gradient k which defined as:

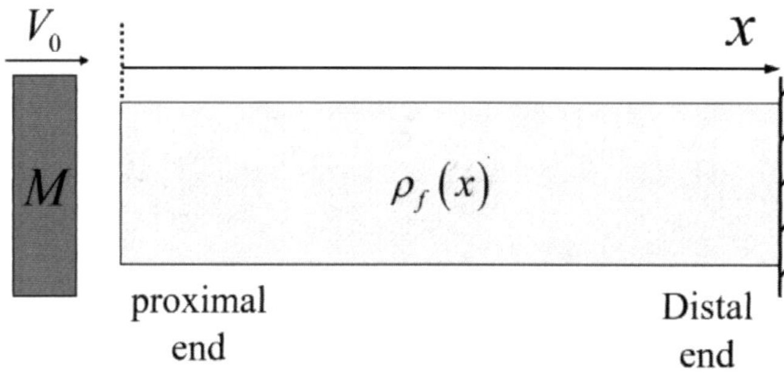

Figure 1. Sketch of cellular rods impact by a rigid mass.

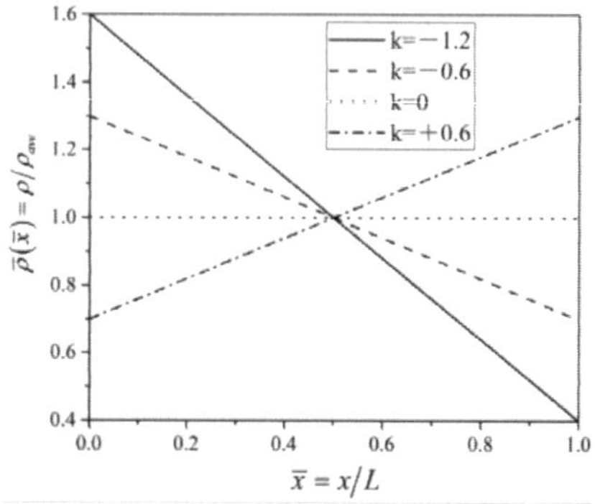

Figure 2. The four types of density distribution with the same mass and thickness.

$$k = \bar{\rho}_f(L) - \bar{\rho}_f(0) \tag{1}$$

Where $\bar{\rho}_f = \rho/\rho_{ave}$. And

$$\bar{\rho}_f(\bar{x}) = 1 + k\left(\bar{x} - \frac{1}{2}\right) \tag{2}$$

Where $\bar{x} = x/L$.

3. Analytical model

3.1. Parameter refined R-PP-L model

The rigid-perfectly-plastic-locking model based on the quasi-static nominal stress-strain curve has been widely adopted to simulate the constitutive relationship of cellular materials, as shown in figure 3. Then the R-PP-L model needs only two parameters: densification strain ε_d and the quasi-static plateau stress σ_{pl}^{qs}. ε_d is obtained through the "efficiency" method and σ_{pl}^{qs} is defined as an average stress value from the initial crushing strain to the densification strain [1].

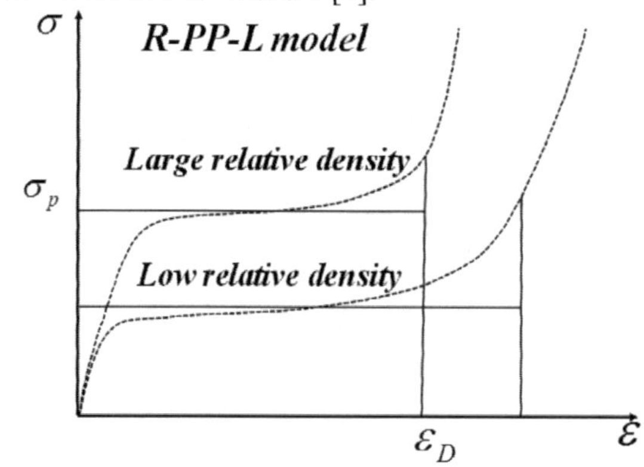

Figure 3. Typical stress-strain relation of different cellular materials and the idealization of constitute relation of R-PP-L model.

However, the crude identifications may cause some inaccuracies. Firstly, the densification strain should be a function of impact velocity and tends to be a constant value with the raise of impact velocity [2]. The "efficiency" method may underestimate the densification strain. So Lopatnikov *et al* replaced ε_d by $\varepsilon_{\max}$ (equation (3) ,where ρ is the foam's density at the shockwave front and ρ_s is the density of base material), duo to the high velocity impact [3].

$$\varepsilon_{\max}=1-\rho/\rho_s \tag{3}$$

It should be noted that the impact velocity (300 m/s) in recent study is high enough to press the rods into fully crushed, then the adoption of $\varepsilon_{\max}$ is suitable. Secondly, the plateau stress may be overestimated. In reality, cellular materials always exhibit strain hardening. Hansen *et al* proposed a materials constitutive and which is modeled as a rate-independent rigid-nonlinear hardening plastic with the quasi-static stress-strain relation in the following form [5].

$$\sigma(\varepsilon)=\sigma_0+\gamma/\varepsilon(1-\rho)-\alpha\ln\left\{1-\left[\varepsilon/(1-\rho)\right]^{\beta}\right\} \tag{4}$$

Where σ_0 is the yield stress, while α , β and γ are coefficients. Zheng *et al* adopted this model to predict the dynamic response of density-gradient cellular materials, which shows good agreement with finite element results [6]. Based on the rigid loading and unloading assumption the yield stress σ_0 can be adopted to be the plateau stress σ_{pl} .date fatting the stress-strain curves of homogenous gives $\sigma_0/\sigma_y=0.4441(\rho/\rho_s)^{2.08}$, where σ_y is the yield stress of base material [6]. Then a parameter refined R-PP-L model with $\varepsilon_D=\varepsilon_{\max}$ and $\sigma_{pl}=\sigma_0$ can be used for analysis later.

3.2. *One dimensional shock wave model*

As shown in figure 4, $\dot{s}=ds/dt$ is the Lagrangian speed of the compress shock front. After the shock front, the cellular material is assumed fully densified while ahead of the shock front, the cellular material remains undeformed with legible velocity and strain. Figure 5 shows the crushing process of one-dimensional shock wave model. The conditions of conservation of mass and momentum flow across the shock front requires that:

$$\rho_s(u-\dot{s})=\rho_f(v-\dot{s}) \tag{5}$$

$$\sigma_d-\sigma_{pl}=\rho_f(v-\dot{s})^2-\rho_s(\dot{u}-\dot{s})^2 \tag{6}$$

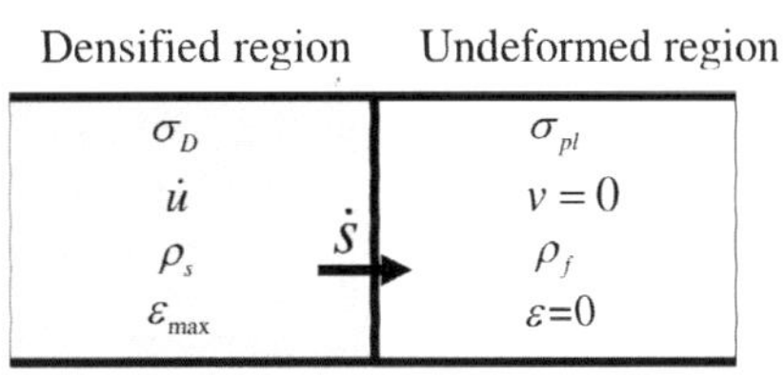

Figure 4.. Illustration of the parameters behind and ahead of the shock front.

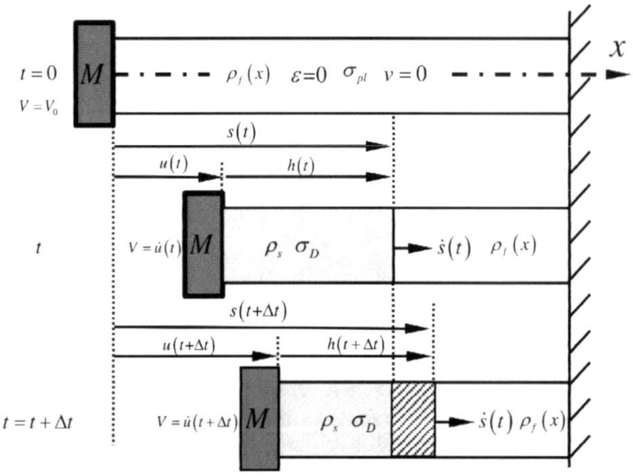

Figure 5. Illustration of the crushing process of the one-dimensional shock wave model.

Combination of equation and equation gives

$$\sigma_D - \sigma_{pl} = \rho_f \dot{s}\dot{u} \tag{7}$$

Conservation of mass across the shock front in a very short time Δt:

$$\rho_f \dot{s}\Delta t = \frac{\rho_f}{1-\varepsilon_{\max}}(\dot{s}-\dot{u})\Delta t \tag{8}$$

It gives that:

$$\dot{s} = \dot{u}/\varepsilon_{\max} \tag{9}$$

Combining equation (7) and (9) leads to:

$$\sigma_D = \sigma_{pl} + \frac{\rho_f}{\varepsilon_{\max}}\dot{u}^2 \tag{10}$$

The fully densified foam gains a velocity same with the rigid mass and decrease with time. The newton's second law gives that:

$$(M/A + m_d)\ddot{u} = -\sigma_D \tag{11}$$

Where M is the mass of the rigid mass, m_d is the mass of densified region and can be given by

$$m_d = \int_u^s \rho_s dx = \rho_s (s-u) \tag{12}$$

The mass behind the shock front gives the relation of s and u

$$u = s - \frac{\rho_{ave}}{\rho_s}\left(s - \frac{k}{2}s + \frac{k}{2L}s^2\right) \tag{13}$$

The initial conditions are set as:

$$u = u(0), \dot{u}(0) = V_0 \tag{14}$$

$$s(0) = 0, \dot{s}(0) = V_0 / \left[1 - \rho_{ave} / \rho_s (1 - k/2)\right] \tag{15}$$

4. Finite element analysis

4.1. 2D random Voronoi technology

To simulate the continuous density gradient, Wang et al. proposed a mathematical method to construct a continuous linear density-gradient in the y-direction by changing the cell size but keeping the cell-wall thickness identical [7]. Following this method, N nuclei are generated in square A_0 and copied to the surrounding neighboring areas by translation. Then voronoi cells are constructed based on these 9 N nuclei and the area out of the square area A_0 is deleted to form a periodic boundary condition. The regularity is defined as $\omega = \delta\sqrt{2A_0/\sqrt{3N}}$, where δ is the minimum distance between any two nuclei.

4.2. Finite element models

In this research, specimen with different density gradients (k=0.6; k=0; k=-0.6; k=-1.2) are constructed in area A_0=100 mm×20 mm, shown in figure 6 (a). And the out-plate thickness is 1 mm. The ω is 0.8. The wall thickness is 0.24 mm. The cell walls are modeled by using the four nodded shell elements. The cell wall material is aluminum, which assumed to be elastic perfectly-plastic with the following properties: the density, Young's modulus, Poisson's ratio, yield stress are 2.7×10^3 kg/m^3, 69 Gpa, 0.3, 170 Mpa.

(a)

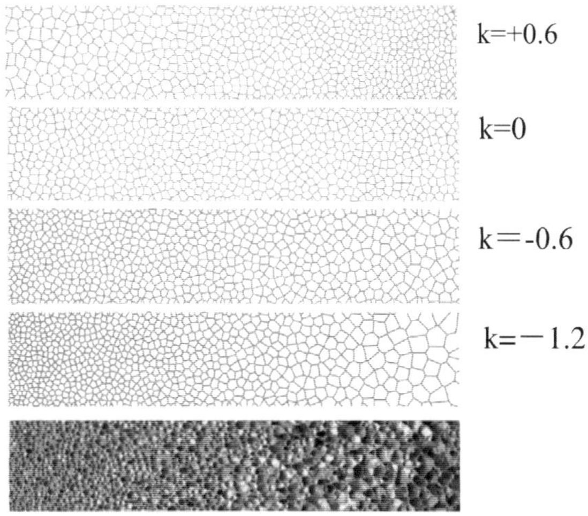

(b)

Figure 6. (a) The schematic diagram of the four density-graded rods. (b) Morphology of functionally graded aluminium foam samples [9].

5. Comparison

Three specimens with different density gradient (k=0.6; k=0; k=-0.6) are used to compare the analytical predictions and FE results as shown in figure 7. The PR-RPPL stands for the parameter refined R-PP-L model. As can be seen from figures, the stress of proximal end predicted by PR-RPPL model shows good agreement with the FEM results. While, the RPPL model gives a higher prediction during the first part of the impact process and tends to the FEM results during the second part of the impact process.

IOP Conf. Series: Materials Science and Engineering **629** (2019) 012036 doi:10.1088/1757-899X/629/1/012036

This phenomenon can be explained by using equation (16). The stress of proximal end difference between the two models can expressed as:

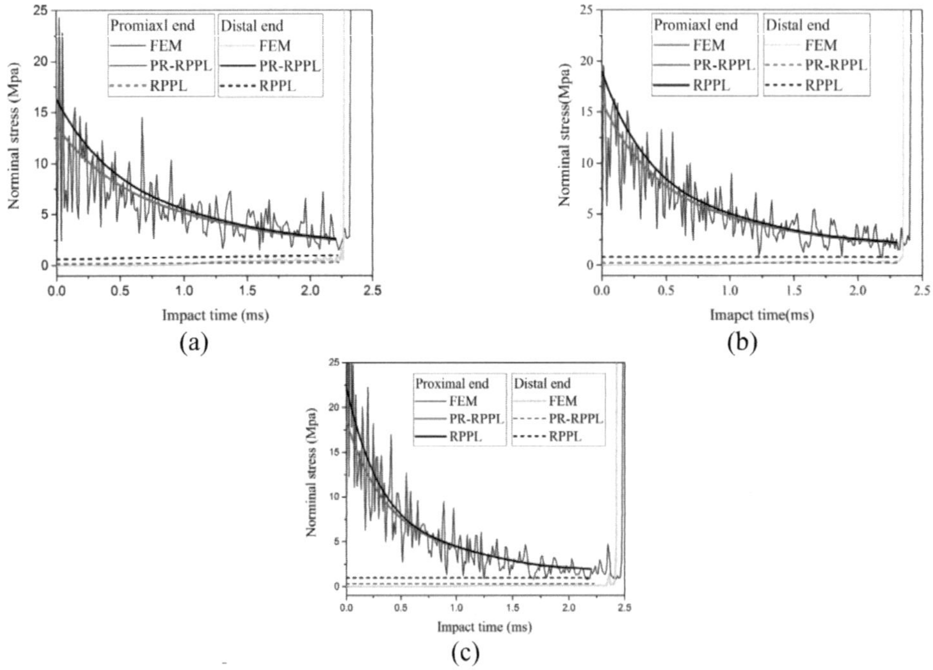

Figure 7. Stress of proximal end and distal end given by FE results and analytical predictions under the initial impact velocity at 300m/s. Density gradient: (a) k=0.6, (b) k=0, (c) k=-0.6.

$$\sigma_{D1} - \sigma_{D2} = \sigma_{pl1} - \sigma_{pl2} + \left(\varepsilon_D^{-1} - \varepsilon_{\max}^{-1}\right)\rho_f \dot{u}^2 \tag{16}$$

Where σ_{D1}, σ_{D2} and σ_{pl1}, σ_{pl2} are stresses of proximal end and stresses of distal end predicted by RPPL model and PR-RPPL model. At the beginning of the impact $\sigma_{pl1} - \sigma_{pl2}$ is a small value compared to $\left(\varepsilon_D^{-1} - \varepsilon_{\max}^{-1}\right)\rho_f \dot{u}^2$ which dominates the difference duo to the high impact velocity and the decrease of impact velocity with time results in the decrease of the difference. Similar to the proximal end, the PR-RPPL model consistent with the FEM results while the RPPL model gives a higher prediction at the distal end. It should be note that the stress's initial time of the distal end is later than the prediction of the two models, the speed of elastic wave should be considered when the impact velocity is extremely high.

6. Discussion

The energy absorbed by cellular rods which converts to internal energy between the time intervals t and $t + \Delta t$ can be expressed as [32]:

$$\Delta E = E(t + \Delta t) - E(t) = \left[\left(\sigma_D + \sigma_{pl}\right)\varepsilon_{\max}\Delta h\right]/2 \tag{17}$$

Then the energy absorbed by cellular rods from the beginning of impact to the time t can be calculated by employing a dummy variable ξ [31]:

$$E=\int_0^S \frac{1}{2}\left[\left(\sigma_D(\xi)+\sigma_{pl}(\xi)\right)\right]\varepsilon_{\max}(\xi)d\xi=\int_0^S\left[\sigma_{pl}(\xi)\varepsilon_{\max}(\xi)+\frac{1}{2}\rho_f(\xi)\dot{u}^2\right]d\xi \tag{18}$$

Where S is the maximum crushed distance at time t shown in figure 6.

Figures 8(a) and 8(b) demonstrate the energy absorbed by cellular rods predicted by analytical model and FE results. Both the two methods show that the lower the density gradient the higher the energy absorption, and with a positive density gradient, the energy absorption is the lowest. At the middle of the crushing process, the difference becomes greatest and decrease with time.

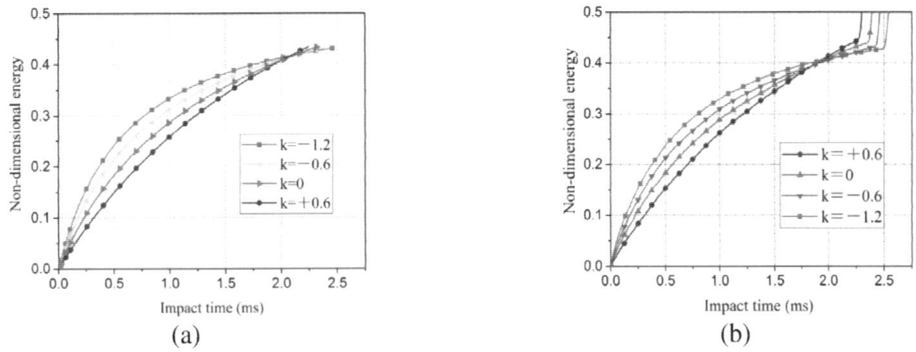

Figure 8. The time history of internal energy of cellular rods. (a) The prediction of PR-RPPL model. (b) The FE results.

Before further discussion, full densification is defined as the one-dimensional shock front reaches the distal end. After that, the rigid mass along with the cellular rod will impact the rigid plate. When a full densification occurs the energy absorbed by cellular rods and the terminal velocities are the same despite the difference of density gradient, which indicates that the density gradient has no effect on the energy absorption and terminal velocity when the cellular rods with the same thickness and mass are fully crushed.

However, when a partial crush happens, the advantages of density gradient can be taken. In this research, partial crush of cellular rods is realized by changing the mass ratio (defined in section 4.2) to a big value (the ratio is set as 15 here) and keeping the other parameters unchanged in the FE models. Figure 9 demonstrates the crushing process of cellular rods with four types of density gradient. The first picture and the last picture in each density gradient stand for the beginning and the end of the collapse process (the maximum collapse distance is reached and the velocity of rigid mass decreased to zero). The deformation occurs at both ends of the rods with a negative density gradient, while the rods with positive density gradient appears continuous collapse. The kinetic energy of the impact system changing with the impact time is shown in figure 10, the cellular rods with a negative density gradient has a larger kinetic energy loss rate than the foam with a positive density gradient during the early time of the impact process (before 2 ms). Duo to the high initial impact velocity, continue crushing happened at the impact end during the early time of the impact process, the energy absorption rate is similar to figure 8, so the lower the density gradient the higher the kinetic energy loss rate.

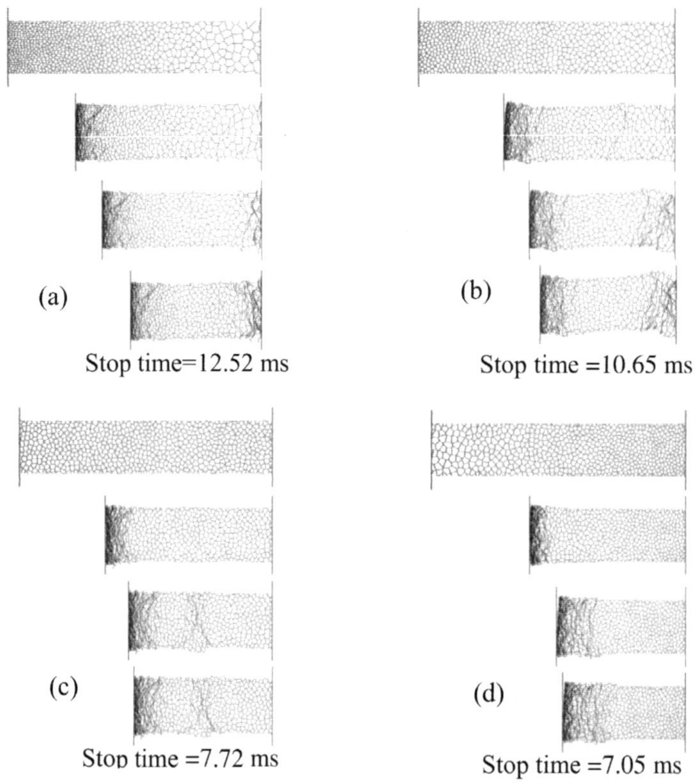

Figure 9. Crushing process of cellular rods. (a) Density gradient k=−1.2. (b) Density gradient k=−0.6. (c) Density gradient k=0. (d) Density gradient k=0.6. Mass ratio=15. V_0 =300 m/s.

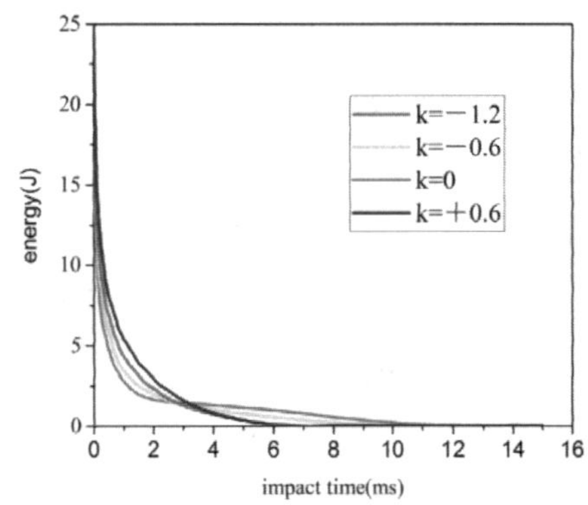

Figure 10. The time history of kinetic energy of the impact system.

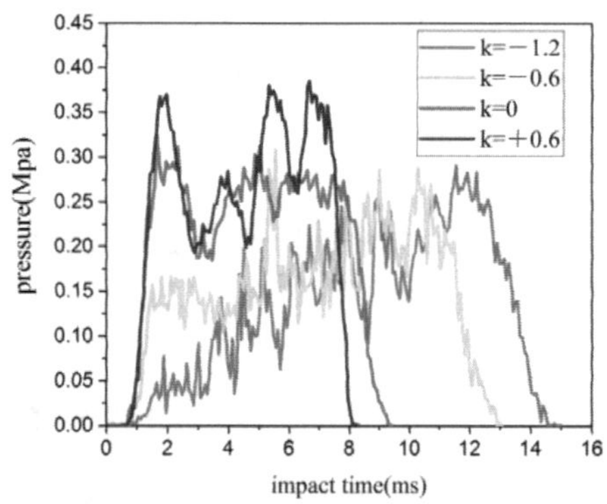

Figure 11. The time history of the pressure of distal end.

With the rapid drop of the velocity of the rigid mass, cellular rods with negative density gradient stop continue crushing of the proximal end and the crushing of the distal end dominates the process, while the positive one remains continue crushing until the velocity of rigid mass decreases to zero. In addition, a crushing band is observed at the middle of the uniform one. The crushing of high-density part of the positive ones causes more kinetic energy loss than the low-density part of the negative ones. Then during the latter time of the impact process (after 3 ms) cellular rods with positive density gradient stop crushing firstly.

Based on the crush paten analyzed before, the crush of distal end can control the pressure of this end to a threshold value related to the density of this end. Figure 11 demonstrates the pressure of distal end, cellular rods with negative density gradient shows lower pressure and longer duration compared with positive one, and this property can be used to benefit the protected object.

7. Conclusion

The dynamic responses of density gradient cellular materials subject to high velocity impact are studied. A parameter-refined R-PP-L model is developed by employing $\varepsilon_{\max}$ and σ_0 which is more suitable for high-speed shock situation. A FE analysis is performed by using the voronoi structure and the FE results achieve good agreement with the analytical prediction.

Both the analytical prediction and the FE results suggest that, compared with the positive one, cellular materials with negative density gradient shows the ability to absorb energy more quickly at the beginning of the high velocity impact while controlling the pressure of the protected structure to a smaller value at the same time. Further numerical studies demonstrate that this advantage benefit the protected structure if a partial crushing happens. When a full densification happens the densified cellular materials along with the rigid mass will impact the protect structure and the advantages of density gradient vanishes and a uniform one should be used for economic reason.

References

[1] Ashby M F, Evans A, Fleck N A, Gibson L J, Hutchinson J W and Wadley H N G 2002 Metal foams: A design guide *Mater Design* **23** 119

[2] Reid S R and Peng C 1997 Dynamic uniaxial crushing of wood *INT J IMPACT ENG* **19** 531-70

[3] Deshpande V S and Fleck N A 2000 High strain rate compressive behaviour of aluminium alloy foams *Int J Impact Eng* **24** 277-98

[4] Tan P J, Reid S R, Harrigan J J, Zou Z and Li S 2005 Dynamic compressive strength properties of aluminium foams Part I - experimental data and observations *J Mech Phys Solids* **53** 2174-205

[5] Radford D D, McShane G J, Deshpande V S and Fleck N A 2006 The response of clamped sandwich plates with metallic foam cores to simulated blast loading *Int J Solids Struct* **43** 2243-59

[6] Shen J, Lu G and Ruan D 2010 Compressive behaviour of closed-cell aluminium foams at high strain rates *Compos Part B-Eng* **41** 678-85

[7] Jing L, Wang Z, Ning J and Zhao L 2011 The dynamic response of sandwich beams with open-cell metal foam cores *Compos Part B-Eng* **42** 1-10

[8] Siegkas P, Tagarielli V L, Petrinic N and Lefebvre L P 2011 The compressive response of a titanium foam at low and high strain rates *J Mater Sci* **46** 2741-7

[9] Wang Z, Jing L, Ning J and Zhao L 2011 The structural response of clamped sandwich beams subjected to impact loading *Compos Struct* **93** 1300-8

[10] Tan P J, Reid S R and Harrigan J J 2012 On the dynamic mechanical properties of open-cell metal foams - A re-assessment of the 'simple-shock theory *Int J Solids Struct* **49** 2744-53

[11] Barnes A T, Ravi-Chandar K, Kyriakides S and Gaitanaros S 2014 Dynamic crushing of aluminum foams: Part I - Experiments *Int J Solids Struct* **51** 1631-45

[12] Hanssen A G, Hopperstad O S, Langseth M and Ilstad H 2002 Validation of constitutive models applicable to aluminium foams *Int J Mech Sci* **44** 359-406

[13] Yahaya M A, Ruan D, Lu G and Dargusch M S 2015 Response of aluminium honeycomb sandwich panels subjected to foam projectile impact - An experimental study *Int J Impact Eng* **75** 100-109

[14] Tan P J, Reid S R, Harrigan J J, Zou Z and Li S 2005 Dynamic compressive strength properties of aluminium foams Part II—'shock' theory and comparison with experimental data and numerical models *J Mech Phys Solids* **53** 2206-30

[15] Karagiozova D, Langdon G S and Nurick G N 2012 Propagation of compaction waves in metal foams exhibiting strain hardening *Int J Solids Struct* **49** 2763-77
[16] Mousanezhad D, Ghosh R, Ajdari A, Hamouda A M S, Nayeb-Hashemi H and Vaziri A 2014 Impact resistance and energy absorption of regular and functionally graded hexagonal honeycombs with cell wall material strain hardening *Int J Mech Sci* **89** 413-22
[17] Zou Z, Reid S R, Tan P J, Li S and Harrigan J J 2009 Dynamic crushing of honeycombs and features of shock fronts *Int J Impact Eng* **36** 165-176
[18] Hohe J, Hardenacke V, Fascio V, Girard Y, Baumeister J, Stoebener K *et al* 2012 Numerical and experimental design of graded cellular sandwich cores for multi-functional aerospace applications *Mater Design* **39** 20-32
[19] Liu Y, Wu H-X and Wang B 2012 Gradient design of metal hollow sphere (MHS) foams with density gradients *Compos Part B-Eng* **43** 1346-52
[20] Fan J, Zhang J, Wang Z, Li Z and Zhao L 2013 Dynamic crushing behavior of random and functionally graded metal hollow sphere foams *Mat Sci Eng A-Struct* **561** 352-61
[21] Liu Y, Wu H-X, Lu G and Wang B 2013 Dynamic properties of density graded thin-walled metal hollow sphere arrays *Mech Adv Mater Struc* **20** 478-88
[22] Gaitanaros S and Kyriakides S 2014 Dynamic crushing of aluminum foams: Part II - Analysis *Int J Solids Struct* **51** 1646-61
[23] Zhu H X, Hobdell J R and Windle A H 2001 Effects of cell irregularity on the elastic properties of 2D Voronoi honeycombs *J Mech Phys Solids* **49** 857-70
[24] Ma G W, Ye Z Q and Shao Z S 2009 Modeling loading rate effect on crushing stress of metallic cellular materials *Int J Impact Eng* **36** 775-82
[25] Zheng Z, Wang C, Yu J, Reid S R and Harrigan J J 2014 Dynamic stress-strain states for metal foams using a 3D cellular model *J Mech Phys Solids* **72** 93-114
[26] Ding Y, Wang S, Zheng Z, Yang L and Yu J 2016 On the stress–strain states of cellular materials under high loading rates *Theoretical and Applied Mechanics Letters* **6** 122-5
[27] Wang X K, Zheng Z J, Yu J L and Wang C F 2011 Impact resistance and energy absorption of functionally graded cellular structures *Advances in Material Engineering and Mechanical Engineering* **69** L Gao ed (Stafa-Zurich: Trans Tech Publications Ltd) 73-8
[28] Zheng J, Qin Q and Wang T J 2016 Impact plastic crushing and design of density-graded cellular materials *Mech Mater* **94** 66-78
[29] He S-Y, Zhang Y, Dai G and Jiang J-Q 2014 Preparation of density-graded aluminum foam *Mat Sci Eng A-Struct* **618** 496-9
[30] Lopatnikov S L, Gama B A, Haque M J, Krauthauser C and Gillespie Jr J W 2004 High-velocity plate impact of metal foams *Int J Impact Eng* **30** 421-45
[31] Zhou H Y, Wang X J and Zhao Z Y 2016 High velocity impact mitigation with gradient cellular solids *Compos Part B-Eng* **85** 93-101
[32] Tan P J, Harrigan J J and Reid S R 2013 Inertia effects in uniaxial dynamic compression of a closed cell aluminium alloy foam *Mater Sci Tech-Lond* **18** 480-8

The equivalent elastic parameters: A lozenge grid structure with negative Poisson's ratio

S Wang[1,3], Z Li[2], K F Zhou[2] and Q W Tan[2]

[1]College of Intelligent Manufacturing, Chongqing Vocational College of Transportation, 7 Xuefu Road, Jiangjin District Chongqing, China
[2]School of Civil Engineering, Chongqing Jiaotong University 66 Xuefu Avenue, Nanan District, Chongqing, China

E-mail: ws15993500427@163.com

Abstract. A new kind of chiral stent was designed by three-dimensional lozenge grid structure structure with four ligaments. Based on the conventional beam theory and energy conservation principle, in-plane Poisson's ratio and equivalent Young's modulus are derived, and finite element analysis was employed to validate the analytical estimates of the elastic module. Then the FEM models are used to investigate the elastic deformation behaviours of lozenge grid structure. The calculated results of equivalent elastic modulus based on beam theory are in consistent with those obtained by finite element method. The theoretical and numerical results show that lozenge grid structure stent proposed in this paper is negative Poisson's ratio material. The presented theory could be used in designing new hierarchical honeycomb structures for multifunctional applications.

1. Introduction

Negative Poisson's ratio (NPR) material has long been widespread concern of scholars due to the unique mechanical features. An auxetic structure with enhanced mechanical properties is designed and manufactured through the tailored negative Poisson's ratio, which aims to treat oesophageal cancer and prevent dysphagia [1]. Two types of innovative chiral stents, lozenge grid structure stent with circular and elliptical nodes and hierarchical lozenge grid structure stents with circular and elliptical nodes, are proposed, which exhibit remarkable radial expansion ability while maintaining axial stability and has a good application prospect in clinical application [2]. Besides, the blood flow in the initial and dilated artery models are simulated by computational fluid dynamics (CFD) [3], which provides personalized design of stent and interventions for the actual situation of patients. Also, a semi-interpenetrating network with better biological activity is developed. The repeated unit cell model is presented based on its repeatability, which is used to research the stent's global deformation [4]. Finite element method is also used to study flexibility and fatigue performance of stent besides its expansion deformation behaviour. Methods for measurement of stent flexibility are proposed based on finite element method [5]. Dumoulin believes that fatigue of stainless steel stent is a small probability event according to the result of simulation [6]. Stolpmann takes the influence of residual stress after inflation and micro-structure on the fatigue of stent into consideration, and thinks general fatigue theory is not suitable for study of stent [7]. Perry [8-9] researches the fatigue of stent by using finite element method and considers finite element method is vital for studying it. Two navel structures are proposed, one is combining re-entrant honeycomb and star-shape honeycomb named as re-entrant star-shaped honeycomb, and another is star-arrowhead honeycomb by combining double arrowhead honeycomb cells and star-shaped honeycomb [10], both of which have enhanced impact resistance.

Based on the research of scholars, a lozenge grid structure with negative passion's ratio aiming to get the large deformation is proposed in this paper.

At the same time, the equivalent Young's modulus and Poisson's ratio of the structure are deduced based on the principle of energy conservation, and the analytical formulas in the linear elastic range are obtained. Finally, the relationship between geometric variables and equivalent elastic parameters is studied.

2. Lozenge grid structure geometry models and linear elastic behaviour

2.1. Lozenge grid structure geometry

Figure 1 shows a typical unit cell of the lozenge grid structure. The parameters *Lx/2, Ly/2, r,* and *t* represent the length along the x and y directions, the circle radius and the rod thickness, respectively. The letter *b* is the cell depth. Structure parameters in detail are shown in table 1.

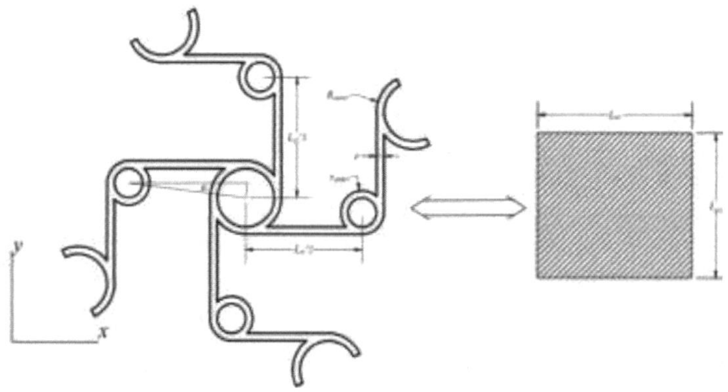

Figure 1. Cell unit.

Table 1. Structural geometry parameters.

Length l (mm)	Angle α (°)	Width t (mm)	Outer Radius R (mm)	Inner Radius r (mm)	Thickness b (mm)
$\pi/6$	arctan(1/2)	0.01	0.05	0.03	0.01

2.2. Analytical formulation

In this paper, the new negative Poisson's ratio structure- lozenge grid structure- is proposed. The schematic diagram of its configuration is shown in figure 1.

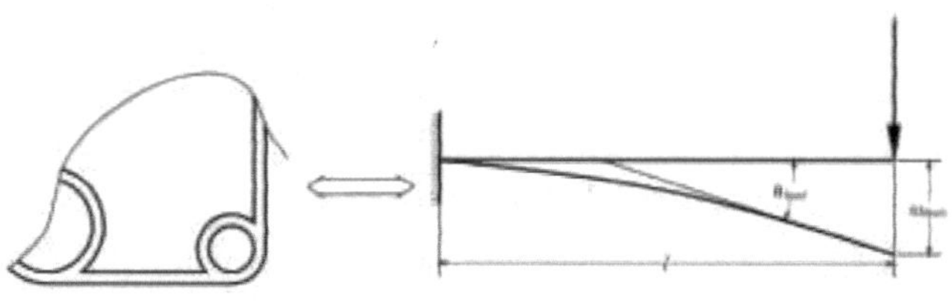

Figure 2. Equivalent beam model.

According to beam theory, the equivalent beam model of structure is established (figure 2). The model uses the cantilever beam calculation method. First, the relationship between curvature and bending moment is obtained. Then, the relationship between curvature and deformation angle is deducted by deflection calculation formula. Finally, the strain and angle formulas are established (equation (1)). The equivalent process is shown in figure 1. The formula is as follows:

$$\varepsilon=\frac{2t}{L}\varphi \tag{1}$$

The equivalent elastic parameters of the proposed structure are deducted in theoretical ways, and the following assumptions are used to derive the model formula.

(1) There are only small deformation structures in the rod, and the nodes in the configuration are completely rigid rotating elements.

(2) The traditional beam theory is employed to derived equivalent Young's modulus and Poisson's ratio of lozenge grid structure.

(3)The circle is considered as rigid body.

Assuming that the unit cell is subjected to an angle along x and y direction and the strain of the unit cell is analysed. Based on assumptions (1) and (2), the x direction strain is ε_x and y direction is ε_y , and they can be expressed as (equations (2) and (3)):

$$\varepsilon_x=\frac{2(R-r-t/2)}{L_i}(i=x,y) \tag{2}$$

$$\varepsilon_y=\frac{2(R-r-t/2)}{L_i}(i=x,y) \tag{3}$$

According to the definition of material Poisson's ratio, the ratio ν_{xy} is expressed as:

$$\nu_{xy}=-\frac{\varepsilon_x}{\varepsilon_y} \tag{4}$$

Meanwhile, substituting equations (2) and (3) into equation (4), the Poisson's ratio is reduced to ν=-1. It's demonstrates that the lozenge grid structure is a negative Poisson's ratio structure.

According to the principle of energy conservation, the linear elastic formula of the lozenge grid is deduced. We assume that the energy generated by the small strain ε_i $(i=x,y)$ is equated to the equivalent beam energy W_i stored in the unit cell (figure 3). Equation (4) is established according to the principle of conservation of energy (equation (4)) and Hooke's law of linear elasticity material (equation (6)). And the volume of equivalent beam is expressed by equation (8).

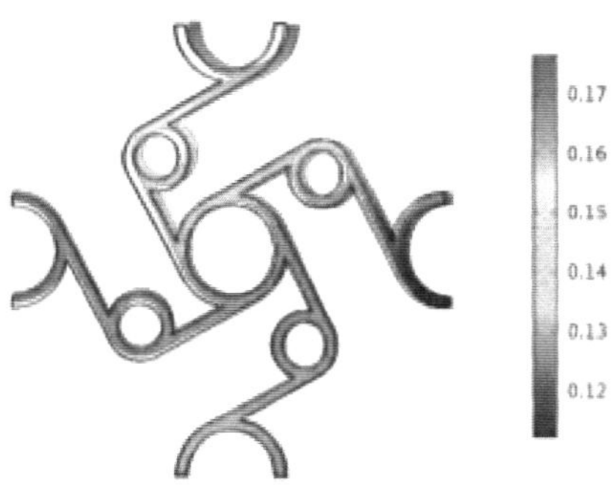

Figure 3. Stent deformation versus radius.

$$\iint_D \frac{1}{2}\sigma_i\varepsilon_i dV = U_e \tag{5}$$

$$\sigma_i = E_i \varepsilon_i \tag{6}$$

$$\frac{1}{2} E_i \varepsilon_i^{\,2} = \frac{4}{V} W_x + \frac{4}{V} W_y \tag{7}$$

$$V = 4 L_x L_y b \tag{8}$$

The edge length of the equivalent beam can be equal to the rod length of lozenge grid structure minus the diameter of the circular hole in the model. The effective length is expressed by *Lex* and *Ley*. According to the geometric relationship of the rod in Figure 1, the effective length *Lex* and *Ley* are equal to:

$$L_{ex} = L_x - 2R - 2r \tag{9}$$

$$L_{ey} = L_x - 2R - 2r \tag{10}$$

According to beam theory, the equal moments *M* which act on each end of the rod will produce an angular deflection θ. The strain energy of the beam can be written as:

$$W = \int_0^{\theta} M d\theta \tag{11}$$

According to the effective length *Lei (i=x,y)* and the equivalent rectangular section of the rod, the rod's second moment of inertia is:

$$\theta = \frac{M L_{ei}}{2 E_c I} \tag{12}$$

$$I = \frac{b t^3}{12} \tag{13}$$

Where, E_c is the material elastic module of equivalent beam. Substituting equations (12) and (13) into equation (11), a new formula for equivalent beam energy W is derived:

$$W = \frac{E_c b t^3 \theta^2}{6 L_{ei}} \tag{14}$$

The small strains ε_x and ε_y are applied to the unit cell. Substituting equations (2), (3), and (14) into equation (7), a new energy expression is obtained by:

$$\frac{1}{2} E_i \varepsilon_i^{\,2} = \frac{4 E_c b t^3 \theta^2}{6 L_{ei} V} \tag{15}$$

Where, E_x and E_y are the Young's moduli of lozenge grid structure. Equation (16) and equation (17) can be rewritten as:

$$E_x = \frac{E_c t^3 L_x^{\,2}}{3 L_{ex}^{\,2} L_{ey} (R - r - \frac{t}{2})^2} \tag{16}$$

$$E_y = \frac{E_c t^3 L_y^{\ 2}}{3L_{ex}^{\ 2} L_{ey}(R - r - \frac{t}{2})^2} \tag{17}$$

3. FEM simulations and discussion

3.1. FEM simulations and material parameters

In this paper, COMSOL Multiphysics computing platform is used for finite element simulation. In order to simulate the real working conditions, the material adopts self-defined material and its parameters are in table 2. And the depth of stent is 0.03 mm,

Table 2. Test parameters.

item	Parameters (mm)				
Radius	0.05	0.08	0.10	0.15	0.20
Length	π/8	π/10	π/12	π/14	π/16

Tetrahedron element is used in the process of mesh generation. The the minimum cell size is 0.01 mm, and the maximum cell growth rate is 1.5. The total number of cells is 1128. Finally, the non-linear large plastic strain solver is used to calculate the plastic strain of the stent during expansion. The data are convergent in the process of solving.

3.2. Comparison between theory and FEM

The pictures are obtained by the FEM results. The elastic variation of the lozenge grid structure along with the cells geometry parameters are showed in Figure 3 and 4. Tensile deformation displacement is obvious and the expansion phenomenon is occurred in the lozenge grid structure.Negative Poisson's ratio structure expands when it is stretched. Figure 3 shows that the bracket expands when it is stretched, this phenomenon shows that the structure is negative Poisson's ratio structure.

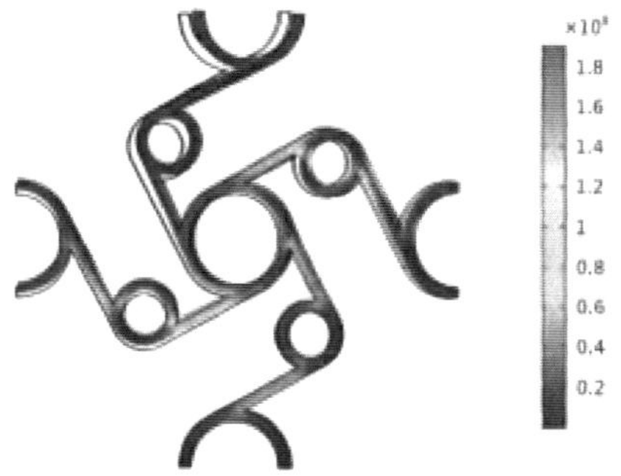

Figure 4. Stent stress versus length.

The Young's modulus errors between FEM and Theory with the geometry parameters are compared in tables 3 and 4. The equivalent Young's modulus is obtained by numerical calculation. Comparing the results in Table 3, the results of equivalent elastic modulus equation based on energy principle are consistent with those of numerical simulation, and the error between them is less than 10% except a few points, and the Young's modulus error raise with the radius growing. The FEM results show a discrepancy of 4.9% over the analytical equation along with the length changing in table 4. The difference between experiment and analysis can be attributed to the use of equivalent elastic modulus *Ec*. The theoretical calculation is basically consistent with the results of the finite element model.

Table 3. The Young's modulus error between FEM and theory with the radius variation.

Radius	0.005	0.008	0.01	0.015	0.018
FEM(GPa)	555	558	505	586	314
Theory(GPa)	595	619	544	458	434
Error(%)	6.72	9.85	7.17	21.84	27.65

Table 4. The young's modulus error between FEM and theory with the length variation.

Length	Pi/8	Pi/10	Pi/12	Pi/14	Pi/16
FEM (GPa)	540	525	505	500	662
Theory(GPa)	511	515	475	583	653
Error(%)	5.37	1.90	5.94	14.23	1.35

4. Conclusions

In this work, a lozenge grid structure is proposed and material equivalent isotropic elastic modulus has been deducted with beam theory. Then structural deformation is carried out by finite element method by COMSOL Multiphysics. By comparing the theoretical equations with the finite element simulation results, it is found that:

- The theoretical equation deduces the Poisson's ratio equation and its value is -1, which shows that the lozenge grid structure is a typical negative Poisson's ratio structure.
- Finite element analysis of the lozenge grid structure tension deformation and the error between the simulation results and theoretical results is less than 10% mostly, the equivalent elastic modulus is good associated with the Finite element results.
- The lozenge grid structure is a new sandwich structures which can be applied in a lot of engineering structures.

References

[1] Wu W W, Song X K and Liang J 2018 *Compos Struct.* **185** 381-92
[2] Bhullar S K, Mawanane Hewage A T, Alderson A, Alderson K and Jun M B G 2013 *Adv. Mater.* **2** 42-7
[3] Gundelwein L, Miró J, Gonzalez Barlatay F, Lapierre C, Rohr K and Duong L 2018 *J Biomech.* **81** 68-75
[4] Petrini L, Migliavacca F, Dubini G, Auricchio F *et al* 2003 *Summer Bioengineering Conf.* (Florida) pp 251-2
[5] Dumoulin C and Cochelin B 2000 *J Biomech.* **33** 1461-70
[6] Stolpmann J, Brauer H, Stracke H J *et al* 2003 *Material Wissensehaft and Werkstofftechnik* **34** 736-45
[7] Perry M, Oktay S and Muskivitch J C 2002 *Min Invas Ther & Allied Technol.* **11** 165-71
[8] Wang H, Lu Z, Yang Z, Li Z Y and Li X 2019 *Compos Struct.* **208** 758-70
[9] Wang H, Lu Z, Yang Z, Li Z Y and Li X 2019 *Int J Mesh Sci.* **151** 746-59
[10] Sharma S, Kushwaha D and Rawal A 2019 *Mater Lett.* **241** 214-18

IOP Conf. Series: Materials Science and Engineering **629** (2019) 012038 doi:10.1088/1757-899X/629/1/012038

Preparation and characterization of catechol-functionalized chitosan thermosensitive hydrogels

D Y Zhang[1], Z Hu[1,3], S T Lu[1], S D Li[1], Z M Yang[2] and P W Li[2,3]

[1]Department of Applied Chemistry, School of Chemistry and Environmental Science, Guangdong Ocean University, Zhanjiang 524088, Guangdong, China
[2]Agricultural Product Processing Research Institute, Chinese Academy of Tropical Agricultural Sciences, Zhanjiang 524001, Guangdong, China

E-mail: huzhangqyx@126.com/puwangli@163.com

Abstract. Catechol functionalization of chitosan was modified by EDC coupling reaction and then catechol-functionalized chitosan thermosensitive hydrogels were prepared by catechol-functionalization of chitosan with β-glycerophosphate disodium as a thermosensitizer. The hydrogels were then characterized by FTIR, ^{1}H-NMR, SEM and rheology analysis. The results demonstrated that the thermosensitive hydrogels were prepared at 37°C within several minutes by mixing 2% catechol-functionalized chitosan with 30% β-glycerophosphate disodium at a ratio of 8:2, and these thermosensitive hydrogels with porous network structure were expected to be developed as biomedical materials with wide applications.

1. Introduction

Hydrogels, novel functional polymeric materials with three-dimensional network structure, are hydrophilic but not soluble in water. Owing to the excellent water adsorption, water retention, and biocompatibility, they have been widely studied and applied [1]. Thermosensitive hydrogels are one of numerous environment-sensitive hydrogels that have been mostly studied, mainly characterized in ability to respond to temperature change and give rise to sol-gel transition. Compared with conventional hydrogels, thermosensitive hydrogels have significant advantages when being applied in tissue engineering, drug delivery, and 3D cell culture.

As tissue engineering materials for wound healing, surgical tissue adhesion, hemostasis in surgical procedure, and local drug delivery, tissue adhesives attract wide attention [2,3]. It's found in many studies that, adhesion behaviors of mussels are mainly affected by 3,4-dihydroxy-L-phenylalanine (DOPA), while a major adhesive component of DOPA is ortho-dihydroxyphenyl functional group that is able to form strong covalent bonds and noncovalent bonds with various organic/inorganic/metallic substrates, which plays an important role in hydrogel-like adhesives [4-6]. Ryu *et al* synthesized temperature-sensitive and adhesive catechol-functionalized Chitosan/Pluronic-SH hydrogel (CHI-C/Plu-SH) after chitosan backbone conjugated with multiple catechol groups was crosslinked with terminally thiolated Pluronic F-127 triblock copolymer [7]. Both in vitro and in vivo assessments indicate that CHI-C/Plu-SH hydrogel has excellent mechanical properties and stability.

Chitosan (CS) is a natural polycationic polysaccharide prepared from deacetylated chitin, and has good bioactivities such as promoting trauma healing, inhibiting scar formation, coagulation, and antimicrobial property. Due to its good biocompatibility, non-immunogenicity and non-irritability, it was approved by FDA as GRAS (generally regarded as safe) substance in 2001 [8]. As chitosan is not readily

IOP Conf. Series: Materials Science and Engineering **629** (2019) 012038 doi:10.1088/1757-899X/629/1/012038

soluble in water but soluble in weak acids, its application is partially limited; moreover, adhesion capabilities of chitosan hydrogels are limited, too, so we conceived of grafting catechol functional groups onto chitosan backbone in order to increase adhesiveness while modifying its water solubility, so that it can work better.

In this paper, we mainly developed thermosensitive tissue adhesive hydrosols based on catechol-functionalized chitosan (CS-C) and β-glycerophosphate disodium(β-GP), which are expected to be applied as bio-adhesive materials.

2. Experimental

2.1. Materials

Chitosan, DD≥90%, MW=100 kDa; 3,4-dihydroxyhydrocinnamic acid (HCA), 1-ethyl-(3-dimethylaminopropyl)-carbodiimide hydrochloride (EDC) and N-hydroxysuccinimide (NHS) were purchased from Shanghai Aladdin Biochemical Technology Co., Ltd; β-glycerophosphate disodium(β-GP), 98%, was purchased from SIGMA.

2.2. Synthesis and characterization of CS-C

The procedure in literatures [6,7] was slightly modified. Briefly, 5 g chitosan was dissolved in 250 mL of 1% acetic acid overnight to result in 2% CS solution for later use. 0.1M HCA was dissolved in 50 mL of deionized water to result in HCA solution for later use. 0.2M EDC and NHS were dissolved in 200 mL of ethanol/water mixture (1:1 by volume). Secondly, HCA solution was added into CS solution and stirred, then EDC/NHS was added dropwise into CS/HCA solution for reaction for 10 hours while pH was kept at 4.5–5.5. After the reaction was completed, the reaction product was dialyzed with pH 5 hydrochloric acid solutions for three days, and then dialyzed with purified water for 4 hours; thereafter the product was freeze-dried to result in CS-C. CS-C was structurally characterized using Spectrum 100 FTIR spectrometer, and AVANCE III NMR spectrometer.

2.3. Preparation of CS-C/β-GP thermosensitive hydrogels

A calculated amount of CS-C was weighed and dissolved in distilled water to result in CS-C solution at a certain concentration; a calculated amount of β-GP was dissolved in distilled water to result in β-GP solution at a certain concentration. At a given ratio by volume, β-GP solution was added dropwise into CS-C solution; thereafter, the mixture was further stirred for 2 min and then put into a thermostat water bath at 37°C to result in CS-C/β-GP-based thermosensitive hydrogel.

2.4. Characterization of CS-C/β-GP thermosensitive hydrogel

2.4.1. Gel time. Gel time was determined by inversion method. 1 mL of the mixture was added into a 3 mL screw neck vial, then the vial was placed into the thermostat water bath at 37°C; 15 min later, gel state was observed every 1 min, and the gel time is the time at which the mixture stops flowing in 30 sec when the vial is inverted.

2.4.2. IR characterization. The freeze-dried CS-C/β-GP hydrogel sample was tabletted with potassium bromide powder, and its IR absorption characteristics were analyzed by Spectrum 100 FTIR spectrometer within a wavenumber range of 4000–400 cm^{-1} with a resolution of 4 cm^{-1} over 16 scans.

2.4.3. Microstructure. A small part of the freeze-dried sample was placed on the stage for gold spray treatment and then placed in S-4800 SEM, observed and photographed at an acceleration voltage of 10 kV.

2.4.4. Rheology analysis. Rheology analysis was performed on HAAKE Rheometer (MARSIII, Haake, Germany) equipped with cone/plate combination (C35/1: φ=35mm, angle=1°) as a measuring system. 1 mL of a hydrogel sample was put into the rheometer, and at a constant frequency of 1 Hz, (1) rheological properties of the sample were determined at a heating rate of 1°C/min within 20°C–60°C, changes in storage modulus G'(Pa) and loss modulus G"(Pa) were recorded as functions of temperature so as to assess gel temperature of the sample; (2) sample rheology was determined at a physiological temperature fixed at 37°C for a longer time, and changes in storage modulus G'(Pa) and loss modulus G"(Pa) were recorded as functions of time so as to assess gel time of the sample. To prevent moisture evaporation, one layer of oil was added around the sample in temperature and time scans.

3. Results and discussion

3.1. Characterization of CS-C

The prepared CS-C was characterized in a number of ways. As can be seen from IR spectrum of CS in figure 1(A), absorption peak at 3433 cm^{-1} is assigned to stretching vibration of O-H and N-H associated via a hydrogen bond, absorption peaks at 1650 cm^{-1} and 1599 cm^{-1} are assigned to stretching vibration of amide group in chitosan, absorption peaks at 1082 cm^{-1} and 1033 cm^{-1} are assigned to stretching vibration of C-O typical of chitosan. Compared with IR absorption of CS, IR spectrum of CS-C varied significantly as a new absorption peak occurred at 1533 cm^{-1}, which is attributed to stretching vibration of aromatic C=C, demonstrating that catechol group has been successfully grafted onto macromolecular chain of chitosan.

CS and CS-C were further structurally analyzed by ^{1}H-NMR (CD_3COOD/D_2O, 500 MHz, 20°C) (figure 1B). In CS spectrum, signal peaks are resolved very well in agreement with what reported in literature [9,10]. In the spectrum of CS-C, the multiplet at chemical shifts of 3.5–4.1 ppm represents absorption peaks of H3–H6 on saccharide ring of CS, the absorption peak at a chemical shift of 4.92 ppm is assigned to H1 on saccharide ring of CS (partially covered by strong solvent peak), the absorption peak at a chemical shift of 3.16 ppm is assigned to H2, absorption peaks at chemical shifts of 2.64 ppm and 2.81 ppm are assigned to two methylene groups (H8,H9) linked to a benzene ring, absorption peaks at chemical shifts of 6.71–6.81 ppm are assigned to hydrogens on aromatic ring (H10, H11, H12), and the ratio of their peak area to methylene peak area was 3:2. Based on relative peak area of catechol functional groups (H10, H11, H12) grafted onto chitosan backbone against peak area of H2 in ^{1}H-NMR, grafting ratio was calculated to be 17.3%.

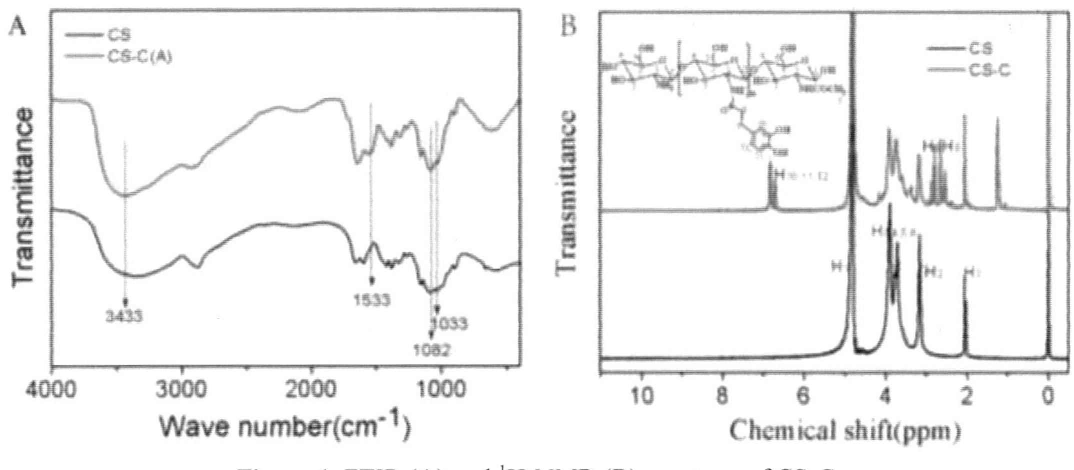

Figure 1. FTIR (A) and ^{1}H-NMR (B) spectrum of CS-C.

3.2. Preparation of CS-C/β-GP-based thermosensitive hydrogels

Based on preliminary work, we selected 2% or 2.5% CS-C, 25% or 30% β-GP, and 7:3 or 8:2 as the ratio of CS-C to β-GP by volume; compositions of thermosensitive gels in various groups are shown in table 1. Sol-gel transition of CS-C/β-GP thermosensitive hydrogels is shown in figure 2.

Table 1. Compositions of CS-C/β-GP thermosensitive hydrogels.

Group	CS-C: β-GP (V/V)	CS-C (%, W/V)	β-GP (%, W/V)	Gel Time (min)
A	8:2	2	25	18
B		2	30	16
C		2.5	25	17
D		2.5	30	20
E	7:3	2	25	28
F		2	30	24
G		2.5	25	30
H		2.5	30	27

Figure 2. Sol and gel of CS-C/β-GP thermosensitive hydrogel.

3.3. IR characterization of CS-C/β-GP thermosensitive hydrogels

Hydrogel samples from groups A and B were selected for IR characterization. As shown in figure 3(A), IR spectra of CS-C, β-GP, and CS-C/β-GP hydrogel samples (A,B) do not differ much, significant blue shift occurred in absorption peaks of CS-C/β-GP hydrogel samples near 3433 cm^{-1} and these absorption peaks became broader, meanwhile symmetric stretching vibration peaks of phosphoric acid at 977 cm^{-1} became weaker, which might be caused by stacking of hydroxyl groups in the modified CS and hydroxyl groups in β-GP, or due to hydrogen bond formation after interactions between amino groups on CS-C molecular chain and hydroxyl groups on β-GP.

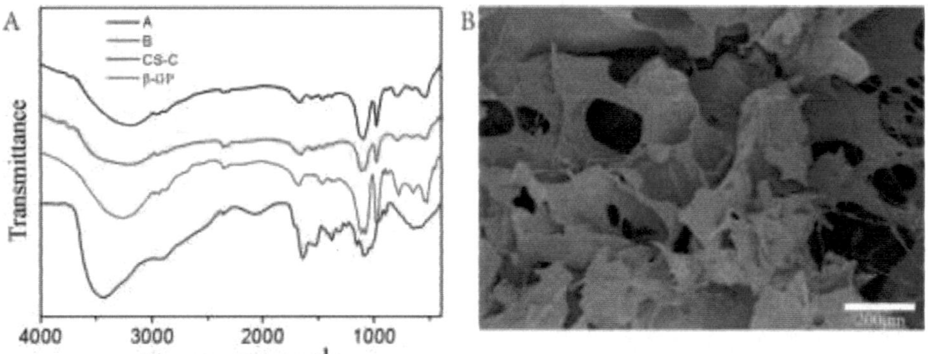

Figure 3. FTIR(A) and SEM(B) of CS-C/β-GP hydrogels.

3.4. SEM examination of CS-C/β-GP thermosensitive hydrogel

The thermosensitive hydrogel sample from Group A was selected and its microscopic morphology was studied using SEM. As shown in figure 3(B), the lyophilized hydrogel prepared from CS-C and β-GP exhibited porous network microstructure that is favorable for water and small molecular drugs to pass through freely, and can be used in wound healing, drug release, etc.

3.5. Rheological study of CS-C/β-GP thermosensitive hydrogels

Hydrogel samples from groups A and B were preferentially selected to evaluate rheological properties, and storage modulus G'(Pa) and loss modulus G"(Pa) were plotted against temperature and time respectively, as shown in figure 4. In general, as temperature rises or time extends, curves of storage modulus G'(Pa) at lower viscosity intersect with curves of loss modulus G"(Pa) at higher viscosity, and the intersection points indicate sol-gel transition [11]. Each intersection point between the curve of storage modulus G'(Pa) and the curve of loss modulus G"(Pa) corresponds to gel temperature or gel time at corresponding temperature, at which the sample transits from sol state to gel state [12]. In the case of group A, gel temperature of the hydrogel was 39°C (figure 4A_1); while in the case of group B, when the amount of β-GP was increased to 30%, its gel temperature was 37.5°C (figure 4B_1) close to body temperature. With increasing temperature and time, storage modulus G'(Pa) increased more quickly while loss modulus G"(Pa) increased more slowly, indicating presence of more networked gel structures in the system, but when G'(Pa)<G"(Pa), only part of the network structure was present in the system; when G'(Pa)=G"(Pa) or at the critical point of sol-gel transition, large-sized physical gel-like network came into being, that is, a hydrogel was formed; in addition, it can be seen after the gel point that G'(Pa)>G"(Pa), in other words, compared with the increase in loss modulus, the increase in storage modulus was higher, indicating that highly elastic gel network was formed during continuous gelation. In the case of group A, gel time was 13 min (figure 4A_2); while in the case of group B, gel time was 12 min (figure 4B_2). Gel times measured by rheology test were shorter than those measured by test tube inversion, as the latter ones were 18 min and 16 min respectively (table 1); such differences should be caused by difference in test methods, and the test tube inversion method would be subjectively affected by observers.

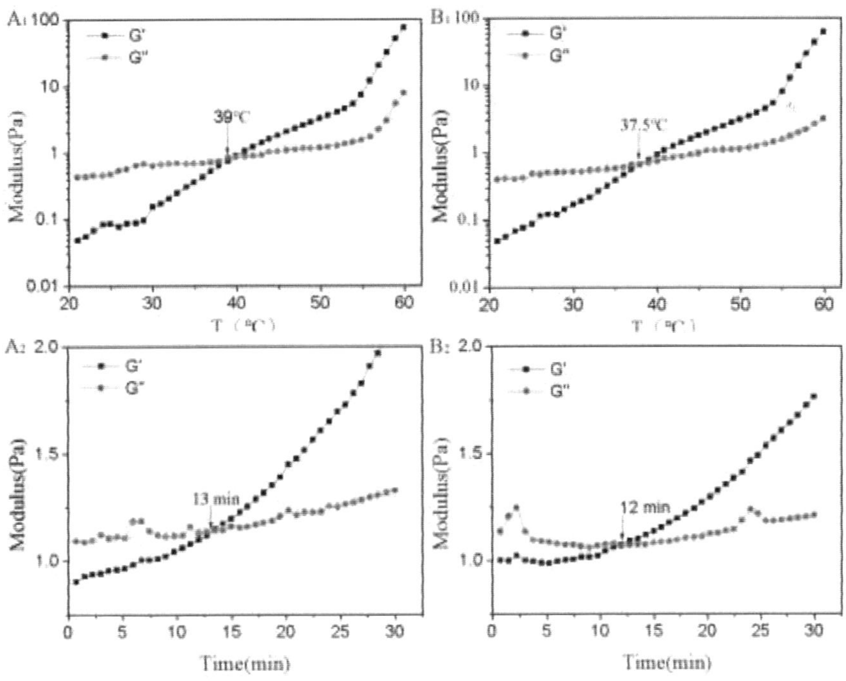

Figure 4. Rheology curves of CS-C/β-GP thermosensitive hydrogels.

4. Conclusion

CS-C/β-GP thermosensitive hydrogels were successfully prepared from catechol-functionalized chitosan by ionic gel reaction. The structural studies indicated that these thermosensitive hydrogels formed porous network structures via intermolecular hydrogen bonds, which were beneficial for adsorption of substantial fluid, and the hydrogels are expected to be developed as biomedical materials.

Acknowledgments
We gratefully acknowledge the financial support by Guangdong Provincial Natural Science Foundation of China (2016A030308009), Project of Science and Technology Plan of Guangdong Province (2015A020216019, 2017A010103023), Project of Enhancing School with Innovation of Guangdong Ocean University (2017KTSCX090) and Postgraduate Education Innovation Project of Guangdong Ocean University (201927).

References
[1] Xia G, Liu Y, Tian M *et al* 2017 Nanoparticles/thermosensitive hydrogel reinforced with chitin whiskers as a wound dressing for treating chronic wounds *J. Mater.Chem.B* **5** 3172-85
[2] Ishihara M, Nakanishi K, Ono K *et al* 2002 Photocrosslinkable chitosan as a dressing for wound occlusion and accelerator in healing process *Biomaterials* **23** 833-40
[3] Kim K, Kim K, Ryu J H *et al* 2015 Chitosan-catechol: A polymer with long-lasting mucoadhesive properties *Biomaterials* **52** 161-70
[4] Lee H, Scherer N F and Messersmith P B 2006 Single-molecule mechanics of mussel adhesion *P. Natl. Acad. Sci. Usa.* **103** 12999-3003
[5] Hight L M and Wilker J J 2007 Synergistic effects of metals and oxidants in the curing of marine mussel adhesive *J. Mater. Sci.* **42** 8934-42
[6] Sparks B J, Hoff E F T, Hayes L P *et al* 2012 Mussel-inspired thiol-ene polymer networks: Influencing network properties and adhesion with catechol functionality *Chem. Mater.* **24** 3633-42
[7] Ryu J H, Lee Y, Kong W H *et al* 2011 Catechol-functionalized chitosan/pluronic hydrogels for tissue adhesives and hemostatic materials *Biomacromolecules* **12** 2653-9
[8] Rao S B and Sharma C P 1997 Use of chitosan as a biomaterial: Studies on its safety and hemostatic potential *J. Biomed. Mater. Res.* **34** 21-8
[9] Perez-Alvarez L, Ruiz-Rubio L and Luis Vilas-Vilela J 2018 Determining the deacetylation degree of chitosan: Opportunities to learn instrumental techniques *J. Chem. Educ.* **95** 1022-8
[10] Fernandez-Megia E, Novoa-Carballal R, Quinoa E *et al* 2005 Optimal routine conditions for the determination of the degree of acetylation of chitosan by H-1-NMR *Carbohyd. Polym.* **61** 155-61
[11] Mirahmadi F, Tafazzoli-Shadpour M, Shokrgozar M A *et al* 2013 Enhanced mechanical properties of thermosensitive chitosan hydrogel by silk fibers for cartilage tissue engineering *Mat. Sci. Eng. C* **33** 4786-94
[12] Wang T, Chen L, Shen T *et al* 2016 Preparation and properties of a novel thermo-sensitive hydrogel based on chitosan/hydroxypropyl methylcellulose/glycerol *Int. J. Biol. Macromol.* **93** 775-82

Institute of Physics
Dirac House, Temple Back
Bristol BS1 6BE UK

ISSN: 1757-8981
ISBN 978-1-5108-9833-2

CURRAN ASSOCIATES INC.
proceedings.com